Physics of Crystal Growth

This text introduces the physical principles of how and why crystals grow.

The first three chapters recall the fundamental properties of crystal surfaces at equilibrium. The next six chapters describe simple models and basic concepts of crystal growth including diffusion, thermal smoothing of a surface, and applications to semiconductors. Following chapters examine more complex topics such as kinetic roughness, growth instabilities, and elastic effects. A brief closing chapter looks back at the crucial contributions of crystal growth in electronics during this century. The book focuses on growth using molecular beam epitaxy. Throughout, the emphasis is on the role played by modern statistical physics. Informative appendices, interesting exercises and an extensive bibliography reinforce the text.

This book will be of interest to graduate students and researchers in statistical physics, materials science, surface physics and solid state physics. It will also be suitable for use as a coursebook at beginning graduate level.

Collection Aléa-Saclay:
Monographs and Texts in Statistical Physics

General editor: Claude Godrèche

C. Godrèche (ed.): *Solids far from equilibrium*
P. Peretto: *An introduction to the modeling of neural networks*
C. Godrèche and P. Manneville (eds.): *Hydrodynamics and nonlinear instabilities*
A. Pimpinelli and J. Villain: *Physics of crystal growth*
D. H. Rothman and S. Zaleski: *Lattice-gas cellular automata*
B. Chopard and M. Droz: *Cellular automata modeling of physical systems*

Physics of Crystal Growth

Alberto Pimpinelli
Université Blaise Pascal – Clermont-Ferrand II

Jacques Villain
Centre d'Études Nucléaires de Grenoble

CAMBRIDGE
UNIVERSITY PRESS

CAMBRIDGE UNIVERSITY PRESS
Cambridge, New York, Melbourne, Madrid, Cape Town, Singapore, São Paulo

Cambridge University Press
The Edinburgh Building, Cambridge CB2 2RU, UK

Published in the United States of America by Cambridge University Press, New York

www.cambridge.org
Information on this title: www.cambridge.org/9780521551984

First published 1998

A catalogue record for this publication is available from the British Library

Library of Congress Cataloguing in Publication data

Pimpinelli, Alberto.
Physics of crystal growth / Alberto Pimpinelli, Jacques Villain.
p. cm. – (Collection Aléa-Saclay: 4)
Includes bibliographical references and index.
ISBN 0 521 55198 6. – ISBN 0 521 55855 7 (pbk.)
1. Solids–Surfaces. 2. Crystal growth. 3. Statistical physics.
I. Villain, Jacques. II. Title. III. Series.
QC176.8.S8P56 1997
548′.5–dc21 96-51772 CIP

ISBN-13 978-0-521-55198-4 hardback
ISBN-10 0-521-55198-6 hardback

ISBN-13 978-0-521-55855-6 paperback
ISBN-10 0-521-55855-7 paperback

Transferred to digital printing 2007

Contents

Preface

In writing a preface, an author is faced with the question: what is this book of mine? Of course, in the end only the reader will decide what it really is. The scope of this preface, as of all prefaces, is to say what it was intended to be.

This book tries to offer a reasonably complete description of the physical phenomena which make solid materials grow in a certain way, homogeneous or not, rough or smooth. These phenomena belong to chemistry, quantum physics, mechanics, statistical mechanics. However, chemistry, mechanics and quantum physics are essentially the same during growth as they are at equilibrium. The statistical aspects are quite different. For this reason, the authors have insisted on statistical mechanics.

Another reason to emphasize the statistical mechanical concepts is that they will probably survive. The concepts developed many years ago by Frank, or more recently by Kardar, Parisi and Zhang are still valid while, for instance, quantum mechanical calculations of the relevant energy parameters will certainly evolve a lot in the next few years. We have not considered it useful to devote too many pages to them, but we have tried to present the frame in which the data can be inserted, as soon as they are known.

However, although emphasis is on statistical mechanics, other aspects are not ignored, even though they may have been treated somewhat superficially. The reader will find more detailed information in an extensive bibliography, where all titles are given in extenso, thus making its use much easier.

This book is mainly devoted to growth, and therefore to non-equilibrium processes. Nevertheless, we have tried to make it self-contained and to incorporate some elements of equilibrium surface physics, for instance the roughening transition and the equilibrium shape. The reader eager to know more will again find the necessary references in the bibliography.

The authors are theorists and their book is mainly devoted to theory. Few details are given on experimental methods, but many experimental pictures (mostly from scanning tunneling microscopy) show how real materials do behave. In this domain, too, an abundant bibliography is available.

Although the responsability for all which is written here–good or bad– is completely ours, we owe a lot to all those who contributed to our understanding of the subject. We wish to thank J.M. Bermond, H. Bonzel, J.-P. Bucher, J. Chevrier, G. Comsa, J. Ernst, J. Frenken, M. Hanbücken, J.C. Heyraud, K. Kern, R. Kern, M. Lagally, J. Lapujoulade, J.J. Métois, B. Mutafschiev and E. Williams, whose experimental works revealed to us all the beauty of Surface Physics and Crystal Growth–and often contributed to the iconographic asset of the book. We are also very grateful to D. Wolf, Ph. Nozières, R. Kern again, J. Krug, P. Jensen, J. Langer, C. Misbah, L. Sander, D. Vvedensky and A. Zangwill, who shared with us some of their secrets. Special thanks are due to P. Politi and M. Schroeder, whose untiring reading of preliminary versions has been a source of most valuable suggestions and improvements. A good share of the chapters on elasticity has much profited from the competence of C. Duport, who corrected all our formulae, and even explained some of them to us!

We thank and beg pardon to all who are omitted here either for space or memory limitations. A final thank is due to the people in Cambridge University Press, and most of all to R. Neal, for waiting patiently for the completion of this work.

List of symbols

DLA: *Diffusion limited aggregation (section 11.2)*
MBE: *Molecular beam epitaxy*
ML: *Monolayer*
RHEED: *Reflection high-energy electron diffraction*

a: *lattice constant or atomic distance (usually taken equal to 1 in this book).*
A_α, $A_{\alpha\gamma}$, A: *kinetic coefficients in sections 13.1 and 13.2*
$A(t)$: *amplitude appearing in section 13.7*
B: *a thermodynamic coefficient in (2.38)*
$B_{\alpha\gamma}$: *kinetic coefficient in chapter 13*
C: *a constant (section 15.7)*
d: *dimension of the space (usually 3)*
d': *surface dimension* $d-1$
$d' = D/D'$, $d'' = D/D''$: *see (6.20)*
d_c^u: *upper critical dimension (section 12.4)*
d_c^ℓ: *lower critical dimension (section 12.4)*
d_f: *fractal dimension of a fractal terrace (section 11.2)*
$\tilde{D}$: *Fick diffusion constant (sections 7.1, 8.4)*
$\tilde{D}_s$, $\tilde{D}$: *surface Fick diffusion constant*
D^*: *tracer diffusion constant (section 7.1)*
D_{int}: *diffusion constant of interstitials*
D_{vac}: *diffusion constant of vacancies in section 7.3.*
D_s^0: *see eq. (7.13)*
D_0: *see eq. (11.28)*
D_2: *diffusion constant of dimers at a surface (sections 7.10 and 11.8)*
D', D'': *step kinetic coefficients (section 6.4)*
D_s *(D when no ambiguity is possible): surface diffusion constant of adatoms*
e : *basis of Napierian logarithms*
E: *Young modulus*
E_2: *binding energy of a dimer (section 11.8)*

$\delta\mathbf{f}^{\text{ext}}$, $\delta\mathbf{F}^{\text{ext}}$: *elementary external forces (chapter 16)*
F_α^{ext}: *External force*
F: *beam intensity in MBE*
$\mathbf{F_R}$: *force acting at point* $\mathbf{R}$
g: *gravity (in chapter 1)*
$G(\mathbf{R})$: *height-height correlation function at equilibrium, eq. (1.4)*
$G(\mathbf{r}, t)$ *height-height correlation function during growth, eq. (12.1)*
h: *amplitude of the modulation of a surface, height of a defect*
i^*: *critical size of a cluster of adatoms (section 11.2)*
$\mathbf{j}_s$: *surface current density of adatoms*
$\mathbf{j}_s^{\text{adat}}$: *the current density of adatoms (section 7.5)*
$\mathbf{j}_s^{\text{advac}}$: *current density of advacancies (section 7.5)*
$\mathbf{j}_k$: *two-dimensional Fourier transform of the surface adatom current density*
$\mathbf{k} = (k_x, k_y)$: *a vector of the two-dimensional reciprocal space*
k_B: *the Boltzmann constant*
K: *bulk modulus (section 16.4)*
K: *kinetic coefficient defined by (12.6)*
L_x, L_y: *sample sizes in the* x *and* y *directions*
L: *linear size of a system*
L: *wavelength of a modulation (chapter 8)*
$m_{\alpha\gamma}$, m: *force dipole moment, eq. (15.2)*
M: *a point at the surface of a solid*
$\mathbf{n}$: *unit vector normal to a surface at a point, directed outward*
N: *number of particles in a physical system*
N: *quantity related to the size of the surface in section 2.6.*
$N(t)$: *cluster density (section 11.6)*
$p_{\alpha\gamma}$, p_0: *stress*
$p_j(\mathbf{r}, t)$: *probability that the j-th impurity is at* $\mathbf{r}$ *at time* t *(section 7.1)*
$\mathbf{q} = (q_x, q_y)$: *a vector of the two-dimensional reciprocal space*
r_1: *characteristic length (chapter 8)*
R: *radius of a terrace (section 8.5)*
R: *radius of curvature of a line or of a surface (chapter 10)*
$\mathbf{R}$, $\mathbf{r}$: *a point of the two-dimensional (chapter 1) or three-dimensional space*
dS: *surface element (chapter 2)*
S: *surface of a crystal (chapters 3 and 4)*
t: *time*
T: *temperature*
T_M: *melting temperature.*
T_R: *roughening transition temperature*
$u(\mathbf{r})$: *atomic displacement at point* $\mathbf{r}$
$\mathbf{v}$, $\mathbf{v}(\mathbf{n})$: *velocity of the surface of a crystal (chapter 5)*
v: *velocity of a step*
$v = V/N$: *volume per atom (chapter 2)*

v_g: *volume per atom in the vapour (chapter 2)*
v_ℓ: *volume per atom in the liquid (chapter 2).*
V: *volume*
δV: *volume element (section 16.1)*
W: *energy or free energy barrier*
W_0: *energy of a kink on a step (section 1.8)*
W_0: *energy barrier (section 15.7)*
w_0: *energy of a chemical bond (section 15.5)*
W_a: *adatom creation energy or free energy, see eq. (7.14a)*
W_{adv}: *see eq. (7.14b)*
W_1: *energy of a step per bond*
W_{sd}: *activation energy for surface diffusion (eq. (7.13) and section 11.8)*
$W_{int}(r)$: *interaction energy between defects at distance r (chapter 15)*
W_{coh}: *cohesive energy*
x, y, z: *coordinates of a point in the three-dimensional space.*
$x_s = \sqrt{D_s \tau_v} = 1/\kappa$: *average adatom diffusion length before desorption*
y': *derivative* dy/dx
$\tilde{z}$: *dynamical "critical exponent" for the correlation length (eq. 12.14)*
$z_q(t)$; *Fourier transform of* $z(\mathbf{r}, t)$
j_s, D_s, Λ_s, $\tilde{D}_s$: *section 8.4*
α, γ: *coordinates* x, y *or* z
α: *kinetic coefficient (eq. 7.8)*
α: *critical exponent for the spatial decay of the correlation function (eq. 12.18)*
β: *critical exponent for the temporal decay of the correlation function (eq. 12.18)*
$\beta = 1/(k_B T)$
γ: *free energy of a step per unit length or per atom (line tension)*
$\tilde{\gamma} \equiv \gamma(\theta) + d^2\gamma/d\theta^2$: *step stiffness (section 2.4)*
γ: *exponent defined in section 11.9*
$\mathbf{\Gamma}(\mathbf{r} - \mathbf{r}')$: *elastic Green function (eq. 15.10)*
Γ: *a step, in chapter 10*
δa: *misfit (section 15.4)*
$\delta \mathscr{F}$, $\delta \mu$, *etc.: increment of* $\mathscr{F}$, μ, *etc. from some reference value*
$\delta \Sigma$: *surface element (section 16.1)*
ΔF: *"supersaturation"* $F - \rho_0/\tau_v$
Δ' *and* Δ'': *kinetic coefficients defined in section 6.4*
ϵ: *Interaction energy between steps (section 1.7)*
$\epsilon_{\alpha\gamma}(\mathbf{r})$: *strain*
$\epsilon(t)$: *amplitude of a modulation in eq. (10.14) (called* η *in eq. (10.7))*
ζ: *Poisson ratio (eq. 6.12)*
ζ: *in section 10.6, a control parameter*
η: *amplitude of a modulation in eq. (10.7) (called* ϵ *in eq. (10.14))*

θ: value of an angle
$\kappa = 1/x_s$: inverse evaporation length of an adatom (eq. 6.6)
λ: capillary length (chapter 1)
λ: Lamé coefficient (chapters 15 and 6).
λ: kinetic coefficient in the KPZ equation (13.4)
λ: a constant (e.g. in section 4.2)
Λ_s, Λ: surface diffusion constant of advacancies (section 6.5)
μ: Lamé coefficient (chapters 15 and 16)
μ: chemical potential (elsewhere)
$v(t)$: number of sites visited by a diffusing adatom in time t (sections 6.3, 11.4)
v: kinetic coefficient defined by (12.6)
$\xi(t)$: correlation length (section 12.6)
(Π): tangent plane to the crystal surface (chapter 3)
$\rho(\mathbf{r},t)$, ρ_1, ρ_s: adatom density (occupied sites/surface sites)
ρ_n: surface density of clusters of n atoms (section 11.4).
ρ_0: equilibrium density of adatoms. (section 6.5)
$\bar{\rho}$: see section 6.3
$\rho_{\text{int}}(\mathbf{r},t)$: concentration of interstitials
σ_0: equilibrium density of advavancies (section 6.5)
σ, $\sigma(\mathbf{n})$: free energy of a surface per unit length or per atom (surface tension)
$\sigma(\mathbf{r})$: density of advacancies
Σ: Wulff's plot (section 3.2)
$\tilde{\sigma}$: surface rigidity, formula (2.5)
$1/\tau_v$: evaporation probability of an adatom per unit time (section 6.1)
$1/\tau_{\text{nuc}}$: nucleation rate of new terraces (section 11.2)
$1/\tau_k$: the rate of emission of "gradatoms" (problem 10.6)
$\phi(z_x, z_y)$: the projected free energy density per unit area (eq. 2.2)
$\varphi\left(\{\epsilon(\mathbf{r})\}\right)$: elastic free energy density (section 16.5)
φ_p: components of an eigenvector of the transfer matrix in Appendix C
Φ: Gibbs free energy or free enthalpy $\mathscr{F} + PV$ (chapters 2 and 16)
ω: decay rate of a modulation, see eq. (10.15)
Ω: atomic area, generally set equal to 1 in this book
Ω: Grand potential $\mathscr{F} + PV - \mu N$
$\Omega^{\xi\zeta}_{\alpha\gamma}$: elastic constants
$\tilde{\Omega}^{\alpha\gamma}_{RR'}$: "discretized elastic constants", eq. (16.25)

ℓ: distance between steps
ℓ_s: adatom diffusion length before nucleation of a terrace (section 11.1)
ℓ_s: length on which diffusion is able to heal the surface (chapter 12)

$\mathrm{d}\mathscr{A}$: elementary projected area $\mathrm{d}x\mathrm{d}y$ (section 16.8)
$\mathscr{A}$: total projected area L_xL_y

$\mathscr{F}$: *free energy*
$\mathscr{G}$: *Gibbs free energy* $\mathscr{G} = \mathscr{F} + PV$.
$\mathscr{N}$: *number of lattice sites on a surface in section 14.6*

$\simeq$: *nearly equal to (e.g.* $\sin x \simeq x$ *for small* x*)*
$\sim$: *proportional to (e.g.* $10x \sim x$*)*
$\approx$: *of the order of magnitude of (e.g.* $10 \approx 1$*)*

cot: *cotangent*
z_x, z_α: *partial derivative* $\partial z/\partial x$, $\partial z/\partial x_\alpha$
v_α, j_α, R_α: *the components of the vectors* $\mathbf{v}$, $\mathbf{j}$, $\mathbf{R}$
$\langle \chi \rangle$: *average value of a quantity* χ
$|z|$: *absolute value of* z
$\dot{z}$, $\dot{\rho}$, *etc.: derivative of* z, ρ, *etc., with respect to time*

$(hk\ell)$: *orientation of a crystal surface. E.g. the (001) face, the (111) orientation*
$[hk\ell]$: *orientation of a crystal axis. E.g.* $[110]$ *steps*
$\{hk\ell\}$: *set of crystal planes which are crystallographically equivalent. E.g. (cubic crystal): the* $\{001\}$ *orientations are the (001), (010) and (100) planes.*

1
Morphology of a crystal surface

Je ne me persuade pas aisément qu'Epicure, Platon et Pythagore nous aient donné pour argent comptant leurs Atomes, leurs Idées et leurs Nombres. Ils étaient trop sages pour établir leurs articles de foi de choses si incertaines et si débattables.

I can hardly convince myself that Epicurus, Plato and Pythagoras have sold us for sound their Atoms, their Ideas and their Numbers. They were too wise to found their belief on such uncertain and questionable things.

Montaigne (*Essais*, II, 12)

Twenty five centuries after Democritus, we are now able to see individual atoms, or at least small groups of them, through a variety of electron microscopy techniques. The visible atoms are of course those at the surface of a solid. Even motions at atomic scales can be observed.

It is thus possible to identify the fundamental elements of the morphology of a surface: terraces, steps, kinks, adatoms, advacancies.

Despite the fantastic progress in surface experimentation between 1970 and 1995, surface physics relies much on theory. The theory of two-dimensional systems (e.g. surfaces) is fascinating. The transition of a surface at equilibrium from its high-temperature, rough state, to its low-temperature, smooth state, has quite unusual features, e.g. no observable specific heat singularity. The roughness of an equilibrium surface is also extraordinarily weak: a few atomic distances over several centimetres.

1.1 A high-symmetry surface observed with a microscope

In 1986, the Nobel Prize for Physics was awarded to Binnig and Rohrer for the invention of scanning tunneling microscopy (STM). This technique (Binnig & Rohrer 1987) allows the observation of the atomic structure of crystal surfaces.

The observation is easier when the orientation of the surface is close to a high-symmetry direction: (001) or (111) in the case of a cubic crystal. Such orientations are mostly used in technological applications and will generally be considered in this book. For such orientations, the microscope 'sees' (Fig. 1.1) *terraces* separated by *steps* of atomic height. These steps are not straight; they contain straight parts separated by *kinks*. On the terraces one can see surface vacancies (or *advacancies)* resulting from

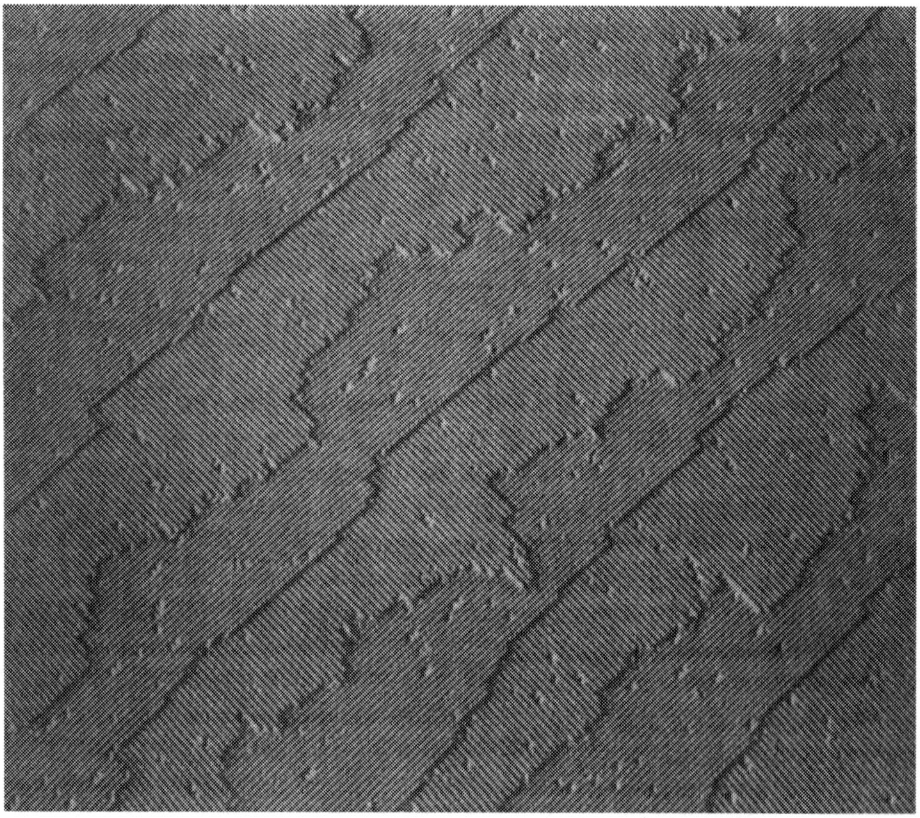

Fig. 1.1. [110] steps on a (001) vicinal silicon face. Terrace width: about 100 Å. Note the alternation of steps of different roughness, which will be explained in chapter 9. Experimental technique: scanning tunneling microscopy (STM) (Lagally *et al.* 1990, with the kind permission of the authors).

missing surface atoms. Certain features (e.g. the alternation of steps with different roughness) are special characteristics of the material which is observed, silicon in this case. They will be discussed in chapter 9.

Technical progress has been fast, and STMs exist that can now be used at very high temperatures (at present, the beginning of 1997, the limiting temperature seems to be around 1500 K). However, varying the temperature without upsetting the instrument is quite a delicate matter, and therefore the STM is not always adequate for the observation of collective motions on a surface. Indeed, the evolution of the morphology of a metal or semiconductor surface at room temperature (still the ideal temperature for most STMs) is very slow, as will be seen in chapter 8. Other methods of electron microscopy, e.g. reflection electron microscopy (REM) (Fig. 1.2) bridge this gap, and also allow the observation of larger scales. Fig. 1.2 is remarkable for the different lengthscales observed in 3 directions: the step height (perpendicular to the surface) is of order a few Ångstrøms, while the length of the steps in the image amounts to hundreds of Ångstrøms and the distance between the steps is of the order of one micron. Moreover, several steps can be seen at the same time, because the electron beam makes a very small angle with the surface

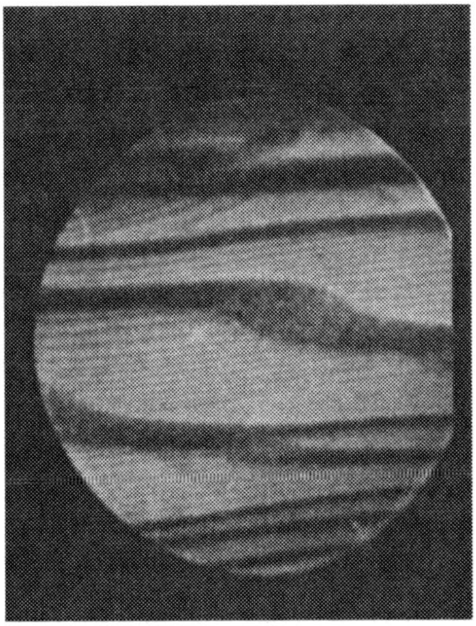

Fig. 1.2. Steps on the silicon (111) face during evaporation by Joule heating. Dark bands are macrosteps or step bunches, while thin lines are monolayer-high steps. The origin of the step bunches is discussed in chapter 9. Width: 1 μm (parallel to the steps). Depth: 35 μm (perpendicular to the steps). Experimental technique: reflection electron microscopy (Alfonso *et al.* 1992, with the kind permission of the authors).

(grazing incidence). The steps are seen to be smoother than on Fig. 1.1. This is partly due to the different scale and surface orientation, but it is also a consequence of the fact that Fig. 1.1 shows the surface of a growing crystal, and a growing surface is far from equilibrium, as will be seen in chapters 11 to 14.

Observation of crystal growth at the atomic scale is not always easy, especially when the crystal grows from the melt or from a solution. Even in growth from the vapour, it is difficult to control the thickness of the deposited atom layer. The type of growth used in the case of Fig. 1.1 was molecular beam epitaxy (MBE). An MBE apparatus is shown in chapter 17. The principle is simple: under ultra-high vacuum conditions, atoms (single or in molecules) are sent onto the surface, where they diffuse until they meet a step where they are incorporated. These diffusing atoms are called *adatoms.* Free adatom diffusion has clearly a meaning only on a high-symmetry terrace (between steps). Adatoms are in general not easily visible by STM, either because their number is too small or because they move too quickly to be resolved by the instrument. Possible instrument-induced perturbations should also be mentioned. A scanning tunneling microscope is essentially a metal tip taken very close to the surface, from which or to which electrons can flow by tunneling. This is a rather disturbing device for the surface, and an isolated atom has a great chance to be strongly perturbed. Isolated atoms can be seen, however, by another technique called field ion microscopy (FIM) (Ehrlich & Hudda 1966, Ehrlich 1977, Wang & Ehrlich 1991). Indeed, field ion microscopy allows the observation of the motion of individual adatoms.

Adatoms are already present at equilibrium. Their equilibrium density (number of occupied sites/number of surface sites) is, according to the Gibbs formula

$$\rho = \exp(-\beta W_a) \tag{1.1}$$

Table 1.1. *Values of some typical energies (in Kelvin) for four fcc metals (Stoltze 1994)*

	Ni	Cu	Ag	Au
Adatom energy $W_a(001)$	8700 K	5900 K	4200 K	3550 K
Step energy $W_1(001)$	2200 K	1450 K	1000 K	750 K
Kink energy W_0	1800 K	1250 K	950 K	800 K
Advacancy energy (001)	8400 K	5500 K	3850 K	2900 K
Adatom energy $W_a(111)$	11650 K	8300 K	6450 K	6600 K
Cohesive energy $-W_{coh}$	51600 K	40800 K	34400 K	44000 K

Fig. 1.3. a) The silicon (001) face after MBE growth at room temperature and annealing at 625 K. Note the terraces (or 'islands') and the steps. Scale: 800 × 560 Å. Experimental technique: STM (Lagally *et al.* 1990, with the kind permission of the authors). b) The silicon (001) face after MBE growth at room temperature, but without annealing. The elongated island shape indicates anisotropy in adatom sticking to steps. Scale: 400 × 400 Å. Experimental technique: STM (Lagally *et al.* 1990, with the kind permission of the authors).

where $1/\beta = k_B T$ and k_B is the Boltzmann constant. W_a is the (free) energy needed to extract an atom from a step and to make it an adatom.

It is one of the many parameters which are required for a quantitative description of surface kinetics. In this book, these parameters will be taken for granted. In principle, they can be obtained from a theory of the electronic structure, which can be *ab initio* (Gross 1990, Ruggerone *et al.* 1997) or make use of simple approximations such as 'tight binding', 'embedded atom', 'effective medium'... The reader will find details in textbooks (Desjonquères & Spanjaard 1993, Lannoo & Friedel 1991, Noguera 1995) and review articles (Stoltze 1994).

Typical values of W_a/k_B according to recent model-potential calculations (Stoltze 1994) are listed in Table 1.1. According to these data, there is at room temperature, on the (111) face of nickel, about one adatom per square centimetre, which is a very low density. On the (001) face of Cu, the proportion of lattice sites occupied by adatoms is about 10^{-9} at room temperature, which is still rather low. This proportion is considerably increased on a growing crystal, as will be seen in chapter 11.

The experimental images displayed in this chapter confirm much older conjectures. We shall not review the twenty five centuries which elapsed between Democritus and the Nobel Prize of 1986, but it is worth mentioning a fundamental article by Burton, Cabrera & Frank (1951) where crystal growth was described as a motion of steps at a time where direct observation of steps (and even less of moving steps) was impossible. This article will be discussed at length in chapter 6.

What happens if the growing crystal is limited by a perfectly oriented high-symmetry surface without steps? The answer is given by Fig. 1.3: new terraces appear, at least temporarily, and therefore steps appear. In practice, a crystal surface always contains steps for two reasons: first, because it is not possible to cut it perfectly straight, and second because a crystal always contains dislocations. A screw dislocation through a

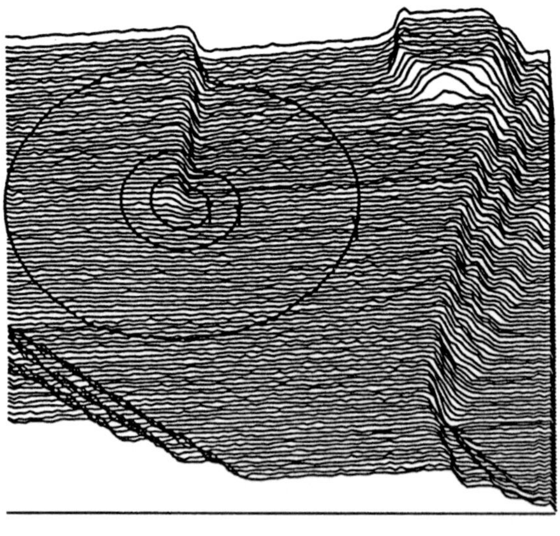

Fig. 1.4. Dislocation and steps on Ag (111) observed by STM (Wolf & Ibach 1991, with the kind permission of the authors.)

surface is necessarily, for geometric reasons, the origin of a step (Fig. 1.4). However, dislocation-free surfaces on macroscopic scales can nowadays be grown for certain materials, e.g. the technologically essential element silicon.

1.2 *In situ* microscopy and diffraction

Now that we have a surface microscope, we would like to look at the surface of a growing crystal as it grows–or, as the jargon demands, *in situ*. It is not that easy, however. On one hand, it is hard to imagine introducing any microscope into a liquid metal or silicon melt! On the other, even though the use of a microscope is in principle easier during vapour phase or ultra-high vacuum growth, each different technique has some specific limitations. Reflection electron microscopy can be used at any temperature, but the images suffer from the strong foreshortening effect seen in Fig. 1.2, which limits the ability of seeing small details. Low-energy electron microscopy (LEEM) does not have this problem, and it has been used for *in situ* studies of growth (see Bauer 1992 for a review), but its resolution is poor (5 to 15 nm) compared to STM. The latter is limited to not-too-high temperatures by technical difficulties which will be certainly overcome in the future, and still suffers from a long–but steadily decreasing–image acquisition time. Some *in situ* observation of growth has started showing up in the literature (Voigtländer & Zinner 1993).

More appropriate for *in situ* observations than microscopy is still diffraction, i.e. scattering of electrons, X-rays or atoms. In a scattering or diffraction experiment (we shall use both words with essentially the same meaning), one measures the scattered radiation as a function of the scattering angle. The analysis as a function of the energy is essential in certain cases, but we shall not care very much about it in this book. The essential difference between microscopy and diffraction is that the diffractionist works in reciprocal space rather than in real space. For a good, three-dimensional crystal, the diffraction spectrum can be interpreted by any student. In the case of a reasonably smooth crystal surface, i.e. a two-dimensional object, the task is already harder. And if the number of imperfections becomes very large, the interpretation is very difficult. On the other hand, pictures from a microscope are understandable (after some image processing, sure...) by the layman, and can in principle be obtained even for a non-crystalline material.

1.3 Step free energy and thermal roughness of a surface

At zero temperature, a surface at equilibrium should contain no step. At low temperature, there are a few adatoms as seen above; there are fewer pairs of adatoms, whose number (or better, whose density) may be obtained by a Gibbs formula similar to (1.1); and smaller yet is the density of larger clusters. Atom clusters are closed terraces bounded by steps. The step density (total step length per unit surface) increases with temperature.

This increase is not easily seen directly by microscopy. Indeed, most of microscopic methods work best at low temperature, when the surface does not easily attain thermal equilibrium. The atom scattering or X-ray diffraction signal does exhibit a dramatic broadening when temperature is increased (Fig. 1.5). Above some temperature, the lineshape, which is lorentzian at low temperature, undergoes a qualitative change. One can for instance measure the 'specular' reflection, i.e. that whose reflection angle almost equals the incidence angle (*speculum* is the Latin for mirror).

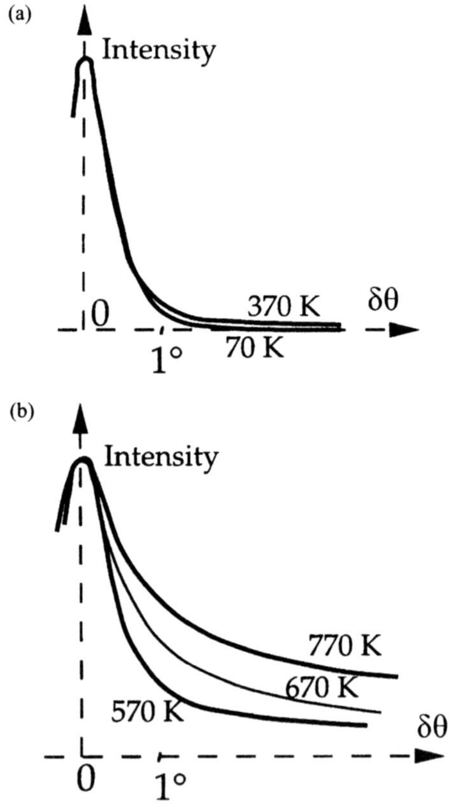

Fig. 1.5. Atom scattering lineshape at temperatures (a) below and (b) above T_R =380 K in Cu(115), after Fabre *et al.* (1987b). The reflected intensity is plotted vs. the deviation $\delta\theta$ with respect to the reflection angle corresponding to the ideally smooth surface.

The specular peak (as well as the Bragg peaks) is narrow for a smooth surface while a rough surface scatters radiation in all directions.

This type of experiment is a quantitative version of everyday observation. Galileo (1632) was perhaps not the first who noticed the importance of surface roughness in light scattering, but he was presumably the first one who realized that light was scattered rather than lost. His problem was to understand the brightness of the moon, which he compared to a wall scattering sunlight:

You see the difference between the reflections occuring on the respective surfaces, that of the wall and that of the mirror: ... Look how the reflection from the wall scatters to all parts of the opposite wall, while that from the mirror goes to a single part, not larger than the mirror itself... If you want to understand all that, you should notice that, for a surface, to be rough means the same thing as to consist of innumerable small surfaces of innumerable orientations, and it necessarily occurs that, among them, many have the appropriate orientation to direct the reflected beams to this place, and many to that place.

The interpretation of diffraction patterns from a hot surface is difficult (Blatter 1984, Levi 1984, Armand & Manson 1988) because the effect of atomic vibrations adds to roughness to broaden the reflected beam. It is however clear, from a quantitative analysis, that the total step length is greatly increased by heating.

The reason is basically the following. Even if the step energy per unit length, W_1, does not change much, its entropy increases, so that the free energy decreases. In order to calculate it, we consider (Fig. 1.6) on the (001)

Fig. 1.6. A step on a surface with a square symmetry, e.g. the (001) face of an fcc metal. For this orientation, the line tension is given by (1.2).

face of a cubic crystal a zig-zag step, whose average direction makes an angle of 45° with bond directions. This step orientation is chosen because it allows a simple, though approximate, calculation. The only configurations which will be considered are those which result from random walks going from the left to the right, each step of the random walker being parallel to a lattice bond, and backward steps being forbidden. This definition may be complicated, but what it means is clearly suggested by the figure. If the width of the system in the direction of the walk is uniformly equal to the bond length multiplied by $L\sqrt{2}$, all configurations have the same energy $2LW_1$. Since there are 2^{2L} configurations, the entropy is $2L \ln 2$ and the free energy per bond is

$$\gamma = W_1 - k_B T \ln 2 . \tag{1.2}$$

The free energy per unit length γ/a is called the line tension of steps. Since a, the lattice parameter, is generally chosen as the length unit in this book, the term line tension will be often employed for γ itself.

When γ is positive, one has to provide mechanical work to introduce a step into the surface. If the total step free energy $L\gamma$ becomes negative, thermodynamics tells us that one should provide mechanical work to remove steps from the surface. Thus, equation (1.2) tells us that the surface undergoes a transition at a temperature T_R approximately given by

$$T_R \approx \frac{W_1}{k_B \ln 2} . \tag{1.3}$$

This transition is called the roughening transition.

1.4 The roughening transition

As seen above, the roughening transition temperature may be defined as the temperature at which the line tension of steps vanishes. According to the experiment (Fig. 1.7) the line tension does vanish at some temperature. The linear temperature dependence predicted by formula (1.2) is in pretty good agreement with experimental observation, at least far from T_R. Near T_R, the experimental curve bends away from the straight line: one can wonder whether it is an instrumental effect or a fundamental one. As will be seen in the next section, it is a fundamental effect. But before discussing that, we would like to make three remarks.

1) The roughening transition temperature depends on the surface orientation, i.e. it is different for a (111) and for a (1,1,19) surface. This point will be addressed in section 1.7.

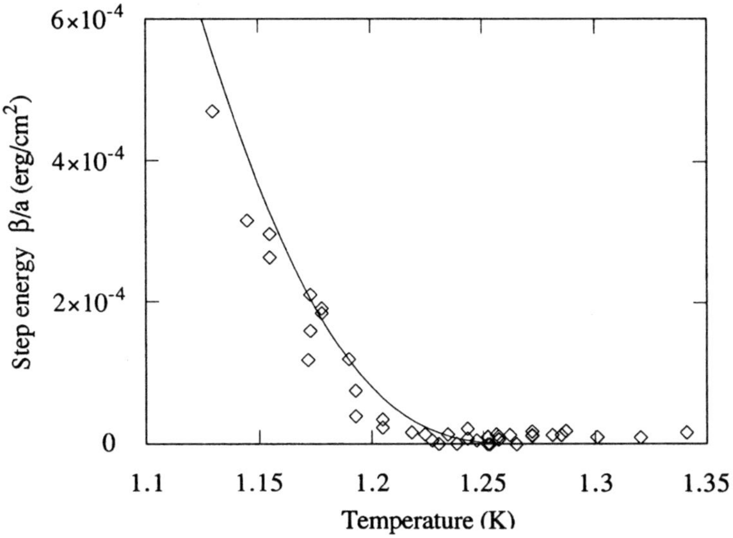

Fig. 1.7. Line tension of a step as a function of temperature, as measured by Gallet *et al.* in helium (Gallet *et al.* 1987, with the kind permission of the authors).

2) For a given surface orientation, the step line tension vanishes at the same temperature for all step orientations, so that T_R is independent of the step orientation. This can easily be proved in the case of a square or hexagonal symmetry. If a step of orientation x has a vanishing free energy, the free energy of a step of any orientation y deduced from x by a lattice symmetry operation also vanishes. But then, a step of average orientation ζ may be obtained as a succession of two consecutive steps of orientations x and y, and has therefore a vanishing free energy, apart from the energy of the kink between the two pieces, which is negligible for a long step.
3) Formula (1.2) suggests that the step line tension becomes negative for $T > T_R$. Actually, it is not so, if a step is defined in a model-independent way. An appropriate definition is to consider a rectangular sample of sizes L_x and L_y in two orthogonal directions x and y, and to fix the surface height at the two ends of the sample at $z(0)$ and $z(L_x) = z(0) + \delta z$, respectively. The step free energy is the difference between the free energies of the system for $\delta z = 0$ and for $\delta z = 1$ in atomic layer units. With this definition, it can be proved that the line tension is identically zero for $T > T_R$. Actually, the very concept of a step is not very meaningful for $T > T_R$. In this case, if $\delta z = 1$ one can say that there is a step in the system, but if one looks at the surface one cannot say where it is.

1.5 Smooth and rough surfaces

The roughening transition is much more complicated than suggested by the discussion of section 1.3. Indeed, thermally excited steps are not isolated objects as suggested by Fig. 1.6, they are closed loops. In this section, we shall try to give an idea of what a rough surface really is. The reader will find more details about the roughening transition in the monographies by Balibar & Castaing (1985), Van Beijeren & Nolden (1987), Lapujoulade (1994), Nozières (1991) and Weeks (1980).

As seen in the previous section, the concept of step is not useful above T_R. It is therefore appropriate to characterize roughness in an alternative way. Consider an infinite surface of average orientation perpendicular to the z axis. Let (x, y, z) be a point of the surface. The 'height' z will be assumed to be a one-valued function of x and y, so that 'overhangs' are excluded. Let $\mathbf{R} = (x, y)$ be a point of the two-dimensional (x, y) space. We define the *height-height correlation function*

$$G(\mathbf{R}) \equiv \left\langle [z(\mathbf{r}) - z(\mathbf{r} + \mathbf{R})]^2 \right\rangle . \tag{1.4}$$

The interest of this function is that it has, if gravity is neglected, two qualitatively different behaviours above and below T_R:

$$\lim_{R \to \infty} G(R) = \begin{cases} \text{finite value for} \quad T < T_R & (1.5a) \\ \infty \qquad\qquad\quad \text{for} \quad T > T_R . & (1.5b) \end{cases}$$

The finiteness of $G(R)$ at low temperature, in agreement with (1.5a), results from a low-temperature expansion, whose lowest order will be derived below. At very low temperature, the surface is flat, except for very few adatoms, whose density is, as in section 1.2, equal to $\exp(-\beta W_a)$. However, there can be advacancies, whose energy is not very far from W_a in usual metals as seen from Table 1.1. The probability that $z(\mathbf{r})$ or $z(\mathbf{r} + \mathbf{R})$ is 1 or -1 is therefore $2 \exp(-\beta W_a)$. The probability that both $z(\mathbf{r})$ and $z(\mathbf{r} + \mathbf{R})$ are equal to 0 is $1 - 2\exp(-\beta W_a)$ and the probability that both $z(\mathbf{r})$ and $z(\mathbf{r} + \mathbf{R})$ are different from 0 and ± 1 is negligible. The first term of the low-temperature expansion is therefore

$$G(\mathbf{R}) \equiv \left\langle [z(\mathbf{r}) - z(\mathbf{r} + \mathbf{R})]^2 \right\rangle \simeq 2 \exp(-\beta W_a) \quad (R \neq 0). \tag{1.6}$$

Of course, $G(0) = 0$.

Thus, we have proved (1.5a). It is more difficult to prove (1.5b), and only a plausibility argument will be given. It is indeed reasonable to assume that, at sufficiently high temperature, say $k_B T \gg W_1$, the discreteness of the crystal lattice becomes negligible, so that the surface height $z(x, y)$ may be regarded as a differentiable function of x and y, as it is in a liquid. Thus, we just forget that we have a crystal, and we write the surface energy as if it were a liquid. The surface free energy $\mathscr{F}_{\text{surf}}$ of a liquid is simply proportional to the surface area, and the proportionality coefficient σ is called the surface tension. Writing $z_\alpha = \partial z/\partial x_\alpha$,

$$\mathscr{F}_{\text{surf}} = \sigma \int\int \mathrm{d}x\mathrm{d}y\sqrt{1 + z_x^2 + z_y^2}$$

or, for small fluctuations,

$$\mathscr{F}_{\text{surf}} = \text{Const} + \frac{\sigma}{2}\int\int \mathrm{d}x\mathrm{d}y \left(z_x^2 + z_y^2\right) . \tag{1.7a}$$

However, in our world subject to gravity, thermal fluctuations of the surface have an additional energy resulting from gravity. The effect of gravity might easily be taken into account in the treatment below T_R and would be found to be irrelevant. It is not so above the transition temperature as will be seen shortly. The energy of a column of matter of cross section dxdy, whose ends are at heights z_1 and z, is

$$\rho g \mathrm{d}x\mathrm{d}y \int_{z_1}^{z} \zeta \mathrm{d}\zeta = \rho g \mathrm{d}x\mathrm{d}y(z^2 - z_1^2)/2$$

where ρ is the specific mass and g the gravity acceleration. The term containing z_1 is constant and will be omitted. The energy excess associated with surface shape fluctuations and resulting from both gravity and surface tension is

$$\delta\mathscr{F} = -\frac{1}{2}\int\int \mathrm{d}x\mathrm{d}y \left[\rho g z^2 + \sigma\left(z_x^2 + z_y^2\right)\right] . \tag{1.7b}$$

This quadratic form is readily diagonalized if one introduces the Fourier transform of z of wavevector $\mathbf{q} = (q_x, q_y)$:

$$z_q = \frac{1}{L_x L_y}\int \mathrm{d}x\mathrm{d}y z(x,y)\exp\left(iq_x x + iq_y y\right) .$$

One obtains then

$$\delta\mathscr{F} = \frac{1}{2}\sum_q \left[\rho g + \tilde{\sigma}(q_x^2 + q_y^2)\right] \mid z_q^2 \mid , \tag{1.8}$$

where $\tilde{\sigma} = \sigma$ for a liquid. The notation $\tilde{\sigma}$ has been introduced because, in the case of a crystal, two different quantities appear, as will be seen in chapter 2. The variance of z_q is easily deduced, and the result (the so-called equipartition theorem) is

$$\left\langle \mid z_q^2 \mid \right\rangle = \frac{k_B T}{\rho g + \tilde{\sigma}(q_x^2 + q_y^2)} . \tag{1.9}$$

Inverting the Fourier transform, the correlation function (1.4) is readily obtained as

$$\left\langle [z(\mathbf{r}) - z(\mathbf{r} + \mathbf{R})]^2 \right\rangle = 2k_B T \int_0^{1/a} \mathrm{d}q_x \mathrm{d}q_y \frac{1 - \cos(\mathbf{q}\cdot\mathbf{R})}{\rho g + \tilde{\sigma}(q_x^2 + q_y^2)} . \tag{1.10}$$

An order of magnitude may be obtained if the lower limit of integration is replaced by an appropriate cutoff, below which the numerator is almost zero. This allows us to replace the cosine by its average value 0, and one obtains

$$\left\langle [z(\mathbf{r}) - z(\mathbf{r}+\mathbf{R})]^2 \right\rangle \approx \int_{1/R}^{1/a} 2\pi q \mathrm{d}q \frac{2k_B T}{\rho g + \tilde{\sigma} q^2} . \tag{1.11}$$

If gravity is neglected, which implies $R \ll \sqrt{\tilde{\sigma}/\rho g}$, (1.11) diverges at large R as:

$$\left\langle [z(\mathbf{r}) - z(\mathbf{r}+\mathbf{R})]^2 \right\rangle \approx (4\pi k_B T/\tilde{\sigma}) \ln R . \tag{1.12}$$

This proves (1.5b). A surface is called *rough* if $\left\langle [z(\mathbf{r}) - z(\mathbf{r}+\mathbf{R})]^2 \right\rangle$ diverges as in (1.12), and *smooth* if there is no divergence, as in (1.6).

Previously, we defined the roughening transition in terms of the vanishing of the step free energy γ. We should worry about the equivalence of the two definitions. In fact, the proof of (1.12) relies on the use of (1.7b). If the step line tension is positive, equation (1.7b) cannot be true. Indeed, since the number of steps per unit length is given by $|z_x|$, we must add to the surface free energy (1.7b) a contribution of the form

$$\int\int \mathrm{d}x\mathrm{d}y \gamma |z_x| ,$$

assuming for simplicity that $g = z_y = 0$. The surface free energy is thus non-analytic in contrast with (1.7b). If $\gamma > 0$, the formation of a terrace of size L requires a free energy which diverges with L. Since this is forbidden, the function (1.4) is finite for $R = \infty$. Therefore, the roughening transition can be defined (at least on a high-symmetry surface) either by (1.5) or by the vanishing of γ.

From (1.11), gravity is seen to kill the roughening transition. It is also seen from (1.11) that gravity is negligible if the length $1/q$ of interest is shorter than the *capillary length*

$$\lambda = \sqrt{\tilde{\sigma}/\rho g} . \tag{1.13}$$

The order of magnitude of $\tilde{\sigma}$ is typically the energy of a chemical bond, i.e. 1 eV per atom. The resulting value of λ is a few centimetres, i.e. much larger than the distance over which equilibrium can be reached at a crystal surface. For this reason, gravity will be neglected everywhere else in this book. Note that in the case of water, λ depends on temperature but is of order 0.5 cm as known from everyday life (see next chapter, Fig. 2.1).

1.6 The SOS model and other models

If one wishes to go beyond (1.6) and to determine the next terms of the low-temperature expansion, one has to know, in addition to the energy of a single adatom, the energy of a pair of adatoms, of a triplet, etc. These energies are not known with precision. On the other hand, the qualitative properties of the system should not depend much on those details. Therefore, the theory is often done for simplified models. One of them, called the 'SOS' (solid on solid) model is described in Fig. 1.8. It belongs to the broad class of 'broken bond models'. In such models, atoms are supposed to be hard spheres and the energy (counted from the completely dissociated state) is assumed to be proportional to the number of bonds between nearest neighbours. If the energy per atom is counted from that of the ideal, infinite crystal, then the energy of a particular configuration is proportional to the number of bonds which should be broken to obtain this configuration from the infinite crystal. The proportionality coefficient is $2W_1$ where W_1 is the step energy per atom introduced in section 1.3. The factor 2 appears because, when breaking an atomic layer into two pieces, one makes two steps.

In a broken bond model, the quantities W_a and W_1 introduced previously in this chapter are linked, on the (001) face of a cubic crystal, by the obvious relation

$$W_a = 4W_1 \,. \tag{1.14}$$

According to Stoltze's review article (1994) this relation is satisfied with an accuracy of order 1% by calculated values for Cu and Ni, while the accuracy is not so good for Au. Other predictions of broken bond models

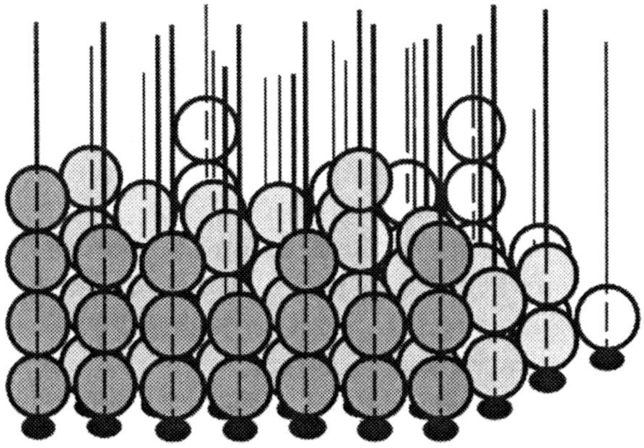

Fig. 1.8. The SOS (solid on solid) model. The adatom positions are fixed on a rigid lattice, and the only degree of freedom is the presence or the absence of an atom, so that there are no overhangs. Each bond has an energy $\epsilon = 2W_1$.

do not fit reality so well. For instance, in the broken bond model of an fcc crystal, the cohesive energy per atom (energy needed to break the material into invidual atoms) is 6 times the bond energy $2W_1$, so that its value is $12W_1$ in broken bond models. The actual ratio predicted by more elaborate theories (Stoltze 1994) and to some extent confirmed by experiment, is much larger than 12 (23 for Ni, 28 for Cu, 35 for Ag, 58 for Au). Another prediction of the broken bond model is that the surface energy per atom (1.14) of a (001) face is $4W_1$. More sophisticated approaches (Stoltze 1994) agree with this simple result within 13%: an astonishingly good success for the broken bond model.

As a general rule, the energy of defects (adatoms, advacancies, steps or the surface itself) is much lower than the value expected from the broken bond model. This is due to the fact that the solid has many continuous degrees of freedom which can be used to reduce the energy. This reduction, however, is often spectacular.

The SOS model (Fig. 1.8) is a broken bond model on a simple cubic lattice. A related model is the discrete Gaussian model defined in appendix A.

Another model, generally used for the theoretical treatment of the roughening transition, employs the free energy (1.7b) modified by an additional term $\eta \cos(2\pi z)$ which mimics the crystal periodicity.

1.7 Roughening transition of a vicinal crystal surface

Let us come back to reality and ask the essential question, whether the roughening temperature of a high-symmetry surface occurs below the melting point T_M. If one accepts the crude estimate (1.3) and the theoretical evaluations of Stoltze (1994) the answer is negative for most materials. This means that the roughening temperature occurs when the crystal is in contact with its melt. In the case of semiconductors, the situation may be further complicated by elastic mechanisms, as will be seen in chapter 15. All this makes experimentation difficult for any material except helium, studied by Balibar *et al.* (1993).

However, experimentation is possible on stepped or 'vicinal' surfaces (Fig. 1.9). A *vicinal* (001) surface, for instance, has an orientation in the *vicinity* of the (001) orientation. We can rightfully wonder what is the roughening temperature for such surfaces.

An example are the stepped (11 2n+1) faces of metals: copper has been studied by Fabre *et al.* (1986, 1987a,b) and Ernst *et al.* (1995) using atom diffraction, and using STM by Giesen-Seibert *et al.* (1993) and Girard *et al.* (1994); nickel has been investigated with X-ray diffraction by Robinson *et al.* (1990). These surfaces can be considered as a succession of (001)

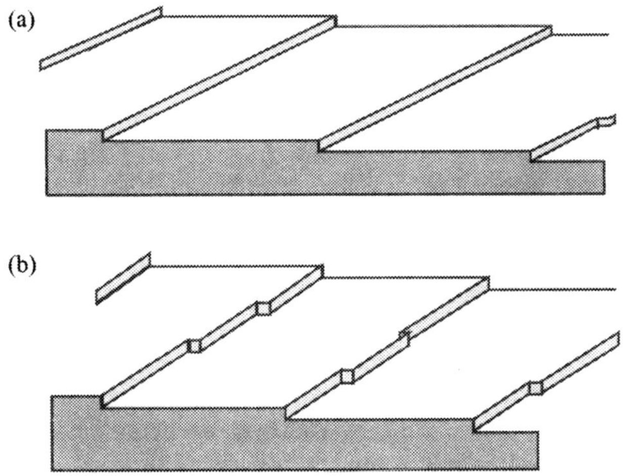

Fig. 1.9. Schematic representation of a stepped (or 'vicinal') surface below (a) and above (b) T_R.

terraces separated by steps (Fig. 1.9). These steps would like to be straight, since each kink costs an energy W_0. On the other hand, the steps would like to be equidistant. If they are closer or farther than the average by one atomic distance, this costs an energy ϵ for each atom of the step. This energy should be small if the average distance between steps is large. The kink energy W_0 is equal to W_1 in the broken bond model, and actually not very different even within the more sophisticated evaluations of Stoltze's (1994) shown in Table 1.1. On the other hand, it is larger than ϵ.

It will now be shown that, if ϵ is small enough, a vicinal surface is rough at fairly low temperatures $T < T_M$, in contrast with high-symmetry surfaces. The average distance between kinks is about $\exp(\beta W_0)$ atomic distances. In other words, and roughly speaking, the independent elements of a step are not atoms, but step pieces of length $\exp(\beta W_0)$ atomic distances. The interaction energy between these pieces is of order $\epsilon \exp(\beta W_0)$. The roughening transition occurs when this energy is of the same order as the temperature T. The roughening transition temperature is therefore given by

$$k_B T_R \approx \epsilon \exp\left(\frac{W_0}{k_B T_R}\right) . \qquad (1.15)$$

This hand-waving argument can be confirmed by an exact calculation on a special model (Villain *et al.* 1985).

1.8 The roughening transition: a very weak transition

In fact, thermal roughness is very 'weak' (see problem 1.1). We mean by that that the surface undulation arising from thermal fluctuations is very shallow in comparison with the waves produced by wind on a pool. This is because the logarithm diverges very slowly in (1.12).

Let us try to imagine people living in a d-dimensional space with d slightly larger than 3. The dimension of a surface would be $d - 1$, in the integrand in (1.11) they would replace qdq by $q^{d-2}dq$, and the integral would no longer diverge! In more than 3 dimensions, there is no phase transition, and d-dimensional scientists would say that a surface at equilibrium is always smooth. On the contrary, in less than 3 dimensions, the divergence of $G(R)$ is faster than a logarithm (see problem 1.2).

It is conceivable that a transition which is absent above 3 dimensions is very weak at the critical dimension 3. We shall see later what this means, but we first want to see how the roughening transition can be studied theoretically. The standard method to study phase transitions is the so-called renormalization group technique. This method was first used by high-energy physicists, and the idea to apply it to phase transitions was rather natural. It was the merit of K.G. Wilson to put this idea into a simple form (Wilson 1983). Wilson's articles generated a cascade of discoveries in the 1970s, and were rewarded by the Nobel prize in 1982. The renormalization group method is outlined in appendix B in the case of the roughening transition, and in appendix J in another instance. In appendix B, we use the equivalence of a model having a roughening transition (the discrete Gaussian model) with a two-dimensional electrolyte. This equivalence (Chui & Weeks 1976) is derived in appendix A. The two-dimensional electrolyte has a transition, whose renormalization group treatment was done for the first time in a celebrated paper by Kosterlitz & Thouless (1973). In appendix B, we present a simpler treatment due to A.P. Young.

We shall only mention two results of the renormalization group treatment:

i) $G(R)$ diverges at T_R as

$$G(\mathbf{R}) \equiv \left\langle [z(\mathbf{r}) - z(\mathbf{r}+\mathbf{R})]^2 \right\rangle \to \frac{2a^2}{\pi^2} \ln R \,. \tag{1.16}$$

This is formula (1.12), but with a well-defined value of the prefactor, in which we have explicitated the lattice constant a (usually taken equal to 1 in this book).

ii) The specific heat does not diverge, and all its derivatives with respect to T are continuous, in contrast with usual transitions (e.g. at the Curie temperature of a ferromagnet). The specific heat has therefore no observable singularity. It has, however a singularity called an essential singularity–a seemingly strange name (coined by mathematicians, probably inspired by philosophic language) for an invisible singularity!

As a matter of fact, the surface specific heat would not be easy to measure, since the surface is but a very small part of a solid. It is easier–at least in some instances–to measure the step line tension γ, which also possesses an essential singularity:

$$\gamma \sim \exp\left[-\sqrt{\frac{T_1}{T_R - T}}\right]$$

where T_1 is a constant. This theoretical result (Nozières & Gallet 1987) can be compared with the experimental curve of Fig. 1.7.

The weakness of this singularity justifies the title of this section.

It is also remarkable that the local aspect of a surface does not change very much when crossing the phase transition (Fig. 1.10).

Then, if there is no observable singularity of the specific heat, it is usually hard to measure the step free energy and if the aspect of the surface does not change much, how is it possible to determine the roughening transition temperature T_R? It is indeed a difficult problem. The usual method is to determine the temperature at which (1.16) is satisfied. An unfortunate situation for an experimentalist, who has to rely entirely on a theoretical formula!

The derivation of formula (1.16) disregards a possibly important feature of solid-fluid interfaces, namely the fact that hills and valleys have not the same energy. Symmetry is known to have a great importance in the theory of phase transition, and one should be particularly careful for a very weak transition as the roughening transition. Indeed, according to Rys (1986) in a model without hill-valley symmetry, the transition is killed and the surface is always smooth. According to Den Nijs (1990) Rys's result is incorrect. Rys considered a special model in which steps are not allowed to meet. He claimed that such a model is never rough if the hill-valley symmetry does not exist. This result is correct although the proof is difficult and only accessible to specialists. Den Nijs claimed that, even in the presence of hill-valley symmetry, Rys's model cannot be rougher than the critical roughness. In other words, at sufficiently high temperature, according to Den Nijs, the correlation function (1.4) has a logarithmic divergence corresponding to (1.12), but the multiplying factor is independent of temperature and equal to the universal value which it has, in well-behaved models, only at the roughening temperature. This value turns out to be $2/\pi^2$, so that

$$G(R) \sim \frac{2}{\pi^2} \ln R \,.$$

Den Nijs concludes plausibly that, since the system can only be critically rough without hill-valley symmetry, it cannot be rough at all with hill-

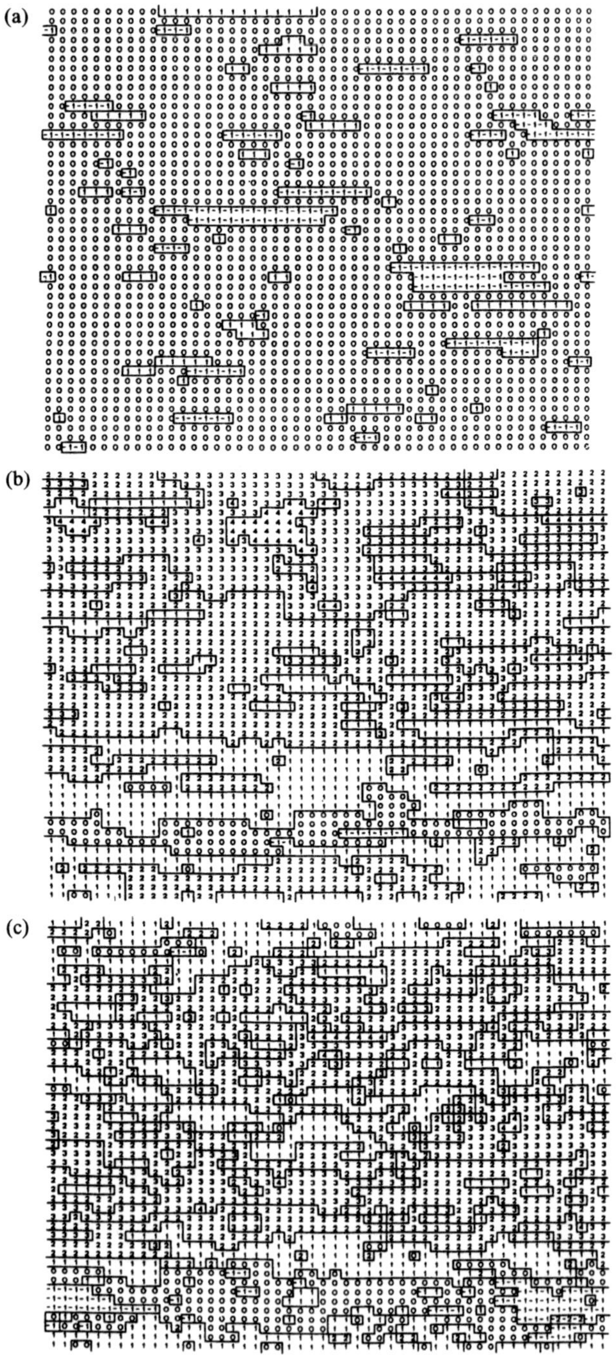

Fig. 1.10. Roughness of an anisotropic surface, from simulations by Lapujoulade *et al.* (1990). The model is an anisotropic SOS model equivalent to that of section 1.7. The value of the ratio of the parameters ϵ and W_0 is $\epsilon/W_0 = 0.1$. The lines correspond to steps and the numbers are the surface heights. a) $T = 0.8T_R$; b) $T = T_R$; c) $T = 1.4T_R$.

valley symmetry. Den Nijs's proof that the system can only be critically rough is not less difficult than Rys's paper (see problem 1.3).

Another unsolved problem is the roughening of a reconstructed surface (Vilfan & Villain 1991, Den Nijs 1991). In that case the roughening transition is accompanied by the disappearance of the reconstruction, which is usually a 'strong' transition, i.e. with a true specific heat anomaly. According to numerical calculations of Den Nijs (1991) the observable specific heat anomaly persists when there is also roughening, but this plausible result is not supported by analytical arguments.

1.9 Conclusion

In this introductory chapter, we have enumerated the fundamental elements (steps, adatoms, advacancies...) which contribute to the morphology of a surface at the microscopic level. Then we have seen how their interaction leads to a phase transition in an infinite, planar surface (or solid-fluid interface) at equilibrium.

We have mainly discussed the case of a metal, because metals have a simpler crystal structure (one atom per unit cell!). The case of semiconductors, which generally have the crystal structure of diamond, is not less important. Chapter 9 will be devoted to this point.

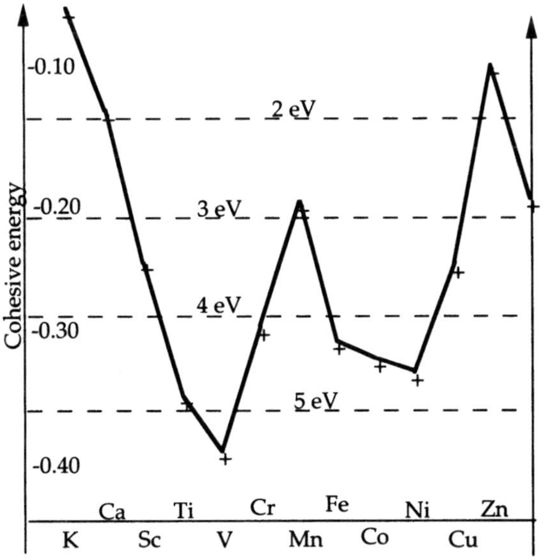

Fig. 1.11. Experimental cohesive energy of some common metals. It is the energy required to separate all the atoms of a solid and take them to infinity. The values on the vertical axis are in Rydberg per atom.

However, crystals are not infinite. The equilibrium shape of a finite crystal will be discussed in chapter 3.

It turns out that crystal surfaces are generally not in equilibrium, except on very small scales. We will therefore investigate the shape of a growing or evaporating crystal (chapters 4, 6 and 8), the roughness of a growing crystal (chapters 11 to 14), its instabilities (chapter 10).

In order to do that, it is often useful to replace the microscopic description in terms of steps and kinks by a thermodynamic description. This will be done in the next chapter.

Thermodynamics introduces a limited number of phenomenological parameters. This is an advantage because most of the microscopic quantities are only known from not quite reliable numerical computations (Table 1.1). Few microscopic parameters have been experimentally measured. The most important one is the cohesive energy (Fig. 1.11).

Problems

1.1. Calculate the roughness function $G(R)$ at 2000 K assuming $\tilde{\sigma} = 0.5$ eV. Consider the values $R = 1$ µm, 1 mm, 1 m, 1 km.

1.2. Replacing the integrand $q\mathrm{d}q$ in (1.11) by $q^{d-2}\mathrm{d}q$, prove that $G(R)$ would diverge at infinity as a power of R in a space of dimension $d < 3$.

1.3. Consider Rys's model introduced in section 1.8, defined as follows on a square lattice. i) All sites have an integer height. ii) If a site has height h, its neighbours and its next nearest neighbours can only have height h, $h+1$ or $h-1$. iii) Each pair of neighbours with different heights costs an energy w_0. Consider then a stripe of $N_x \times N_y$ sites with $N_y \gg N_x$ and $h(0,0) = 0$. Let $Z^+(1)$ be the partial partition function of the states satisfying the boundary condition $h(x,y) = h(x+N_x, y) + 1$ and $Z^-(1)$ the partial partition function of the states satisfying the condition $h(x,y) = -h(x+N_x, y) + 1$. Prove that $Z^+(1) = Z^-(1)$. This property is a step of Den Nijs's difficult proof that Rys's model cannot be more than critically rough. Den Nijs associates this 'hidden symmetry' with the 'hard-core interaction between steps', i.e. the fact that steps cannot have common points. Discuss the case of Van Beijeren's model (Van Beijeren 1977) which has a roughening transition of the standard type although steps cannot have common points either.

1.4 A Pb atom is mobile in a finite area of a Si(111) crystal surface, say a triangle of side about 3 nm. Discuss qualitatively the STM picture obtained when this atom is alone and when it is paired with another Pb atom which reduces its mobility (Gómez-Rodríguez *et al.* 1996).

2

Surface free energy, step free energy, and chemical potential

No pienso yo que la cultura, y mucho menos la sabiduria, haya de ser necesariamente alegre y cosa de juego.

I do not think that culture, and even less science, have to be necessarily matters of fun.

Antonio Machado (*Juan de Mairena*)

According to thermodynamics, physical properties can be deduced from the knowledge of the free energy. In this chapter, the surface free energy is introduced in a simplified way (disregarding, in particular, elasticity, for the moment). The morphology of a surface at equilibrium, in particular its roughness, may be related to the behaviour of the surface free energy per unit area (or 'surface tension') as a function of the orientation. The chemical potential which is essentially the free energy per atom, will also prove to be an useful concept, since it is the same everywhere at equilibrium.

The surface tension of a crystal depends on the surface orientation. As a function of the orientation, it may show singularities which are related to roughness and faceting.

When steps of atomic height are present on a surface, it is also of interest to define their free energy per unit length, or *step line tension*. This concept is not adequate when steps are too close to one another, and it is necessary in that case to take into account the interactions between steps. As seen in the preceding chapter, the line tension of steps vanishes at the roughening transition, if the latter occurs below the melting point.

2.1 Surface tension

In order to create a surface, one has to break chemical bonds, and this costs energy. At finite temperature, it is more appropriate to consider the free energy. It is easy to give a precise definition of the surface free energy: to break a crystal along a plane, it requires a work W. If S is the area of the crystal section, the surface free energy for unit area or *surface tension* is $\sigma = W/(2S)$ (because two surfaces are created this way).

The surface tension of a fluid is accessible to everyday experience through rather striking phenomena (Fig. 2.1). A water drop does not spread on glass, despite gravity, because spreading would increase its surface. For the same reason a water drop at the tip of one's finger does not fall. Water climbs into a narrow glass pipe because the interfacial free energy between water and glass is smaller than that between air and glass (in a more microscopic language, because of an attractive interaction between water molecules and glass). If R is the internal radius of the pipe, the force responsible for water climbing is proportional to R and has to compete with the weight of the water column, proportional to hR^2. Consequentely, the height h of the liquid column is proportional to $1/R$, so that it can only be large in thin pipes or 'capillaries' (from the latin *capillus*: hair). The word *capillary* is often used as an adjective to qualify anything related to interfacial energies.

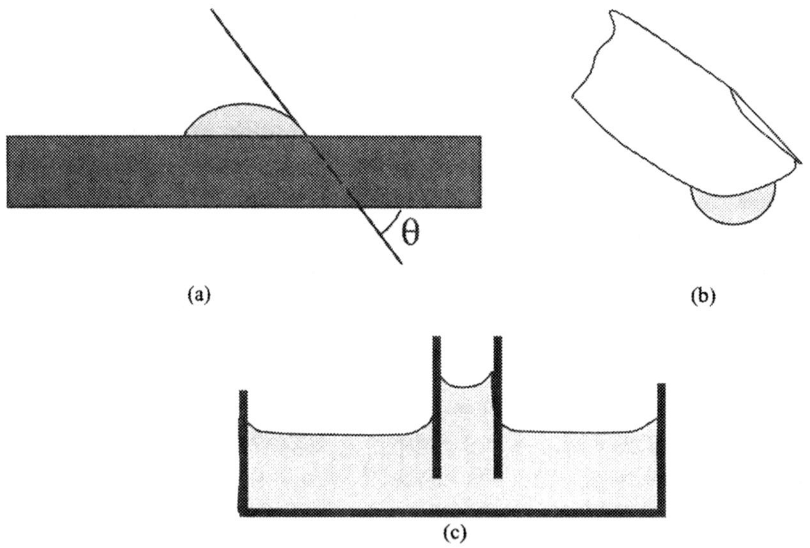

Fig. 2.1. Some familiar situations which illustrate the concept of surface and interface free energy. a) A water droplet on a table does not spread, because spreading would increase the surface energy. b) A water droplet on my fingertip does not fall down. c) Water rises in a glass capillary because the water-glass interface energy is less than that of the glass-air interface. The lengthscales involved are of some millimetres, which is the order of the capillary length (1.13).

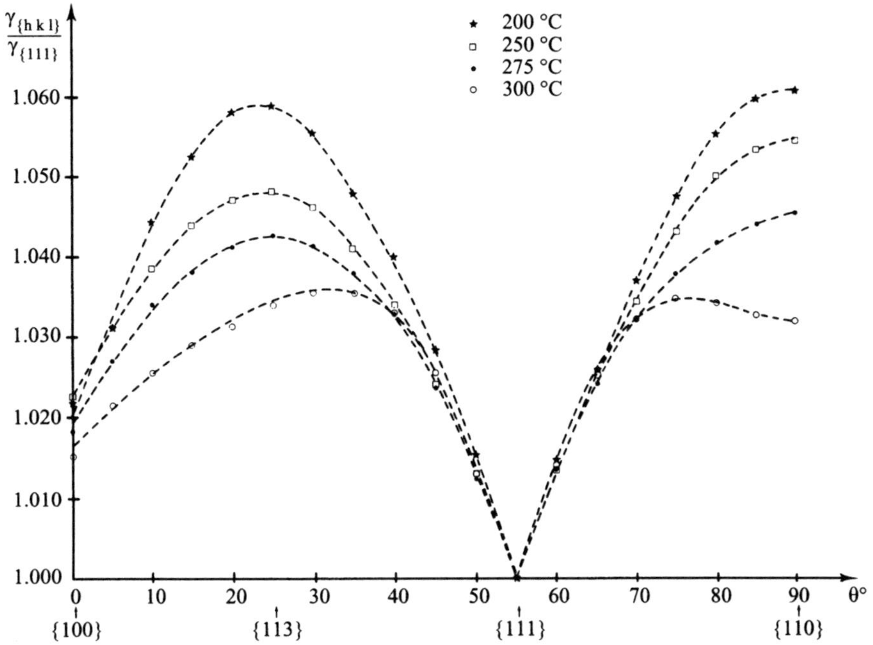

Fig. 2.2. The surface tension of indium at various temperatures, for the orientations $\{11k\}$, normalized to the value for $\{111\}$. Here, k is larger than 1 to the left and smaller than 1 to the right of the cusp. Note the weak variation (visually amplified by the choice of the normalization) and the roughening singularities (cusps) in the directions $\{100\}$ and $\{111\}$. The function ϕ has similar singularities (Heyraud & Métois 1986, with permission by the authors).

As seen from its definition, the surface tension σ of a crystal depends on the orientation of the surface (though in general not tremendously much, as seen from Fig. 2.2). If the surface is not a plane, the total surface energy may be expected to be the integral of the energies of all surface elements. This is only true, as it will be seen later, for an *incompressible solid* of large enough size. If x, y and z are Cartesian coordinates and $z(x, y)$ is the height of the surface over the xy plane, the local surface tension is then a function $\sigma(z_x, z_y)$ of the partial derivatives $z_x = \partial z/\partial x$ and $z_y = \partial z/\partial y$. The total free energy of a large, incompressible solid body containing N atoms is then:

$$\mathscr{F} = (\mu_0 - Pv)N + \int\int \sigma(z_x, z_y)\mathrm{d}S \tag{2.1}$$

where the second term on the right-hand side is the surface free energy. The integration is made over the surface and dS the surface element. The first term on the right-hand side has been written as a function of the equilibrium pressure P and the equilibrium chemical potential μ_0, and $v = V/N$ is the volume per atom.

For an incompressible solid, one can equivalently use the (Helmholtz) free energy $\mathscr{F}(T,V)$ or the free enthalpy (Gibbs free energy) $\Phi(T,P) = \mathscr{F} + PV$, which is known from thermodynamics to be equal to $\mu_0 N$ (in the thermodynamic limit $N \to \infty$ and in the absence of a surface). For a compressible solid, PV is not a constant, and processes at constant pressure are more adequately described if Φ is used. In the future, we shall mainly use the free energy $\mathscr{F}$ because, in the absence of external forces, it corresponds to the interactions between the molecules of the solid. Sometimes, we will drop the word 'free', especially at low temperature where energy and free energy are not very different.

In formula (2.1), the free energy is assumed to be insensitive to the local curvature. This is only true for a large sample, since the curvature scales as the reciprocal linear size. Another shortcoming is that the surface itself may be difficult to define: its location may be rather imprecise. This point has been addressed by Gibbs (see Caroli *et al.* 1991). Also, in formula (2.1) an analytic function $z(x,y)$ is assumed, which implies that fluctuations on an atomic scale are ignored.

We close this introductory section on the surface tension with a remark on membranes. In membrane theory the integral in (2.1) is disregarded and only curvature-dependent terms are taken into account. They are responsible for the shape of a red blood cell, or of a deflated balloon, because the surface area is nearly a constant, as well as σ and hence the integral in (2.1) (Miao *et al.* 1991, Seifert *et al.* 1991, Fourcade *et al.* 1992).

2.2 Small surface fluctuations and surface rigidity

As long as σ is not known, formula (2.1) is not very useful. However, one can assume for $\sigma(z_x, z_y)$ an analytic expansion, whose use is comparable with that of the Landau expansion in the theory of critical phenomena (Landau & Lifshitz 1959a). It will be seen, however, that an analytic expansion does not exist below the roughening transition temperature.

It is convenient to transform the surface integral in (2.1) into an integral over Cartesian coordinates x and y. For a closed surface, it is only possible if the surface is cut into a few pieces (e.g. 6) and different projection planes are used for the various pieces (e.g. xy, yz and zx, each twice). For a given piece, it is convenient to introduce the function ϕ (the *projected free energy*),

$$\phi(z_x, z_y) = \sigma(z_x, z_y)\sqrt{1 + z_x^2 + z_y^2} \tag{2.2}$$

so that we can rewrite (2.1) as follows:

$$\mathcal{F} = (\frac{\mu_0}{v} - P) \int\int z(x, y) \mathrm{d}x\mathrm{d}y + \int\int \phi(z_x, z_y) \mathrm{d}x\mathrm{d}y \,. \qquad (2.3)$$

Take now an infinite crystal surface globally parallel (within local fluctuations) to the plane $z = 0$. Then formula (1.7) can be deduced from (2.1). Expanding ϕ, (2.1) and (2.3) become

$$\mathcal{F} = (\mu_0 - Pv)N +$$
$$\int\int \mathrm{d}x\mathrm{d}y (\sigma + \sigma_1 z_x + \sigma_2 z_y + \frac{1}{2}\sigma_{11} z_x^2 + \sigma_{12} z_x z_y + \frac{1}{2}\sigma_{22} z_y^2)$$
$$\times (1 + \frac{1}{2} z_x^2 + \frac{1}{2} z_y^2)$$

where $\sigma_1 = \partial\sigma/\partial z_x$, $\sigma_{11} = \partial^2\sigma/\partial z_x^2$, etc. The mixed second derivatives $\sigma_{12} = \sigma_{21} = \partial^2\sigma/\partial z_x \partial z_y$ may be made to vanish by an appropriate choice of the coordinate axes. Also, the integrals of z_x and z_y vanish. In fact,

$$\int_{-L}^{+L} z_x(x, y) \mathrm{d}x = z(L, y) - z(-L, y)$$

is either a negligible fluctuation for a surface of linear size $2L$ large enough, or exactly 0 for periodic boundary conditions.

If $\sigma_{11} = \sigma_{22} = \sigma''$, one obtains (within an additive constant):

$$\phi(z_x, z_y) \simeq \frac{1}{2}\tilde{\sigma}(z_x^2 + z_y^2) \qquad (2.4)$$

where we introduced the *surface stiffness* (or *rigidity*)

$$\tilde{\sigma} = \sigma + \sigma'' \,. \qquad (2.5)$$

Relation (2.4) is a generalization of (1.7a), as mentioned. The preceding calculation only makes sense if the various partial derivatives exist. Intuitively, we expect them to exist at $T > T_R$, because there the discreteness of the crystal lattice should not play any role. We will shortly show that they do not exist, however, if $T < T_R$.

2.3 Singularities of the surface tension

Let us suppose that the plane $z = 0$ is parallel to a high-symmetry orientation, for instance (001), and that the temperature is lower than the roughening temperature for that orientation. We want to know σ or equivalently ϕ for a vicinal surface (see Fig. 1.9) making a small angle θ with that orientation. If γ is the free energy per unit step length, also called *line tension*, and if step-step interactions are neglected, the contribution to

ϕ due to the steps is simply equal to the step free energy multiplied by the step density $|\tan\theta|$:

$$\phi(\theta) \simeq \phi(0) + \gamma|\tan\theta| = \phi(0) + \gamma\sqrt{z_x^2 + z_y^2}\,. \tag{2.6}$$

The line tension γ is a function of the step orientation $y' = dy/dx = -z_x/z_y$. We will see that it is an analytic function at all T. The right-hand side of (2.6) has infinite second derivatives at $\theta = 0$ ($\nabla z = 0$). More generally, if a given surface orientation is below its roughening temperature, the surface tension is a non-analytic function of z_x, z_y in the neighbourhood of that orientation (Fig. 2.2). We say that the surface tension exibits a *roughness singularity* for that orientation. The reason why the argument fails above T_R is that γ vanishes right at T_R as we saw in Fig. 1.7. Above T_R a surface cannot be smooth because its free energy is minimal at a *non-vanishing* step density.

Summarizing, the expansion of ϕ for small deviations from a high-symmetry orientation $z = 0$ is given by (2.4) above T_R, and by (2.6) below. The generalization for non-symmetric orientations (other than (001) or (111)) or to non-cubic crystals is straightforward.

2.4 Chemical potential

The chemical potential is the Gibbs free energy per particle (here, per atom):

$$\mu = \left[\frac{\partial}{\partial N}(\mathscr{F} + PV)\right]_{T,P}\,. \tag{2.7}$$

The chemical potential is very useful for determining the equilibrium crystal shape, as we will see in the next chapter. It will be also used in chapter 8 for investigating the time evolution of the crystal shape. In fact, since each particle seeks to minimize its free energy, the chemical potential is the key to surface dynamics as well as to statics. In macroscopic systems at equilibrium, the chemical potential of each species of particles is uniform. If it were not so, one would lower the free energy by taking a particle from a point with a high chemical potential and putting it in a place where the chemical potential is low. This argument does not work for a small system, because this operation might modify μ.

Note that, for an infinite incompressible solid, $\mathscr{F}/N$ does not depend on the pressure P, while the chemical potential does. We will now compute μ for a surface of given profile $z(x, y)$.

Let us locally modify the surface (Fig. 2.3) so as to change the particle number by a quantity δN. Let $\delta\mathscr{F}$ be the corresponding variation of the free energy. If the transformation is isothermal, this variation is

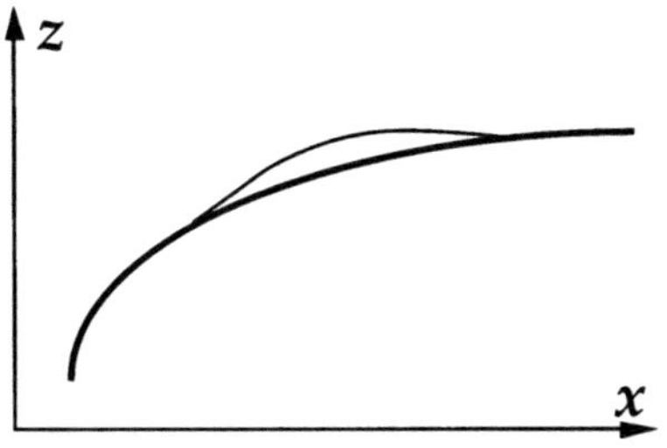

Fig. 2.3. Small deformation of a surface.

$\delta\mathcal{F} = -P\delta V + \mu\delta N$. It is enough to compute $\delta\mathcal{F}$ at fixed T to deduce the chemical potential μ. If $z(x, y)$ varies by δz, the corresponding variation of (2.3) is:

$$\delta\mathcal{F} = \left(\frac{\mu_0}{v} - P\right)\int\int \delta z(x, y)\mathrm{d}x\mathrm{d}y + \int\int\left(\frac{\partial\phi}{\partial z_x}\delta z_x + \frac{\partial\phi}{\partial z_y}\delta z_y\right)\mathrm{d}x\mathrm{d}y \,. \qquad (2.8)$$

Each of the terms within brackets can be integrated by parts, provided ϕ is at least twice differentiable. Thus:

$$\int\int\left(\frac{\partial\phi}{\partial z_x}\delta z_x\right)\mathrm{d}x\mathrm{d}y = \int\int\left(\frac{\partial\phi}{\partial z_x}\frac{\partial}{\partial x}(\delta z)\right)\mathrm{d}x\mathrm{d}y$$
$$= \int\int\frac{\partial}{\partial x}\left(\frac{\partial\phi}{\partial z_x}\delta z\right)\mathrm{d}x\mathrm{d}y - \int\int\delta z\frac{\partial}{\partial x}\left(\frac{\partial\phi}{\partial z_x}\right)\mathrm{d}x\mathrm{d}y \,.$$

We only consider local variations δz, so that the first term vanishes. The second term in brackets in (2.8) can be handled in the same way, and we find:

$$\delta\mathcal{F} = \int\int\delta z(x, y)\left[\frac{\mu_0}{v} - P - \frac{\partial}{\partial x}\left(\frac{\partial\phi}{\partial z_x}\right) - \frac{\partial}{\partial y}\left(\frac{\partial\phi}{\partial z_y}\right)\right]\mathrm{d}x\mathrm{d}y \,.$$

Since δz is localized in a small domain, the factor within brackets must be constant and $\mathcal{F}$ must be equal to this constant multiplied by the integral of δz, which is equal to δV or to $v\delta N$ as well. Since $\delta\mathcal{F} = -P\delta V + \mu\delta N$, it follows (Herring 1953, Mullins 1963):

$$\mu = \mu_0 - v\left[\frac{\partial}{\partial x}\left(\frac{\partial\phi}{\partial z_x}\right) + \frac{\partial}{\partial y}\left(\frac{\partial\phi}{\partial z_y}\right)\right] \,. \qquad (2.9)$$

In the following we will often choose our units in such a way that the molecular volume $v = V/N$ is 1. If the solid is infinite, μ_0 is the chemical potential inside the solid, according to (2.3). At equilibrium, the chemical potential at the surface must also be equal to μ_0, and from (2.9) the surface must be flat.

The Herring-Mullins formula (2.9) defines a *local chemical potential.* It only needs local, not global, equilibrium, to be valid. Local equilibrium implies that slowly-relaxing thermodynamic variables (such as the energy or matter density) need not possess their equilibrium distribution, while fast-relaxing variables (e.g. the single particle velocity) have the distribution they would possess at equilibrium, for the given local value of the slow variables.

Indeed, it is often convenient to give names to formulae, but the choice of the name is not always clear. For instance, the 'Herring-Mullins' formula (2.9) was written by Herring and later by Mullins in a different form, which uses the radii of curvature or the surface. We have written it in a particularly simple way, which has the drawback that the function ϕ depends on the choice of the coordinate system.

If (2.4) holds, the chemical potential is $\mu_0 + \delta\mu$, where $\delta\mu$ is, according to (2.9), given by

$$\delta\mu = -v\tilde{\sigma}(z_{xx} + z_{yy}) \tag{2.10}$$

where $\tilde{\sigma}$ is the surface stiffness (2.5). In the case of an isotropic surface tension, $\tilde{\sigma} = \sigma$. Assume now a convex surface $z(x, y)$. Choosing any point as the origin, it is possible to choose the axes in such a way that

$$z(x, y) \approx z_0 - \frac{1}{2}\left(\frac{x^2}{R_1} + \frac{y^2}{R_2}\right), \tag{2.11}$$

where R_1 and R_2 are the two principal radii of curvature of the surface. Inserting (2.11) into (2.10) we find the Gibbs-Thomson relation

$$\delta\mu = v\tilde{\sigma}\left(\frac{1}{R_1} + \frac{1}{R_2}\right). \tag{2.12}$$

In particular, for a sphere of radius R

$$\delta\mu = \frac{2v\sigma}{R}. \tag{2.13}$$

A similar expression yields the excess chemical potential in the vicinity of a curved step on the surface, as a function of the *step stiffness* $\tilde{\gamma} = \gamma(\theta) + \mathrm{d}^2\gamma/\mathrm{d}\theta^2$, where γ is the step line tension. Writing the atomic area a^2 (usually equal to unity in this book), and letting R be the local radius of curvature of the step, the relation reads

$$\delta\mu = a^2\frac{\tilde{\gamma}}{R}. \tag{2.14}$$

Note the factor 2 in (2.13)–a consequence of the 2 terms in (2.9).

The equation (2.13) is reminiscent of, but quite different from, the Laplace relation giving the pressure difference $P_\ell - P_g$ between the inside and the outside of a liquid droplet. In order to derive the Laplace relation, we consider the mechanical work done in increasing by δR the droplet radius. It is $(P_\ell - P_g) \times 4\pi R^2 \delta R$. It must be equal to the surface free energy increase $\sigma \times 8\pi R \delta R$. The result is the Laplace relation

$$P_\ell - P_g = \frac{2\sigma}{R} . \tag{2.15}$$

The pressure is higher inside the droplet because the surface wants to shrink in order to reduce its energy. A similar effect exists in a solid, as we will see in chapter 16. The surface tension exerts a force (the surface stress) which adds to the external pressure and needs a stronger pressure from inside to be balanced (Fig. 2.4b). Note the pressure in (2.15) is the *total* pressure, and corresponds to a *mechanical* effect. Formula (2.15) gives the difference between the pressure inside and outside the droplet. It is possible, however, to know the two separately by considering how the thermodynamic equilibrium equation $\mu_\ell(P_0, T) = \mu_g(P_0, T)$ for a planar interface, is modified by the presence of a curved interface.

The chemical potential and the temperature must still be the same everywhere; the pressure, on the other hand, is changed according to (2.15), so that at equilibrium:

$$\mu_\ell(P_\ell) = \mu_g(P_\ell - \frac{2\sigma}{R}) . \tag{2.16}$$

Writing $P_\ell = P_0 + \delta P_\ell$, $P_g = P_0 + \delta P_g$, which define the *excess pressures* δP_ℓ and δP_g, equation (2.16) yields

$$\mu_\ell(P_0 + \delta P_\ell) = \mu_g(P_0 + \delta P_\ell - \frac{2\sigma}{R}) . \tag{2.17}$$

Expanding (2.17) near P_0 and using the formula

$$v = \partial\mu/\partial P \tag{2.18}$$

we find

$$\delta P_g = \frac{v_\ell}{v_g - v_\ell} \frac{2\sigma}{R} \tag{2.19a}$$

$$\delta P_\ell = \frac{v_g}{v_g - v_\ell} \frac{2\sigma}{R} \tag{2.19b}$$

where we also used (2.15) in the form $\delta P_\ell = \delta P_g + 2\sigma/R$.

(a)

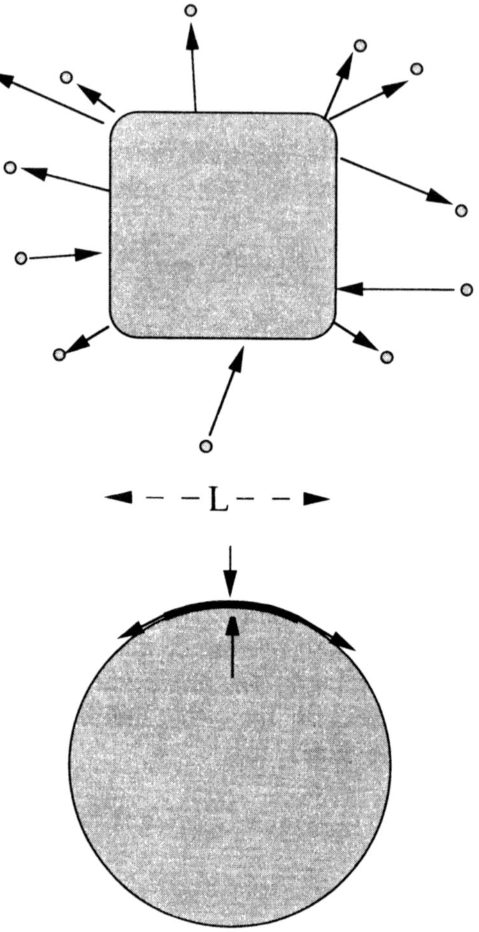

(b)

Fig. 2.4. a) Statistical interpretation of the Gibbs-Thomson relation. The arrows represent the trajectories of atoms emitted by and impinging on the crystal (shaded area). Atoms at corners are less bound to the crystal and evaporate more easily, thus increasing the vapour pressure by an amount δP with respect to the equilibrium value for a planar surface. The relative number of corner atoms is proportional to $1/L$, and thus $\delta P \sim 1/L$. This is a global form of the Gibbs-Thomson relation, analogous to the relations (2.19) which also hold for a spherical solid particle. b) Dynamical interpretation of the Laplace relation. The arrows represent the forces acting on a surface element (thick line) of a liquid droplet (shaded area). The forces tangent to the sphere are due to the surface tension σ, since shrinking the surface would reduce its energy (because $\sigma > 0$). The sum of all forces should vanish at equilibrium. An analogous effect exists for a compressible solid, and the density of tangential force per unit length is called surface stress (see chapter 16).

Equations (2.19a) and (2.19b) are the Gibbs-Thomson relations giving the excess pressures inside and outside a liquid droplet in equilibrium with its vapour.

Various forms of the Gibbs-Thomson relation exist, relating the local values of excess thermodynamic quantities (pressure, chemical potential, temperature) to the local curvature of an interface. In the case of a liquid droplet or a solid particle in equilibrium with its vapour, the excess pressure δP is related to the excess chemical potential $\delta\mu = \mu(P_0 + \delta P) - \mu(P_0)$. Using (2.18), $\delta P_g \approx \delta\mu_g / v_g$, and the Gibbs-Thomson relation (2.19a) reads then

$$\delta\mu_g \approx \frac{v_\ell v_g}{v_g - v_\ell} \frac{2\sigma}{R} \approx v_\ell \frac{2\sigma}{R} \tag{2.20}$$

where we considered that $v_g \gg v_\ell$. Equation (2.20) coincides with (2.13). Note also that $\delta\mu_g = \delta\mu_\ell$.

Also in the case of a crystal particle (which is in general not spherical) the excess pressure in the saturating vapour is proportional to the reciprocal of the linear size of the particle (Fig. 2.4a).

The pressure in (2.19b) in the case for instance of a water droplet in the atmosphere, is the partial water pressure. More generally, it depends on the *chemical* nature of the molecules. In this sense, it corresponds to a *chemical* effect.

We remind the reader that throughout all this chapter we have considered only an incompressible solid body. According to Nozières (1991), the concept of a chemical potential in a compressible solid body does not even make sense! In fact, the usefulness of the Gibbs free energy, and consequently of the concept of the chemical potential, disappears in non-hydrostatic conditions, when the pressure is not constant everywhere. In this cases, it is more appropriate to use the Helmholtz free energy; an example of equilibrium conditions obtained using the latter will be given in section 2.8 below.

Formula (2.9) strictly applies only to non-singular orientations, i.e. when all derivatives are well defined. If the orientation is singular, it is better described as a system of steps. For this reason we turn now to the investigation of surface steps, using a two-dimensional version of what we have just seen.

2.5 Step line tension and step chemical potential

Formula (2.1) gives the free energy of the surface of a solid body as an integral over the surface itself of a function $\sigma(\mathbf{n})$, known in principle. Analogously, we assume that the free energy of a one-dimensional object, such as the step limiting a terrace, is the line integral of a function γ of the in-plane normal $\mathbf{n} = (\cos\theta, \sin\theta)$ to the step:

$$\mathscr{F} = \int_{\text{step}} \mathrm{d}\ell \; \gamma(\theta) \,. \tag{2.21}$$

This contribution is added to the free energy of what is inside the step, which equals $(\mu_0/v - P)S$, if S is the area bounded by the step. The function γ is called *step free energy*, *step line tension*, or simply *line tension.*

Formula (2.21) is the one-dimensional version of (2.3). It is only valid for long lengthscales, otherwise the step curvature must be included.

The chemical potential along the step is given by the following formula, which consists of writing (2.9) for a one-dimensional interface (the step) on a two-dimensional space (the surface); $y = y(x)$ describes the step profile. We let $y' = \mathrm{d}y/\mathrm{d}x = \tan\theta$, $\gamma(\theta) = f(y')$ and $\delta\mu = \mu - \mu_0$:

$$\delta\mu = -\frac{\mathrm{d}}{\mathrm{d}x}\left[\frac{\mathrm{d}}{\mathrm{d}y'}\left(f(y')\sqrt{1+y'^2}\right)\right] \,. \tag{2.22}$$

Writing all differentiations explicitly, after some lengthy algebra (2.22) reads:

$$\delta\mu = -y''\left\{f''(y')(1+y'^2)^{\frac{1}{2}} + \frac{2y'f'(y')}{(1+y'^2)^{\frac{1}{2}}} + \frac{f(y')}{(1+y'^2)^{\frac{3}{2}}}\right\} \,.$$

It is useful to rewrite this formula introducing the radius of curvature ρ of the step explicitly, $1/\rho = -y''(1+y'^2)^{-\frac{3}{2}}$:

$$\delta\mu = \frac{1}{\rho}\left\{f(y') + 2y'f'(y')(1+y'^2) + f''(y')(1+y'^2)^2\right\} \,. \tag{2.23}$$

Anticipating the next chapter, we will ask here about the equilibrium shape of a closed terrace (Fig. 2.5). It is found requiring μ =constant.

Fig. 2.5. Equilibrium shape of a terrace, realization of a two-dimensional crystal. Note that the shape contains *no* straight parts, as shown in section 3.7.

Applying (2.22) and integrating once, we find:

$$\frac{\mathrm{d}}{\mathrm{d}y'}\left(f(y')\sqrt{1+y'^2}\right) = -\delta\mu\ x \tag{2.24}$$

for the integration constant can be made to vanish by an appropriate choice of the origin on the x axis. A second integration can be performed by parts and gives:

$$f(y')\sqrt{1+y'^2} = -\delta\mu \int x\ \mathrm{d}y' = -\delta\mu \int [\mathrm{d}(xy') - \mathrm{d}y] = \delta\mu(y - xy')\,.$$

Finally, appropriately choosing the origin on the y axis we find:

$$f(y')\sqrt{1+y'^2} + \delta\mu\ x\ y' - \delta\mu\ y = 0\,. \tag{2.25}$$

In the case of a square symmetry it is possible, using (2.25), the result (2.28) of section 2.6 or problem 1.1, to show that the equilibrium shape of a large closed terrace at low temperature is that portrayed in Fig. 2.5.

It is important to stress the difference between the description adopted in this paragraph–where a step appears as a continuous object–and the microscopic description of Fig. 1.6. This difference is a translation of the experimental reality: a step appears as continuous in low-resolution electronic microscopy, and at high temperature (Fig. 1.2), while the low-temperature STM images (Figs. 1.1 and 1.3) look more like Fig. 1.6. Mathematically, one goes from a discretized (microscopic) to a continuum description through a coarse-graining procedure, which is at the basis of the renormalization group technique that we will employ in chapter 13 and in appendix J. We do not need such complications for the moment, since we limit ourselves to a computation of γ within a simple model. The calculation can be pushed to its end if one assumes a step of uniform average orientation. This is what we did in section 1.3, and we repeat it here in a different case.

2.6 Step line tension close to a high-symmetry direction at low T

In section 1.3 we have approximately computed the line tension of a step within a simple broken bonds model and for a direction forming an angle of 45° with the bonds. We investigate here the step free energy within the same model, but for a step direction close to that of the bonds, that we denote by x. The step goes from point A to point B, and each elementary (of atomic length) step segment can be perpendicular or parallel to x, but in the latter case only in the positive direction (Fig. 2.6). We also assume that two consecutive elementary step segments perpendicular to x are never allowed. Let θ be the angle of AB with the x axis, and $L = N/\cos\theta$

the length of AB. Following (2.21), the partial free energy in the direction θ is:

$$\mathcal{F}_N(\theta) = L\gamma(\theta) = N\frac{\gamma(\theta)}{\cos\theta} = Nf(y')\sqrt{1+y'^2}\,.$$

Within a factor $-k_BT$, this is the logarithm of the partial partition function $Z_N(y)$, that we compute in problem 2.1 below by the following method. We assume that the step is the path of a random walker directed from A to B, according to the following rules: the walker can only go right, up-and-right or down-and-right, with probabilities $\epsilon = \exp(-\beta W_0)$, ϵ^2 and ϵ^2 respectively (where W_0 is the bond energy). Treating N as a discrete time, the (approximate) $Z_N(y)$ satisfies the recurrence relation

$$Z_N(y) = \epsilon^2 Z_{N-1}(y-1) + \epsilon Z_{N-1}(y) + \epsilon^2 Z_{N-1}(y+1)\,. \tag{2.26}$$

The result, after taking the logarithm of $Z_N(y)$, neglecting terms of lower order than N, letting $y' = \tan\theta = y/N$, and multiplying by $-k_BT$, reads (in the large N, small ϵ limit):

$$\mathcal{F}_N = N\left(W_0 - k_BT\sqrt{4\epsilon^2+y'^2} + k_BTy'\,\mathrm{Arsh}\frac{y'}{2\epsilon}\right)\,. \tag{2.27}$$

We distinguish two cases.

i) If $|y'| \gg \epsilon$, then $\xi = \mathrm{Arsh}(y'/2\epsilon)$ is large, and thus $y'/2\epsilon \simeq e^{\xi}/2$, $\xi \simeq \ln(y/N\epsilon)$. Hence,

$$\mathcal{F}_N = N\left(W_0 - k_BTy' + k_BTy'\ln\frac{y'}{\epsilon}\right)$$

and recalling the definition of ϵ,

$$\begin{aligned} f(y')\sqrt{1+y'^2} &= \mathcal{F}_N/N \\ &= [W_0 + W_0y' + k_BTy'(\ln y' - 1)]\,. \end{aligned} \tag{2.28a}$$

ii) If $|y'| \ll \epsilon$, from (2.27) we have

$$f(y')\sqrt{1+y'^2} = \mathcal{F}_N/N = (W_0 - 2k_BT\epsilon + k_BT\frac{y'^2}{2\epsilon})\,. \tag{2.28b}$$

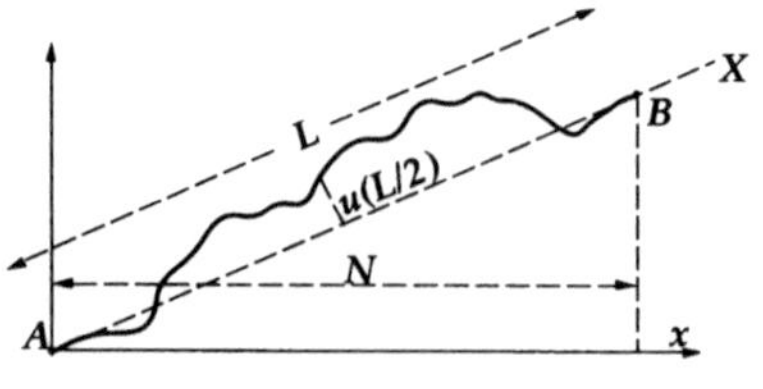

Fig. 2.6. Schematical representation of a step element as a smooth continuous line. One can compare this with the microscopic rough appearance of Fig. 1.6.

The approach of the present section differs slightly from that of section 1.3. In section 1.3 we computed the average energy of a step of given length and average direction. Here, we have forced the step to start at a point A and to end in a point B, and we fixed the length and the direction of the segment AB. In fact, the free energy per unit length turns out to be the same in the two situations if AB is long enough. That is, the free energy depends only weakly on the boundary conditions: we can fix the ends or allow them to fluctuate a little, it does not matter very much. The result is not surprising: it is a consequence of the independence of the free energy of a one-dimensional system with short-range interaction on boundary conditions in the thermodynamic limit.

We leave to the reader the task of verifying this property in the framework of the broken bonds model previously studied (see problem 2.2).

2.7 Thermal fluctuations at the midpoint of a line of tension γ

Consider a line (Fig. 2.6) of line tension γ, whose ends A and B are fixed. Let L be their distance. Define an X axis as passing through these fixed points. Let $u(X)$ the ordinate of the line at the abscissa X. What is $\langle u^2(L/2)\rangle$?

The interesting point in asking this question is that this quantity is measurable, when the line is a surface step (Alfonso *et al.* 1992); γ can thence be deduced.

Let us define the Fourier components u_q of $u(X)$ by

$$u(X) = \frac{1}{\sqrt{L}} \sum_q u_q \sin(qX) . \tag{2.29}$$

With this definition, only positive q are to be kept. The boundary conditions $u(L) = u(0) = 0$ impose $q = \pi m/L$, with m an integer. For small deviations, the step free energy (2.21) may be written as $\mathscr{F} = \mathscr{F}_0 + \delta\mathscr{F}$, where $\delta\mathscr{F}$ represents the contribution of fluctuations, and has the following form, analogous to (1.7):

$$\begin{aligned}
\delta\mathscr{F} &= \frac{1}{2}\tilde{\gamma} \int_0^L \mathrm{d}X\, u'^{\,2}(X) \\
&= \frac{\tilde{\gamma}}{2L} \sum_{qq'} qq'\, u_q u_{q'} \int_0^L \mathrm{d}X \sin(qX)\sin(q'X) \\
&= \frac{1}{4}\tilde{\gamma} \sum_q q^2\, u_q^2
\end{aligned} \tag{2.30}$$

where the *line stiffness* $\tilde{\gamma}$ is:

$$\tilde{\gamma} = \gamma + \gamma'' = \gamma + \frac{\mathrm{d}^2\gamma}{\mathrm{d}\theta^2} . \tag{2.31}$$

The equipartition theorem applied to (2.30) states that

$$\langle |u_q|^2 \rangle = 2k_B T/(\tilde{\gamma} q^2).$$

Whence:

$$\langle u^2(X) \rangle = \frac{1}{L} \sum_q \langle u_q^2 \rangle \sin^2(qX) = \frac{k_B T}{\tilde{\gamma} L} \sum_q \frac{1-\cos(2qX)}{q^2} . \tag{2.32}$$

It might be tempting at this point to replace the sum by an integral, but then the result would be twice as large as the correct one. The correct way is to use the fact that $q = \pi m/L$ and $X = L/2$, so that

$$\begin{aligned} \langle u^2(\frac{1}{2}L) \rangle &= \frac{k_B T}{\pi^2 \tilde{\gamma}} L \sum_m \frac{1-(-1)^m}{m^2} \\ &= \frac{2k_B T}{\pi^2 \tilde{\gamma}} L \left\{ 1 + \frac{1}{9} + \frac{1}{25} + \frac{1}{49} + \cdots \right\} = \frac{k_B T}{4\tilde{\gamma}} L , \end{aligned} \tag{2.33}$$

where we used formula (O.234.2) in Gradshteyn & Ryzhik (1980).

The measurement of $\langle u^2(\frac{1}{2}L) \rangle$ allows thus to deduce $\tilde{\gamma}$. Alfonso *et al.* (1992) found $\tilde{\gamma}/k_B T \approx 3$ at about 900 °C for steps on Si(111). The order of magnitude of the line tension γ is probably not very different.

2.8 The case of a compressible solid body

We are now going to briefly address the question of elasticity, neglected up to this point. The extension of formula (2.1) to a compressible solid requires the intervention of the *deformation* (or *strain*) *tensor* $\epsilon = \{\epsilon\}_{\alpha\gamma}$. We recall that it is defined starting from the *displacements* $u_\alpha(\mathbf{r})$ through the relation $\epsilon_{\alpha\gamma} = (\partial_\alpha u_\gamma + \partial_\gamma u_\alpha)$. The free energy is, under hydrostatic pressure, and for small deformations (*harmonic approximation*),

$$\mathscr{F} = \mathscr{F}_0 + \frac{1}{2} \int\int\int \mathrm{d}^3\mathbf{r} \sum_{\alpha\gamma\xi\zeta} \Omega_{\alpha\gamma}^{\xi\zeta} \epsilon_{\alpha\gamma}(\mathbf{r}) \epsilon_{\xi\zeta}(\mathbf{r}) + \int\int \sigma(\mathbf{n}, \epsilon) \mathrm{d}^2\Sigma \tag{2.34}$$

where $\mathbf{\Omega}$ is a known tensor and σ a known function, and the double integral is taken over the surface of the solid. This formula only applies when ϵ are the deformations counted from the zero pressure state, which is not necessarily compatible with the harmonic approximation. We will come back to this point in chapter 16.

To understand the relation between $\mathscr{F}_0$ and μ_0, we can neglect the surface. The volume variation due to the pressure is $\delta V = V_0 \sum_\alpha \epsilon_{\alpha\alpha}$. Its value is given as the minimum of the Gibbs thermodynamic potential, which is, to second order in ϵ:

$$\Phi = \mathscr{F} + PV = \mathscr{F}_0 + \frac{1}{2} V_0 \sum_{\alpha\gamma\xi\zeta} \Omega_{\alpha\gamma}^{\xi\zeta} \epsilon_{\alpha\gamma} \epsilon_{\xi\zeta} + PV_0(1 + \sum_\alpha \epsilon_{\alpha\alpha}) . \tag{2.35}$$

The minimization with respect to $\epsilon_{\alpha\gamma}$ yields

$$\sum_{\xi\zeta} \Omega_{\alpha\gamma}^{\xi\zeta} \epsilon_{\xi\zeta} = -\delta_{\alpha\gamma} P \tag{2.36}$$

and thus Φ at equilibrium is equal to:

$$\Phi_0 = \mu_0 N = \mathscr{F}_0 + P_0 V_0 + \frac{1}{2} P_0 \delta V \,. \tag{2.37}$$

This is the relation between $\mathscr{F}_0$, the saturating vapour pressure P_0, and the chemical potential of this saturating vapour. A uniaxial deformation can be equally accounted for by imposing appropriate boundary conditions.

2.9 Step-step interactions

Formula (2.6) only applies to noninteracting steps. How is it going to be modified by step interactions?

The most evident interaction between steps is an electrostatic one: symmetry considerations suggest that steps carry an electric dipole moment (Thompson & Huntington 1982). It is in fact a corollary of the electrostatic interaction between adatoms, roughly evaluated by Kohn & Lau (1976) and by Lannoo & Allan (1978). Indeed, even a flat surface carries an electric moment: it is the correction to this electric moment due to the step that we are considering here. As a consequence, two steps parallel to the y axis at mutual distance ℓ must feel an interaction varying as $1/\ell^2$: this results from integrating over y the dipole-dipole interaction which varies as $(\ell^2 + y^2)^{-3/2}$. One should at this point beware of the temptation of adding to (2.6) a term proportional to $1/\ell^2 = z_x^2 + z_y^2$: in fact, one should remember that the step density is itself equal to $1/\ell$. Formula (2.6) must then be modified in the following way:

$$\phi(z_x, z_y) = \phi(0,0) + \gamma(z_x^2 + z_y^2)^{\frac{1}{2}} + B(z_x^2 + z_y^2)^{\frac{3}{2}} \tag{2.38}$$

where the constant B needs not be positive. The derivatives z_α must be small.

For certain metal surfaces, the electronic theory of metals predicts an oscillating adatom-adatom interaction proportional to $r^{-2}\cos(2k_F r)$ (Lau & Kohn 1978), where k_F is the Fermi wavevector. Integrating along an adatom row generates an interaction between parallel rows proportional to $r^{-1}\cos(2k_F r)$ (problem: prove this). Presumably, a similar interaction is present between steps on a clean surface. Its effect may be of making certain surface orientations unstable, if it turns out to be attractive (see section 3.6). However, the oscillations are washed out when r becomes longer than the electron mean free path, and therefore the interactions between parallel steps on a clean surface are likely to decay as $1/r^2$ at long distances in all cases. The oscillating interaction appears when there is a partially unfilled electronic surface band, and corresponds to a Friedel oscillation.

Other types of interactions exist, most of which vary as $1/\ell^2$, and thus do not modify anything else than B in (2.38). We will see in chapter 15 that one type is of elastic origin (Shiflet & Van der Merwe 1991, Van der Merwe & Shiflet 1991). Another one is of 'entropic' origin (Gruber & Mullins 1967, Pokrovsky & Talapov 1979, Haldane & Villain 1981). It is due to the following effect: two steps of identical sign (i.e., both ascending or descending) cannot cross each other when they meet, and this limits their degrees of freedom. Contrary to the other ones, this type of interaction is always repulsive, and, as we show in appendix C, also varies as $1/\ell^2$.

A noteworthy exception to the $1/\ell^2$ rule is the elastic interaction between steps in the case of an adsorbed epitaxial layer which has a different lattice parameter from that of the substrate. In this case, the (elastic) dipole-dipole interaction has to be integrated over the whole misfitting terrace, not only over the step edge: the interaction becomes thus logarithmic! The same behaviour is found for steps on a clean Si(001) surface, as will be seen in chapter 15.

Problems

2.1. Prove the relation (2.27) by the random-walker method described in section 2.6.

Solution – We can seek a solution of (2.26) in the form of a Fourier integral

$$Z_N(y) = \int_{-\pi}^{\pi} \mathrm{d}k \; \zeta_N(k) \cos(ky) \tag{1}$$

where the integration is limited to $[-\pi, \pi]$ because y is an integer quantity. Inserting (1) into (2.26) yields

$$\zeta_N(k) = \epsilon(1 + 2\epsilon \cos k)\zeta_{N-1}(k) = [\epsilon(1 + 2\epsilon \cos k)]^N \zeta_0 \tag{2}$$

where $\zeta_0 = \zeta_0(k) = 1/2\pi$ is independent of k because $Z_0(y)$ vanishes for $y \neq 0$; and since $Z_0(0) = 1$, one finds $\zeta_0 = 1/2\pi$. Inserting (2) into (1) now gives

$$Z_N(y) = \frac{\epsilon^N}{2\pi} \int_{-\pi}^{\pi} \mathrm{d}k \; (1 + 2\epsilon \cos k)^N \cos(ky)$$

which, for small ϵ, may be replaced by

$$Z_N(y) = \frac{\epsilon^N}{2\pi} \int_{-\pi}^{\pi} \mathrm{d}k \ \exp(2N\epsilon \cos k) \cos(ky). \tag{3}$$

We are now faced with a dilemma: to be or not to be lazy? The lazy attitude consists of extending the integral in (3) from $-\infty$ to ∞ and replacing $\cos k$ by its small-k expansion. We would only obtain the small slopes limit ($y' \ll 1$), formula (2.28b). Alas, no, we choose to be bold and brave, invoking the modified Bessel functions of imaginary argument I_y (Gradshteyn & Ryzhik 1980, p. 958, eq. 8.431.5):

$$Z_N(y) = \epsilon^N I_y(2N\epsilon).$$

Asymptotically, for large y (Bateman *et al.* 1953, p. 86, eq. 7.13.2.8), we find

$$Z_N(y) \sim \frac{1}{\pi\sqrt{2}} \epsilon^N (y^2 + 4N^2\epsilon^2)^{-1/4} \exp\left\{ \sqrt{y^2 + 4N^2\epsilon^2} - y \mathrm{Arsh} \frac{y}{2N\epsilon} \right\}$$

which gives (2.27).

2.2. Show that the method of section 2.6 for computing the step free energy gives the correct result if applied to the problem of section 1.3.

Solution – In section 2.6, the total partition function, which could be obtained by integrating (2.27) over y, is equal to $\epsilon^N(1 + 2\epsilon)^N$. The corresponding total free energy is thus:

$$-k_B T N \ln[\epsilon(1 + 2\epsilon)] = N W_0 - 2N k_B T \epsilon.$$

This is exactly the expression obtained by making $y' = 0$ in (2.27).

Consider now the case of section 1.3. We want to compute the partial partition function Z_{AB} of a step element AB, the segment AB being at 45° with respect to the bond direction (Fig. 1.6). Excluding the possibility of stepping backwards, as in section 1.3, each possible configuration is associated to a different configuration of 'right' (r) and 'up' (u) steps, for instance *uururrruuururu*..., thus to a different term of the polynomial $(u+r)^N$. The result is therefore the coefficient of the term $u^N r^N$ in this polynomial. The partition function is thus (making use of Stirling's formula):

$$\begin{aligned} Z_{AB} &= \frac{(2N)!}{(N!)^2} \exp(-2\beta N W_1) \\ &= \frac{(2N)^{2N} \mathrm{e}^{-2N} \sqrt{4\pi N}}{(N^N \mathrm{e}^{-N} \sqrt{2\pi N})^2} \exp(-2\beta N W_1) \\ &= \frac{(2)^{2N}}{\sqrt{\pi N}} \exp(-2\beta N W_1). \end{aligned}$$

Taking the logarithm, multiplying by $-k_B T$ and dividing by $2N$, we find that the free energy per bond is equal to (1.2), within an additive term

$k_B T/(4N)\ln(\pi N)$ which is negligible for large N with respect to the main term, which is of order N.

2.3. Calculate the mean square fluctuation (2.32) of a line of stiffness $\tilde{\gamma}$ for $X << L$.

Solution – Replacing the sum (2.32) by an integral, one obtains:

$$\langle u^2(X) \rangle = \frac{k_B T}{\tilde{\gamma}} X \,.$$

3
The equilibrium crystal shape

You boil it in sawdust: you salt it in glue:
You condense it with locusts and tape:
Still keeping one principal object in view–
To preserve its symmetrical shape

Lewis Carrol (*The hunting of the Snark*)

The equilibrium crystal shape results from the minimization of the surface free energy. This is quite striking, since the latter is proportional to $V^{2/3}$, and one would be tempted to neglect it for a large volume V. Real surfaces, however, are practically never at equilibrium beyond some short lengthscale (≤ 10 μm not too far from the melting point).

Experimentalists know–since approximately 1980–how to make equilibrium crystals and how to observe them (under ultra-high vacuum conditions) with electron microscopy. This is the best technique for measuring the surface tension anisotropy and for determining the roughening transition.

Theoreticians, on the other hand, have known since 1900 how to predict the crystal shape by means of a geometrical construction which would have delighted the students of times now past–and formally related to the Legendre transformation known from thermodynamics. However, only since 1980 have the singularities (facets and angular points) generated by the famous *Wulff construction*, been studied in detail.

3.1 Equilibrium shape: general principles

In all equilibrium systems the chemical potential (or the chemical potentials) must be constant everywhere. In fact, if one takes an atom from a position where $\mu = \mu_1$ to a position where $\mu = \mu_2$, the corresponding free energy variation at constant volume and temperature must be $\Delta\mathcal{F} = \mu_2 - \mu_1$. Since $\mathcal{F}$ is minimum at equilibrium, this variation must vanish. The same result is found at constant pressure by reasoning with the Gibbs free energy $\mathcal{G} = \mathcal{F} + PV$.

In particular, the equilibrium shape of any macroscopic object is determined by the requirement *that the chemical potential* μ *be a constant.*

We have seen that μ is given by formula (2.9). Therefore we know, in principle, the equation of the crystal surface once we know the function $\mathcal{F}(z_x, z_y)$, or $\phi(z_x, z_y)$. Such an equation is

$$\frac{\partial}{\partial x}\frac{\partial \phi}{\partial z_x} + \frac{\partial}{\partial y}\frac{\partial \phi}{\partial z_y} = -\mu = \text{Constant}\,. \tag{3.1}$$

Note that this is a partial differential equation, and thus it is strictly valid only if the derivatives of ϕ exist. In fact, we already know since section 2.3 that they do not exist for high-symmetry orientations below the roughening temperature. We will show later that such singularities imply that the crystal has *facets*. However, even in these cases equation (3.1) may be useful if taken with care.

• *Remark*

At equilibrium the differential of the grand potential $\Phi = \mathcal{F} + PV - \mu N$ is

$$\delta\Phi = (\frac{\partial\mathcal{F}}{\partial V})_{T,N}\delta V + (\frac{\partial\mathcal{F}}{\partial N})_{T,V}\delta N + P\delta V - \mu\delta N = 0\,.$$

The equilibrium crystal shape (Fig. 3.1) – sometimes called crystalline habit – is an extremal of the grand potential at constant pressure. It is not, however, a minimum: if one varies the particle number N at fixed pressure and chemical potential, this number decreases if it is too small

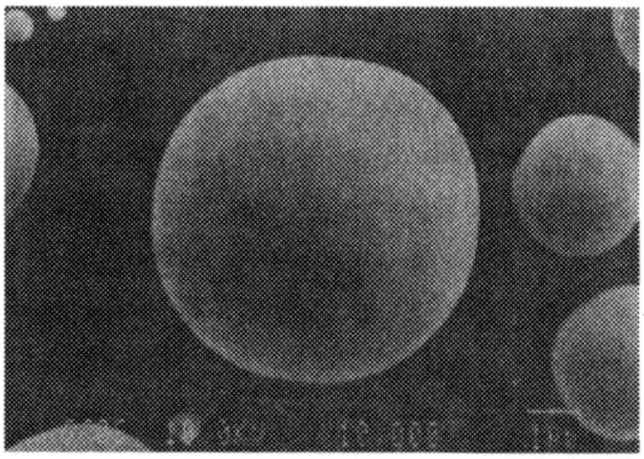

Fig. 3.1. Equilibrium shape of a tin crystal at 192 °C, which means 40 K below the melting point. One can see small facets oriented (111) and (001). The biggest crystal in the middle has a diameter of some 10 μm (with the kind permission of A. Pawlowska).

and increases if it is too large. The value of N which corresponds to a fixed value of μ is a saddle point for the grand potential. At fixed shape, e.g. spherical, it is a maximum: this is the phenomenon of nucleation.

We have shown in section 2.4 that equation (3.1) can be derived by a variational principle, which applies to rounded crystal shapes. We will now investigate the case of a completely faceted crystal (Fig. 3.2), as one would find at zero temperature. In this case (3.1) has no meaning at all, since, strictly speaking, the surface free energy is only defined for a discrete set of orientations–corresponding to the facets in the equilibrium shape.

For a faceted crystal (Fig. 3.2), the surface free energy can be written in the form $\mathscr{F}_{\text{surf}} = \sum_f \sigma_f A_f$ where σ_f is the value of $\sigma(\mathbf{n})$ at $\mathbf{n} = \mathbf{n}_f$, and A_f is the area of the corresponding facet. We wish to find the shape which minimizes $\mathscr{F}_{\text{surf}}$ at given volume. The volume of the crystal is $V = 1/2 \sum_f A_f h_f$, where $h_f = \max_{\mathbf{R}} \{\mathbf{R} \cdot \mathbf{n}_f\}$, $\mathbf{R}$ being any point on the crystal surface. We perform a constrained minimization with a Lagrange multiplier λ to enforce the fixed volume condition

$$\delta(\lambda V + \mathscr{F}_{\text{surf}}) = \sum_f \left(\frac{\lambda}{2} h_f + \sigma_f\right) \delta A_f = 0$$

which gives

$$\frac{\sigma_f}{h_f} = -\frac{\lambda}{2} \,. \tag{3.2}$$

For each facet, the ratio of the surface energy to the distance h_f from the origin is a constant. The resulting set of $\{h_f\}$ gives the equilibrium shape. Rewriting the variational equation as $\delta \mathscr{F}_{\text{surf}} = \mu \delta N = -v \delta N \lambda$,

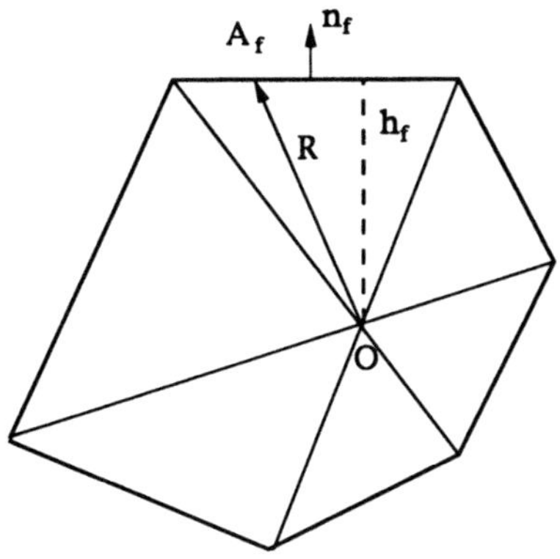

Fig. 3.2. The equilibrium shape of a faceted 2-D crystal. The 'area' of the f-th facet is A_f, and $h_f = \mathbf{R} \cdot \mathbf{n}_f$ its distance from the centre of symmetry.

where we introduced the volume per particle $v = V/N$, it follows that $\lambda = -\mu/v$. The Lagrange multiplier λ turns out to be proportional to the chemical potential, which is thus responsible for setting the overall size of the crystal. Equation (3.2) can be cast in the form

$$\mu = \frac{2v\sigma_f}{h_f} .$$

Comparing with (2.13) shows that (3.2) is the same as the Gibbs-Thomson relation for a solid body with anisotropic surface tension.

3.2 The Wulff construction

Consider now a rounded crystal (Fig. 3.3). Take the tangent plane Π to the crystal surface S in a point M where the normal is $\mathbf{n}_M$. Let $OH = \sigma_M$ be the corresponding value of the surface free energy $\sigma(\mathbf{n}_M)$; H lies by construction on the geometrical locus Σ which defines the *polar plot* of σ as a function of $\mathbf{n}$. It is then clear that $OH = \sigma(\mathbf{n}_M) = \mathbf{OM} \cdot \mathbf{n}_M$. This is the same as equation (3.2).

This result can be phrased as a geometrical construction which allows to find the equilibrium crystal shape when the polar plot of the surface free energy is known: the Wulff construction (Wulff 1901).

Wulff's construction: the equilibrium shape of a crystal is found by taking, for each direction $\mathbf{n}$, the point H defined by $OH = \sigma(\mathbf{n})$, lying on the polar plot Σ of the surface free energy, and the plane Π orthogonal to OH and cutting Σ in H. The crystal surface S is then the inner envelope of the planes such as Π. The envelope of Π is called the *pedal* of the surface Σ spanned by H.

We will prove in appendix D, following Taylor (1974), that the Wulff construction actually yields the unique solution to the problem of finding the convex crystal shape minimizing the surface free energy. It is straightforward to verify (and we invite the reader to do it) that Wulff's construction applies also to the faceted crystal of Fig. 3.2.

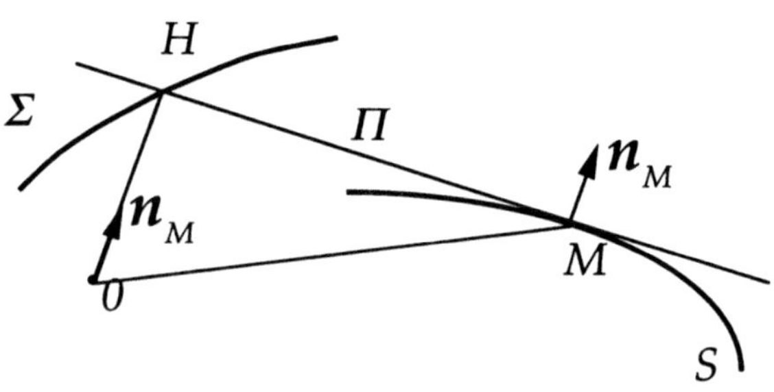

Fig. 3.3. The Wulff construction for a rounded crystal.

We go back now to Fig. 3.3 and assume that the crystal surface is (locally) described by an analytic function $z(x, y)$. The components of $\mathbf{n}$ can be explicitly written by noting that any displacement $(\mathrm{d}x, \mathrm{d}y, \mathrm{d}z)$ must verify $\mathrm{d}z - z_x\mathrm{d}x - z_y\mathrm{d}y = 0$ in order to be tangent to the surface $z = z(x, y)$. Hence,

$$\mathbf{n} = \frac{1}{\sqrt{1 + z_x^2 + z_y^2}}(-z_x, -z_y, 1)\,. \tag{3.3}$$

The vector $\mathbf{OM}$ is then given by

$$\mathbf{OM} = (x, y, z(x, y)) \tag{3.4}$$

and the value OH of the surface free energy in the direction $\mathbf{n}$ is thus proportional (not simply equal for dimensional reasons) to the projection of the vector $\mathbf{OM}$ along $\mathbf{n}$:

$$OH = \sigma(\mathbf{n}) = \lambda\mathbf{n}\cdot\mathbf{OM} = \lambda\frac{z - xz_x - yz_y}{\sqrt{1 + z_x^2 + z_y^2}}\,. \tag{3.5}$$

Formula (3.5) states that if x, y and z are the coordinates of a point on the crystal surface, and we scale variables as $\tilde{x} = \lambda x$, $\tilde{y} = \lambda y$, $\tilde{z} = \lambda z$, they satisfy the equation

$$z - xz_x - yz_y = \phi(z_x, z_y) \tag{3.6}$$

where we used the projected surface free energy defined in equation (2.2) and we omitted the tildas for simplicity.

How should we read (3.6)? Since the right-hand side is only a function of the surface slopes z_x, z_y, equation (3.6) has a nice interpretation as a *Legendre transformation*: the projected surface free energy ϕ is the Legendre transform of the surface profile z (Landau & Lifshitz 1959a, Wortis 1988).

3.3 The Legendre transformation

Legendre transformations are well-known from thermodynamics, and in fact we ourselves used them many times in the preceding chapters.

The idea behind Legendre transformations is the following geometrical property: any convex function f can be described in term of its graph $y = f(x)$, or, equivalently, in terms of the graph $p = g(f')$ of a function of the slope of f. In fact, take a function $f(x)$ (Fig. 3.4), and consider the tangent line to its graph in the point $(x_1, f(x_1))$. The equation of the tangent is $y = m_1x + n_1$, where $m_1 = f'(x_1)$, and $n_1 = f(x_1) - m_1x_1$.

We can thus associate to the function $f(x)$ of x, the function $g(m)$ of the slope m of f in x, according to the following prescription: for each x, take

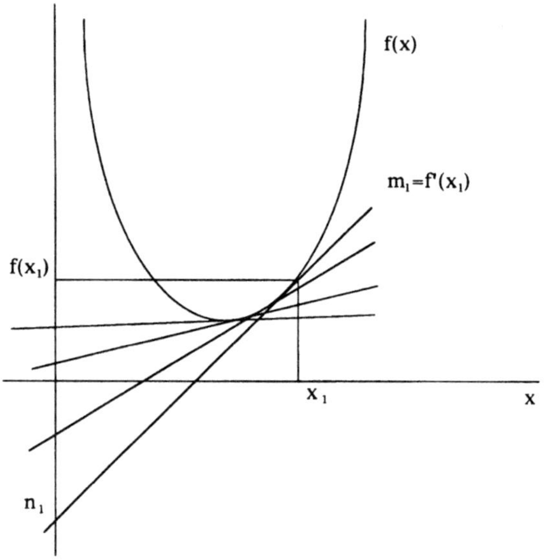

Fig. 3.4. Geometric meaning of the Legendre transformation of a function $f(x)$.

the slope $m = f'(x)$ of f. Then invert this relation to get $x = x(m)$ (which is a well-defined function if f is convex). Finally, define the function $g(m)$ as

$$g(m) = f(x(m)) - x(m)m = f(x(m)) - x(m)f'(x(m)) . \tag{3.7}$$

This works well in thermodynamics because convexity is a necessary condition constraining free energies. It is well-known, however, that statistical mechanics models (e.g. mean field models) often produce non-convex free energies; these are for instance responsible for the van der Waals loops in the equation of state of model gases. In this case, it is an easy exercise to realize that the Legendre transformed function is multivalued (see Fig 3.8 below). A fact that shall not in any case plunge us into despair, for no other reason than that we already know the cure for it: it is the double-tangent construction, which gives the equal-area Maxwell rule. We will see later how this applies to equilibrium shapes. Indeed, all properties of Legendre transformations apply lock, stock and barrel to the relation between the free energy and the equilibrium crystal shape.

3.4 Legendre transform and crystal shape

First of all, note that (3.6) justifies in some sense the existence of Wulff's construction: the surface free energy and the crystal shape carry the same content of geometrical information, since they are related by a Legendre transformation. Knowing the shape of the crystal is completely equivalent to knowing the graph of the projected surface free energy.

This has the interesting practical consequence that it is possible to measure (up to a proportionality factor) the surface free energy of a crystal by observing its equilibrium shape; Fig. 2.2 was obtained this way. Clearly, the difficult part is just obtaining the equilibrium shape.

From (3.6) we immediately prove

$$\frac{\partial \phi}{\partial z_x} = -x \, , \quad \frac{\partial \phi}{\partial z_y} = -y \tag{3.8}$$

and therefore

$$\frac{\mathrm{d}}{\mathrm{d}x}\frac{\partial \phi}{\partial z_x} + \frac{\mathrm{d}}{\mathrm{d}y}\frac{\partial \phi}{\partial z_y} = -2 \, . \tag{3.9}$$

If we explicitate again the scale factor, we recover equation (3.1). Formula (3.6) gives thus a solution of (3.1).

We already stressed that (3.1) is not applicable when the surface free energy–and thus the function ϕ–has singularities. We saw in section 2.3 that such singularities exist: they are commonly angular points for ϕ as a function of the surface slope, or cusps in the polar plot of σ as a function of the crystalline directions $\mathbf{n}$.

It is easy to see that these singularities correspond to facets in the equilibrium shape: a proof based on Legendre transforms can be found in problem 3.1. Here we prefer to give a physical argument.

Suppose that the crystal surface has a singular orientation in a neighbourhood of the point P. This point must be surrounded by steps. Let R be the radius of the smallest of such steps. Along it, as well as inside it, the chemical potential $\delta\mu$ is given by (2.12). It is thus equal to $1/R$ times a positive constant. In other words, $1/R$ has a macroscopic value proportional to $\delta\mu$, and thus, according to (2.9), proportional to the reciprocal of the crystal radius. This reasoning is only approximate, since it neglects the effect of step-step interactions which may modify the formula (2.12). We leave to the reader the task of investigating this problem.

Thus, the chemical potential on a facet is *not* fixed by formula (2.9). The latter is in fact not defined on a facet, since $\phi(z_x, z_y)$ is not differentiable at that point, according to (2.6). If the facet is at equilibrium, the chemical potential is uniform all over it, and equal to the value it takes at the facet edge, where (2.9) can be given a meaning. Therefore the chemical potential on a facet is not fixed only by the orientation of the facet itself, as (2.9) seems to imply; it also depends on the size and the shape of the facet.

Our definition of the equilibrium shape in terms of Wulff's construction is also valid for facets. As we have just discussed, the chemical potential on a facet at equilibrium is fixed by its value at the facet edge. The Wulff construction guarantees that the chemical potential is constant on the

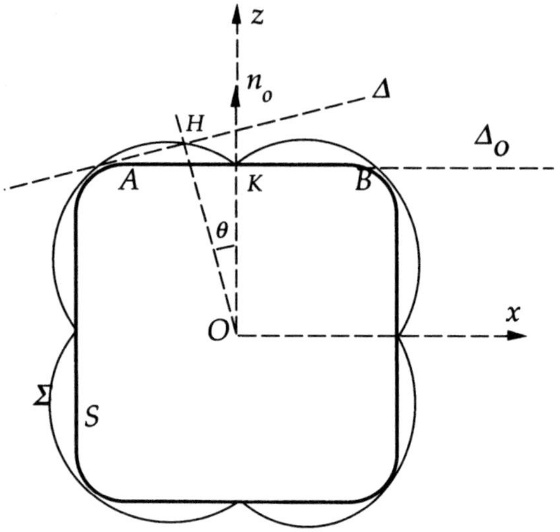

Fig. 3.5. Wulff's construction for a crystal with rough and non-rough orientations. The facet AB joins smoothly to the neighbouring rounded parts.

facet edge (which is necessary for the facet to be at equilibrium) and that it has the required value. For instance, in the–unrealistic–case of a step line tension γ independent of the step orientation, the facet must be circular. Its radius R fixes the chemical potential over the facet according to the Gibbs-Thomson formula $\mu = \mu_0 + \gamma/R$ (chapter 2). Wulff's construction now warrants that the chemical potential takes precisely this value near the facet–as well as everywhere else on the crystal surface.

For the sake of visualization, let us take a two-dimensional crystal (Fig. 3.5). The point H defined by (3.5) spans a curve (Σ) which has an angular point at $K = (0, \sigma_0)$. When H tends to K from the right (or the left, respectively) side, the point of intersection of (Δ) and (Δ_0), normal to, respectively, OH and OK, tends to the point A (resp. B). Considering the angular point of (Σ) as the limit of an arc of very large curvature, one easily realizes that the curve (S) resulting from Wulff's construction (*i.e.* the crystal surface) contains the segment AB.

We recall that cusps in polar plots of σ are called *roughness singularities* because a crystal can only possess facets with a given orientation if the temperature is lower than the roughening temperature relative to that orientation.

3.5 Surface shape near a facet

Two cases are possible: the facet may join smoothly with the neighbouring surface, or with a sharp edge. Thus, the equilibrium shape may possess or not slope discontinuities. We will first consider the case of a smooth junction. The problem is: how does the slope tend to zero?

Let us take a cylindrical symmetry, and choose the xy plane parallel to the facet. The lines z = constant are parallel to the y axis. While the size (in the x direction) of the facet depends on the step line tension (see problem 3.1), the shape of the rounded part near a facet depends crucially on the step-step interactions. Several mechanisms, as we saw in section 2.9, predict interactions decaying as $1/\ell^2 = (z')^2$, where ℓ is the step spacing, and z' is the step density; the projected surface free energy ϕ is then, as in (2.38):

$$\phi(z') = \gamma|z'| + B|z'|^3 , \tag{3.10}$$

whose Legendre transform is easily obtained (see problem 3.2):

$$z(x)/\mu = \begin{cases} -2/(27B)^{\frac{1}{2}}(x-\gamma)^{\frac{3}{2}}, & x \geq \gamma \\ 0, & -\gamma \leq x \leq \gamma \; . \\ -2/(27B)^{\frac{1}{2}}(x+\gamma)^{\frac{3}{2}} & x \leq -\gamma \end{cases} \tag{3.11}$$

The chemical potential μ appears as dictated by (2.9), $\mu = 6B|z'z''|$.

If the origin is chosen at the facet edge, the crystal profile is found to behave as

$$|\delta z| = \frac{2\mu|\delta x|^{\frac{3}{2}}}{3B} . \tag{3.12}$$

We can see that the shape has a singularity, first noted by Jayaprakash *et al.* (1983). Attempts at an experimental verification of this form, namely by Métois & Heyraud (1987), were not always succesful. Experiments by Andreeva's group on helium seem to agree with a longer range interaction than (3.10) (which, we recall, implies $1/r^2$ interactions). According to Carmi *et al.* (1987), however, formula (3.12) would be well established experimentally. A recent paper by Balibar *et al.* (1993) may put an end to this long-lasting controversy; equation (3.12) would be correct only for very small slopes z', and it still awaits experimental verification.

One could wonder whether such singularity also exists in the equilibrium shape, which does not possess a cylindrical symmetry. The answer is yes. In fact, one simply needs to add to (3.10) a term proportional to the step radius of curvature, which hardly varies near a facet. Its role is thus negligible, and the result is the same.

3.6 The double tangent construction

The alternative to a tangential joining of the facet to the rounded part is a sharp edge. This can form if the curvature of the projected free energy ϕ changes sign at small z'. Actually, the curvature of the free energy *cannot* change sign; if it did, the equilibrium crystal shape should contain some non-convex parts (the reader should prove this by using Wulff's construction).

However, as a result, for instance, of a mean field model calculation one could find such a sign change; the situation is the exact analogue of Landau's theory of phase transitions (Landau & Lifshitz 1959a). As in that case, a construction exists which allows discarding of non-equilibrium structures. We describe it in two dimensions.

Let us consider the curve (G) representing the function

$$g(y') = f(y')\sqrt{1+y'^2}$$

with $y' = \mathrm{d}y/\mathrm{d}x$ (Fig. 3.6). The tangent to this curve at the point M cuts the coordinate axis in A. Point M corresponds to a point on the crystal surface of coordinates x_M, y_M, where the slope is y'_M. According to (2.24), the slope of MA is equal to $-x_M\delta\mu$. The equation of the staight line MA is thus

$$\frac{g_M - g_A}{y'_M - y'_A} = -x_M\delta\mu\,.$$

From the definition of g and from $y'_A = 0$, it follows

$$g_A = g_M + x_M y'_M \delta\mu = f(y'_M)\sqrt{1+y'^2_M} + x_M y'_M \delta\mu$$

and this is equal to $y_M\delta\mu$ according to (2.25). We can conclude that, if the straight line MA is a double tangent to the curve (G) at the points M and N, the two points (x_M, y_M) and (x_N, y_N) on the crystal surface corresponding to M and N coincide. In fact, $x_M\delta\mu$ and $x_N\delta\mu$ are both equal to the slope of MA, and $y_M = y_N = g_A$ (Fig. 3.6). On the contrary, the crystal slopes y'_M and y'_N, are different. The double tangent

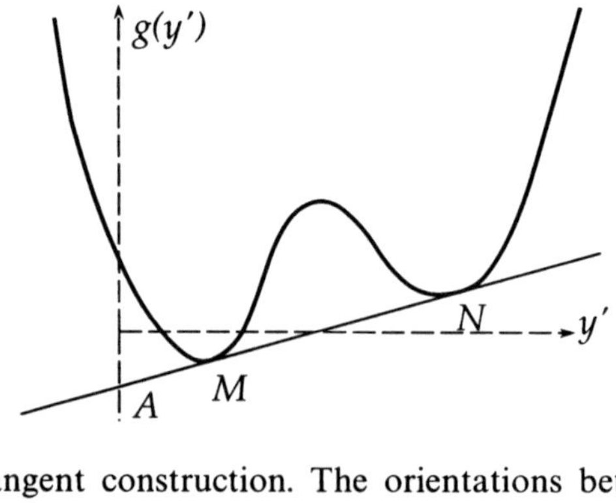

Fig. 3.6. Double-tangent construction. The orientations between M and N do not appear in the equilibrium shape of the crystal.

therefore corresponds to an angular point. In particular, if one of the coexisting orientation is a high-symmetry one (a facet), the double tangent corresponds to a sharp edge joining of the facet with the neighbouring rounded part. One should note the connection between sharp edges and surface melting first investigated by Nozières (1991).

The analogy with phase transitions is in fact complete: if one prepares a vicinal surface cut along a forbidden orientation (which does not appear in the equilibrium shape), the surface *facets*, that is it spontaneously decomposes into the stable, coexisting orientations, just as in a liquid-gas phase separation in the unstable region (Mullins 1961). This has been observed, for instance, for vicinals neighbouring the $\{311\}$ facet of silicon (Song & Mochrie 1994). The equilibrium shape shows indeed a sharp-edge joining of the (113) facet with the crystal (Suzuki *et al.* 1995).

3.7 The shape of a two-dimensional crystal

Two-dimensional crystals are very easy to draw, hence their popularity. One can draw them with facets and angular points. However, real two-dimensional crystals also exist, namely the atomic height islands or terraces which can be found on the surface below T_R, and slightly out of equilibrium (see e.g. Michely *et al.* 1993). They show neither facets nor angular points.

The absence of facets results from the fact that a line, and especially a step, has $T_R = 0$ independently of its orientation. In fact, a step always has kinks at any finite temperature. Thus, its line tension $\gamma(\theta)$ is analytic at any finite temperature. We proved this in special cases (*cf.* section 2.5).

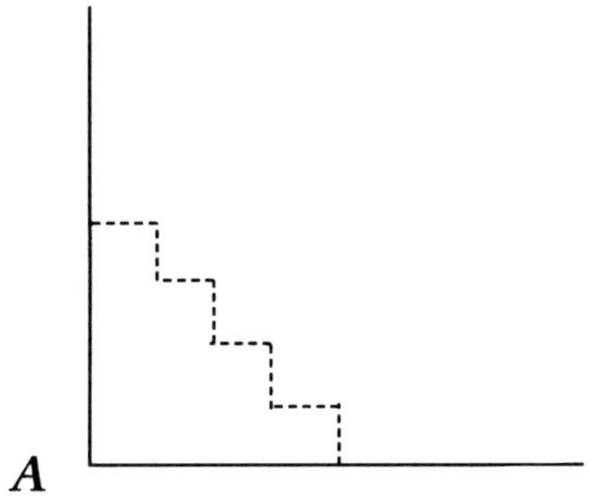

Fig. 3.7. Impossibility of an angular point in a two-dimensional crystal. If one assumes that the crystal has an angular point in A, one shows that the shape given by the dotted line has a lower free energy, which contradicts the starting assumption.

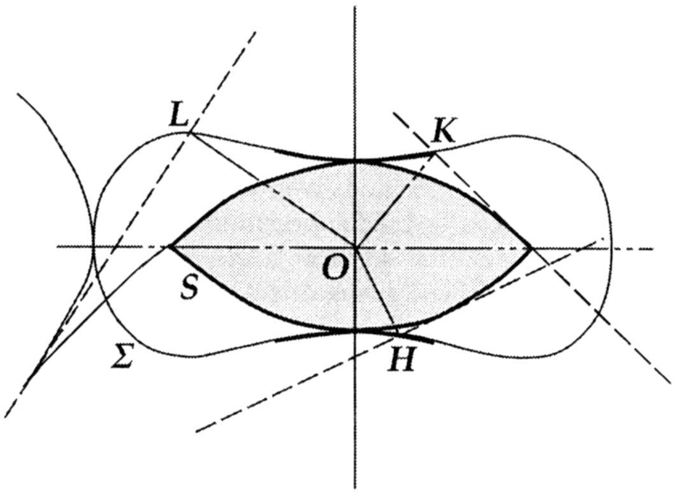

Fig. 3.8. Appearance of an angular line in the equilibrium shape of a crystal (shaded). The intermediate orientations (e.g. the vertical one) are unstable, and do not appear in the equilibrium shape (S). The pedal of Wulff's plot (Σ) is multivalued. The physically meaningful sheet is represented by a bold line. One of the non-physical sheets is also shown on the left (thin line).

The impossibility of an angular point for a large two-dimensional crystal has been shown by Van Beijeren & Nolden (1987), whose reasoning we will reproduce in somewhat greater detail. These authors assume that an angular point exists separating two orientations D and D' (Fig. 3.7). Then they show that this is impossible by considering a different crystal shape, bounded by a broken line consisting of pieces of the two orientations D, D', D, D' repeated a large number of times. The energy is the same, since the two orientations are coexisting at the corner, but one gains an entropy proportional to the length ϵ of the broken line. It is true that now the crystal area is less, but only of a quantity proportional to ϵ^2. We only need to perform a homothety transformation to regain the primitive area, and this only costs an increase in the perimeter, and thus in the surface tension, proportional to ϵ^2. The final crystal has a lower free energy than the initial one, which could not have had its equilibrium shape.

An example of such an impossible crystal shape has been drawn nevertheless in Fig. 3.8: its purpose is to illustrate how the double-tangent construction looks like in a polar plot of the surface tension σ. The angular point in the figure, indeed, means that all orientations between those of the edges are forbidden; and the equilibrium branch of the surface free energy is defined only for those orientations which are in the crystal shape. Its polar plot consists thus of two disconnected pieces (bold lines in the figure). The dashed-line part of the plot, were Wulff's construction applied to it, would provide our crystal of non-convex parts, which would give it quite a fishy look...

The analog of the double-tangent construction consists of replacing the dashed line by the locus of points which are mapped by the Wulff construction into the angular point itself. It is not difficult to see that this locus is a circular arc passing through the angular point. This is the only physically meaningful way to draw the polar plot of σ near unstable orientations; as a corollary we can state that any rounding of the sharp edge must correspond to a surface tension curve lying entirely inside this circular arc.

3.8 Adsorption of impurities and the equilibrium shape

The experimental realization of equilibrium shape is no easy matter. One obvious obstacle is represented by metastable states, in which the crystal may be trapped during growth. Escape from these states may not be easy except at very high temperature. Another, more subtle, hindrance to making equilibrium shapes is impurities.

Impurities are always present during growth. Some of them diffuse into the bulk, and are not dangerous. Some others segregate to the surface, where they can alter the surface free energy and thus the equilibrium shape. Gaseous impurities are present in the observation chamber, and if they are adsorbed to the crystal surface they may also have an effect on surface properties.

In order to investigate the action of adsorbed atoms on a crystal surface, let us consider a two-dimensional crystal with nearest and next-nearest neighbour two-body interactions ϕ_1 and ϕ_2 (Fig. 3.9). Consider also an adsorbed layer of atoms whose binding energy is W_1 on the (01) face (Fig. 3.9a) and W_2 on the (11) face (Fig. 3.9b). For simplicity, we will limit ourselves to $T = 0$.

Removing the atom C at the corner in Fig. 3.9a costs an energy $2\phi_1+\phi_2+2W_1$, since two impurity atoms must also be removed. Replacing it by an impurity (Fig. 3.9c), gains an energy $-W_2$. The total energy balance is thus $\Delta E = 2\phi_1 + \phi_2 + 2W_1 - W_2$. Assume now that $2\phi_1 + \phi_2 > 0$: this implies that the face (11) does not belong to the equilibrium shape (at $T = 0$). On the other hand, if $2W_1 < W_2$ and $|2W_1 - W_2| > 2\phi_1 + \phi_2$, in the presence of adsorbed impurities the face (11) will appear in the 'equilibrium' shape!

Impurities have thus a 'morphactant' effect–a term proposed by Eaglesham *et al.* (1993a) from the Greek μόρφὴ for 'shape, form', and in analogy with 'surfactants'–on equilibrium crystal shapes. They may also affect kinetics, as discussed in sections 6.7 and 11.12.

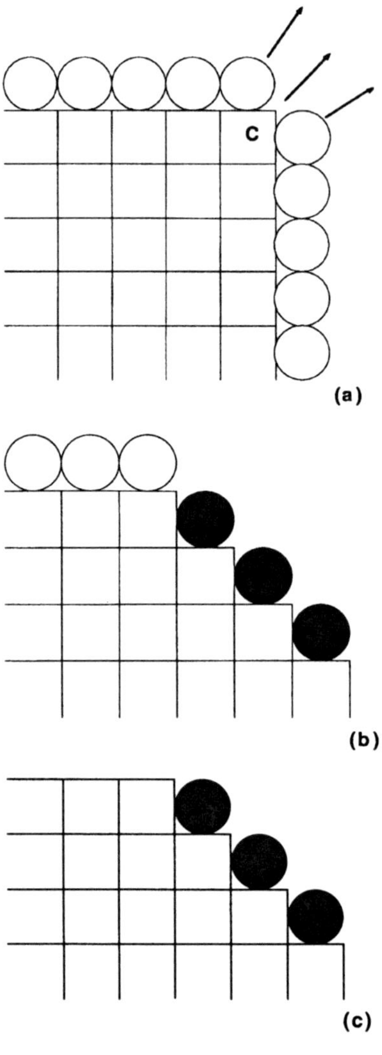

Fig. 3.9. a) Adsorbed layer of impurity atoms on a cubic crystal. b) Absorbed layers of impurity on (10) (open circles) and (11) (closed circles) orientations. c) Stabilization of (11) faces by adsorption. The corner atom C in (a) has been replaced by an impurity (closed circle). Next-nearest neighbours interaction can make this configuration favourable, and lead to the appearance of (11) facets.

3.9 Pedal, *taupins* and Legendre transform

We tried to keep up with fashion by investigating the relation between Wulff's theorem and the Legendre transform: however, knowing what a Legendre transform is, does not allow us to ignore what a pedal is, and how to construct it!

To the French, in fact, the pedal ('*podaire*') recalls the times when the candidates to the École Polytechnique ('*taupins*') had to draw horribly intricate geometrical plans under the control of municipal guards ('*cipaux*') whose goal was to prevent them from cheating ('*crassusser*').

From time to time the *taupins* used to chant a song, suggesting that the guards should not leave their post, lest the *taupins* should start cheating. We write the lyrics below: however, in doing so we comply to the advice of the Red Queen to Alice, and we speak French since we cannot think of the English for it!

Petits cipaux, paux, paux
Qui guardez le taupins, pins, pins,
Petits cipaux, ne vous en allez pas, pas, pas,
Car si vous vous en allez, lez, lez,
Petits cipaux, le taupin crassuss'ra, ra, ra.

3.10 Conclusion

Our proof of Wulff's construction is incomplete since we have not thoroughly discussed the stability and the occurrence of the various singularities one may meet. The stability of the shapes defined by (3.1) would also deserve a deeper analysis, as well as the–related–question of angular points.

We mention in concluding that many beautiful theorems exist–at least in the case of short range interatomic forces–treating crystal shapes, which all represent real mathematical *tours de force* (Abraham 1983, Akutsu *et al.* 1988, Yamamoto *et al.* 1990, Mikheev & Pokrovsky 1991). Also, fluctuations around the equilibrium shape have been investigated by Wolf & Villain (1990a).

Problems

3.1. Using the definition of the Legendre transform of a function f, show that the transform of an angular point function $f(x) = \alpha + \beta|x|$ is a line segment of length 2β.

Solution – Define:

$$f_\epsilon(x) = \begin{cases} \alpha + \beta x & x \geq \epsilon \\ \alpha + \dfrac{\beta x^2}{2\epsilon} + \dfrac{\beta\epsilon}{2} & -\epsilon \leq x \leq \epsilon \\ \alpha - \beta x & x \leq -\epsilon. \end{cases}$$

The conjugate variable $p_\epsilon = f'_\epsilon$ is equal to

$$p_\epsilon(x) = \begin{cases} \beta, & x \geq \epsilon \\ \dfrac{\beta x}{\epsilon}, & -\epsilon \leq x \leq \epsilon \\ -\beta, & x \leq -\epsilon \end{cases}$$

and the transformed function $\hat{f}_\epsilon$ reads

$$\hat{f}_\epsilon(p_\epsilon) = \begin{cases} \alpha, & p_\epsilon = \beta \\ \alpha - \dfrac{\beta p_\epsilon^2}{2\epsilon} + \dfrac{\beta\epsilon}{2}, & -\beta \leq p_\epsilon \leq \beta \\ \alpha, & p_\epsilon = -\beta. \end{cases}$$

Letting now $\epsilon \to 0$, we find

$$\hat{f}(p) = \lim_{\epsilon \to 0} \hat{f}_\epsilon(p_\epsilon) = \alpha, \quad -\beta \leq p \leq \beta$$

which is the equation of a segment of length 2β, as wanted.

3.2. Using the definition of the Legendre transform of a function f, show that the crystal shape near a singular orientation where the projected free energy is $\phi(z') = \gamma|z'| + B|z'|^3$ contains a facet of size 2γ joining the curved part according to (3.10).

Solution – According to the definition, the x coordinate along the crystal surface must satisfy (due to symmetry, we only consider $z' \geq 0$):

$$x = \phi(z')' = \gamma + 3B(z')^2 \geq \gamma$$

hence, inverting this relation, one finds

$$z'(x) = \left(\frac{x-\gamma}{3B}\right)^{\frac{1}{2}}$$

and

$$\begin{aligned} z(x) &= \phi(z'(x)) - z'(x)\ \phi'(z'(x)) \\ &= \gamma z'(x) + Bz'(x)^3 - \gamma z'(x) - 3Bz'(x)^3 \\ &= -\frac{2}{(27B)^{\frac{1}{2}}}(x-\gamma)^{\frac{3}{2}} \quad x \geq \gamma. \end{aligned}$$

The result of the preceding problem shows that the complete profile is as in equation (3.9).

In the same way it can be shown that a free energy ϕ which has a $(z')^2$ term, with a positive coefficient, gives rise to a smooth crystal profile where the rounded part joins the facet as $(\delta x)^2$. We remind that a quadratic term might come, for instance, from $1/r$ step–step interactions. Fitting the equilibrium crystal shape allows, in principle, such interactions to be determined.

3.3. Explain why the equilibrium crystal shape originated by the two-dimensional free energy function

$$\sigma(\theta) = A\cos\theta + B \tag{1}$$

with $0 < A \leq B$, is a circle, even though σ is not isotropic (Lee *et al.* 1975).

Solution – (Arbel & Cahn 1975). It is quite easy to see that the equilibrium shape is actually a circle not centred in the origin of the coordinate system. This paradoxical result is a consequence of an invariance property of the surface free energy: we have seen in chapter 2, equation (2.3), that the equilibrium shape minimizes the function $\phi(y') = \sigma\sqrt{1+y'^2}$, where $y' = \tan\theta$. It is straightforward to see that σ as in (1) yields

$$\phi(y') = A + B\sqrt{1+y'^2}\,.$$

In other words, adding $A\cos\theta$ to the surface free energy only adds a constant to the *projected* free energy, which is the quantity to be minimized. The addition of a constant to ϕ cannot alter the equilibrium properties of the crystal, and in particular it leads to the same equilibrium shape.

We should be anyway worried by the fact that the term $A\cos\theta$ is not in general invariant under crystal symmetries–it might for instance lead to different values of σ for equivalent crystallographic orientations.

Symmetry considerations thus forbid the addition of such a term for most crystals. Arbel & Cahn (1975) in fact show that such a term vanishes for all crystals belonging to the 11 point groups that possess a centre of symmetry. It is forbidden as well for all crystals belonging to the 11 additional symmetry groups which have two rotation axes, or one rotation axis with a mirror plane not through this axis but no centre of symmetry.

There are 10 other point groups, however, for which symmetry is not able to prevent this addition and whose σ can never be given an absolute value.

4
Growth and dissolution crystal shapes: Frank's model

Les matières de géométrie
sont si sérieuses d'elles même
qu'il est advantageux qu'il s'offre
quelque occasion pour les rendre
un peu divertissantes.

Blaise Pascal (*Suite de l'histoire de la roulette*)

The subject of geometry
is so turgid that
any opportunity
to render it entertaining
is welcome.

Crystal growers know that growth shapes possess more pronounced facets than dissolution shapes. Conversely, dissolution shapes appear less faceted. The simplest idea is to look for an equivalent of Wulff's construction. It is in fact possible to find it: one simply needs to replace (in equation (3.5) for the curve S of Fig. 3.3) the surface tension $\sigma(\mathbf{n})$ by $v(\mathbf{n})$, where v is the growth velocity. In general, $v(\mathbf{n})$ has very sharp minima, and very shallow maxima (Fig. 4.5), whilst $\sigma(\mathbf{n})$ does not vary very much (Fig. 2.2). These characteristics, together with Wulff's construction, explain the aforementioned experimental observations.

However, Wulff's construction is not enough during dissolution. Frank proved a very general theorem which is only known to the initiates. We hope to make it better known.

Experimental techniques, temperatures, lengthscales, are the same as for equilibrium shapes. The best suited observational method is still *in situ* electron microscopy. Temperatures must not be too much below the melting point ($2T_M/3 < T < T_M$, say). Typical lengths are of some microns.

A very recommendable reading are the review papers by Taylor *et al.* (1992) and Taylor (1992).

4.1 Frank's model

The physics of growth and dissolution of crystals (Fig. 4.1) is a complex subject. It is tempting to treat first a model which is more related to geometry than to physics: a model in which the velocity of growth or dissolution depends only on the surface orientation. In other words, the rules of the game are as follows.

Take a crystal bounded by a surface S, and let $\mathbf{n}$ be the normal to S at a point M on the surface; then, the velocity of propagation of the surface at M is $\mathbf{v}(\mathbf{n}) = \mathbf{n}\, v(\mathbf{n})$, where $v(\mathbf{n})$ is a given function of the local orientation of the surface.

This very simplified model has been investigated in great detail by Frank (1958). In particular, he has been able to prove a wonderful theorem, which is published, alas, in a paper quite hard to find. We will call the model *Frank's model.* The reader will also find Frank & Ives (1960) and Ives (1961) as useful references.

The essential physical ingredient which is absent from Frank's model is diffusion near the surface: during growth, atoms, energy, impurities need to diffuse up to the surface. During dissolution atoms must be carried away. The real physics is thus non-local. In any case, Ives (1961) has applied Frank's model to definitely realistic situations, as the dissolution of lithium fluoride with several solvents. In the case of growth, we will see in chapter 10 that the most spectacular effect of diffusion near the surface consists of the appearance of instabilities, which force the solidification front to take complicated shapes. The same kind of instability–in this case driven by diffusion of atoms on the surface–also affects the growth shape of two-dimensional 'crystals' (islands) on a substrate (Michely *et al.* 1993).

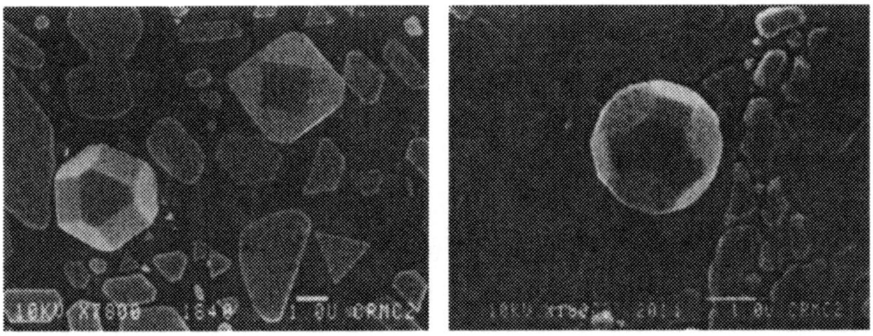

Fig. 4.1. Growth shapes of lead crystals, observed by electron microscopy by Heyraud & Métois near 120 °C. Facets are larger, at a given temperature, than at equilibrium. With permission of the authors.

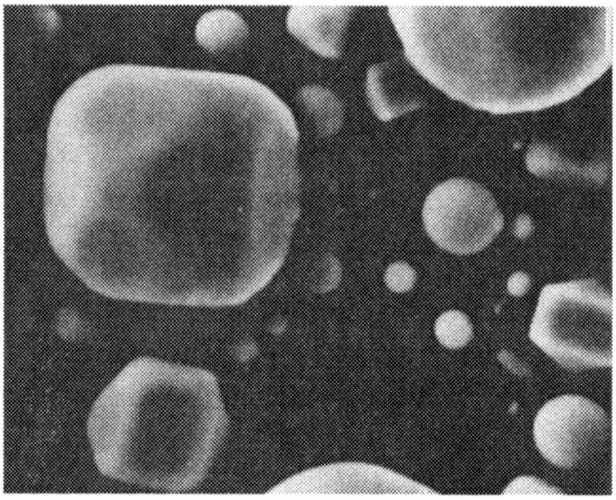

Fig. 4.2. Growth shape at $T_M - 10$ K. Notwithstanding the higher temperature, the growth shape has sharper facets than the equilibrium shape of Fig. 3.1 (with the kind permission of A. Pawlowska).

Nevertheless, it may also happen that growth shapes are observed which are absolutely regular and compatible with Frank's model.

These shapes are most often characterized by sharper facets (at a given temperature) than those seen in equilibrium shapes (Figs. 4.1 and 4.2).

It is clear that the dominant orientations in a growing crystal are those whose growth velocity is the lowest. They thus slow down the growth of the whole crystal. Slow-growing orientations are those which form facets at equilibrium. Since the velocity difference may be fairly large, all other orientations tend to disappear, whence the very sharp facets observed.

Although growth is, in Frank's model, ruled by a very simple automaton, it is tempting to try to simplify the matter further by discovering some geometric properties, for instance:

a) invariant shapes,
b) point trajectories possessing given properties.

We will in the next section discuss the question of invariant shapes. Its limits will be pointed out in section 4.3. Frank's work (point (b) above) will be discussed in section 4.4.

4.2 Kinematic Wulff's construction

In the framework of the geometric growth model we have just defined, a variant of Wulff's construction exists, which allows to find growth or dissolution shapes, under the requirement that they are *self-similar*. That is, the crystal should always stay homothetic to its initial shape during its evolution. In fact, the original formulation of Wulff's in 1901 concerned the solution of this kinematic problem (*cf.* Krug & Spohn 1991). More recently, it has been exploited by Jaccodine (1962) and by Shaw (1979).

Let us see how the construction works: let M be a point on the surface $S(t)$ at time t and $\mathbf{n}$ the normal to S at M. At time $t + \mathrm{d}t$, the point $N = M + \mathbf{v}(\mathbf{n})\mathrm{d}t$ lies on the surface $S(t + \mathrm{d}t)$.

By assumption, the surface $S(t)$ transforms homothetically with respect to a center O as time goes by. If the straight line OM cuts at point N' the plane tangent in N to $S(t + \mathrm{d}t)$, we must have $MN' = \lambda OM\mathrm{d}t$, the coefficient λ being the same for all points M on the surface. Consider now the projection H of O on the plane tangent in M to $S(t)$: it results (Fig. 4.3) in

$$OH = OM\frac{MN}{MN'} = OM\frac{v(\mathbf{n})\mathrm{d}t}{\lambda OM\mathrm{d}t} = \frac{v(\mathbf{n})}{\lambda}\,. \tag{4.1}$$

This condition corresponds precisely to Wulff's construction, if one replaces the surface tension with the velocity of front propagation. Whence we have the following.

Kinematic Wulff's construction – A self-similar crystal growth shape is obtained by considering for each vector $\mathbf{n}$ the point H defined by $\mathbf{OH} = \mathbf{n}v(\mathbf{n})$ and the plane (Π) perpendicular to $\mathbf{OH}$. The crystal surface is the interior envelope of (Π).

It is not difficult to understand the origin of this construction: we have seen in chapter 3 that the Wulff construction can be expressed locally (in Cartesian coordinates) as a Legendre transform of the projected free energy $\phi(\mathbf{n}) = \sigma(\mathbf{n})\sqrt{1 + z_x^2 + z_y^2}$. The transformed function is the equilibrium shape $z(x, y)$, up to a scale factor $2/\mu$, which is fixed by fixing the crystal volume. In other words, the Wulff construction–as well as the Legendre transformation–gives a 1-parameter family of surfaces, $S(\tau)$,

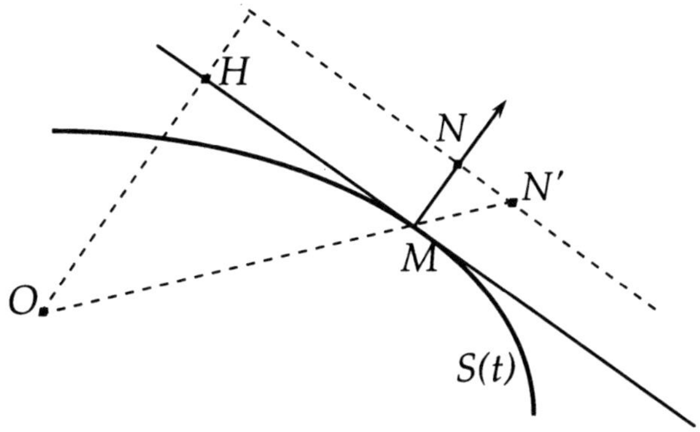

Fig. 4.3. Growth of a crystal (the case of dissolution is analogous).

which can be written locally as a function $z = z(x, y; \tau)$. Differentiating z, we find

$$dz = z_x dx + z_y dy + z_\tau d\tau \,. \tag{4.2}$$

Since we require all surfaces to be self-similar, the shape function z must satisfy the scaling relation

$$z(x, y; \tau) = \tau \zeta(x/\tau, y/\tau) \,.$$

The parameter τ being the scale factor, it follows $dx/d\tau = x/\tau$, and the same for y and z, so that (4.2) reads

$$z = z_x \; x + z_y \; y + z_\tau \; \tau \,. \tag{4.3}$$

Let $e^t \equiv \tau$ and interpret t as the physical time: let also $Z(x, y; t) = z\,(x, y; \tau(t))$. We can write $\tau z_\tau = Z_t$, where $Z_t = \partial Z/\partial t$ is interpreted as the growth or dissolution velocity of the crystal profile Z parallel to the z axis. The growth (or dissolution) velocity is therefore

$$v(\mathbf{N}) = \frac{Z_t}{\sqrt{1 + Z_x^2 + Z_y^2}}$$

where $\mathbf{N}$ is the normal to Z. From (4.3) we find

$$Z_t = Z - Z_x \; x - Z_y \; y$$

and thus

$$v(\mathbf{N}) = \frac{Z - Z_x \; x - Z_y \; y}{\sqrt{1 + Z_x^2 + Z_y^2}}$$

which is the analog of (3.5), the analytic form of Wulff's construction, the only difference being the velocity at the place of the surface tension.

Unfortunately, this kinematic construction is at best useful for growth situations. Indeed, as we are going to see, there is no reason for dissolution to lead to self-similar shapes.

4.3 Stability of self-similar shapes

For the sake of simplicity, we will content ourselves with investigating the case (Fig. 4.4) of a centrosymmetric initial surface $S(0)$. Let $M(0)$ and $P(0)$ be two points on $S(0)$ such that the respective normal directions $\mathbf{n}_M$ and $\mathbf{n}_P$ intersect in a point O. Let v_M and v_P be the corresponding values of the surface velocity. The two points $M(t) = M(0) + v(\mathbf{n}_M)t$ and

$P(t) = P(0) + v(\mathbf{n}_P)t$ belong to the surface $S(t)$. Therefore (see Fig. 4.4):

$$\frac{OM(t)}{OP(t)} = \frac{OM(0) + v_M t}{OP(0) + v_P t} . \tag{4.4}$$

This relation rests on the hypothesis that no angular point forms. We will see in section 4.5 that this hypothesis is false. Nevertheless it is implied by the assumption of self-similarity.

The case of *growth* corresponds to $v_M, v_P > 0$. In this case, (4.4) agrees with self-similarity at long times. One can easily convince himself that the shape of a growing crystal becomes increasingly self-similar as growth proceeds.

On the contrary, in the case of *dissolution* ($v_M, v_P < 0$), (4.4) cannot in general be satisfied for an arbitrary initial shape. The crystal profile becomes flatter and flatter or thinner and thinner before disappearing, except for exceptional cases where $OM(0)/OP(0) = v_M/v_P$. We will examine this question further in section 4.5.

The problem essentially arose because we started from an initial shape which did not coincide with the self-similar one selected by Wulff's construction. Clearly, nothing compels the initial shape to be the Wulff shape!

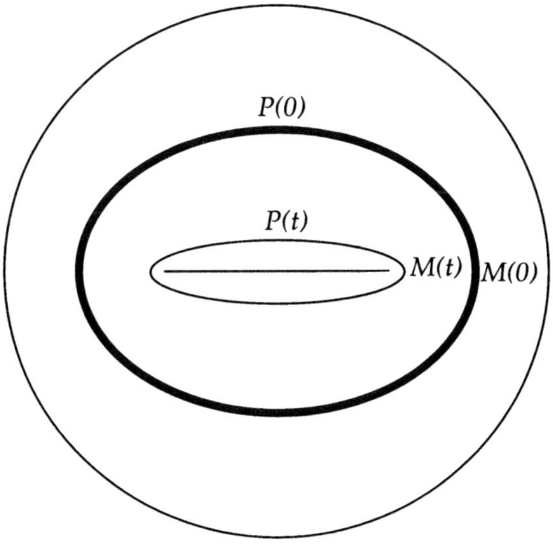

Fig. 4.4. Crystal evolution. Bold line: initial surface. External thin curve: growth shape. Internal thin curves: dissolution shapes. The line segment shows the crystal shape just before complete dissolution, if one assumes (4.2) to be correct (which is not, as we shall show in section 4.5). The value of the velocity ratio $v_M/v_P = 2$ has been assumed.

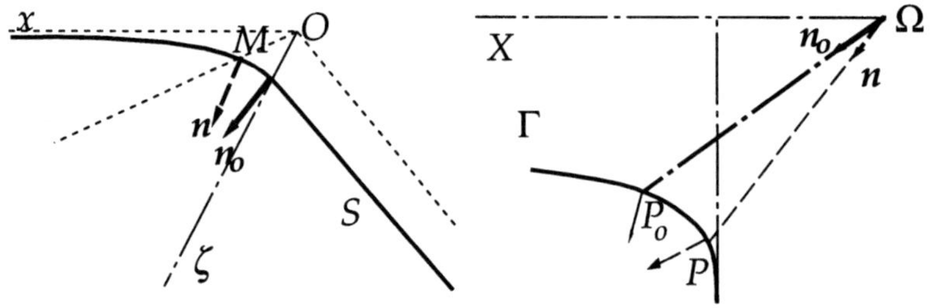

Fig. 4.5. Frank's theorem. The point on the crystal surface where the normal is $\mathbf{n}$ follows a straight line trajectory parallel to the normal in P to the surface Γ. The latter being the locus of points P such that $\mathbf{\Omega P} = \mathbf{n}/v(\mathbf{n})$. The example chosen here has two characteristic properties: a) the growth velocity v vanishes in the direction Ox, so that the section of Γ shown has an asymptote orthogonal to Ox; b) the surface S of the crystal is initially planar to the right of some plane $O\zeta$, and normal to a given direction $\mathbf{n}_0$. Thus, it always retains this orientation to the right of the plane $O\zeta$.

Fortunately, the geometric growth (or dissolution) model allows to determine the evolution of any initial shape once the function $v(\mathbf{n})$ is known.

4.4 Frank's theorem

Frank (1958) discovered a remarkable property of the time evolution of crystal shapes in the geometric model. The trajectory of all points of the surface $S(t)$ where the tangent plane has a given orientation, is a straight line (or a straight half-line, or a segment). Also, this line can be found by a simple geometric construction. We will prove in appendix E the theorem that we are going to state in the following.

Consider (as in Fig. 4.5) the point P defined from an origin Ω by the relation $\Omega P = \mathbf{n}/v(\mathbf{n})$, where $\mathbf{n}$ is the normal to the crystal surface S. Frank's theorem states that all points of the surface where the normal is $\mathbf{n}$, move along a straight line parallel to the normal in P to the surface Γ spanned by P. An equivalent statement is obtained if we call $1/v$ the *reluctance* and the surface Γ spanned by P its *polar plot*.

Frank's theorem – The trajectory of all points where the crystal surface is oriented normally to a given vector $\mathbf{n}$ is a straight line normal to the polar plot of the reluctance at the point corresponding to $\mathbf{n}$.

4.5 Examples

1) A first example (borrowed from Frank) is given by Fig. 4.6.

2) A second application of Frank's construction concerns the ellipsoid seen in section 4.3. We are going to show that the reasoning of that section

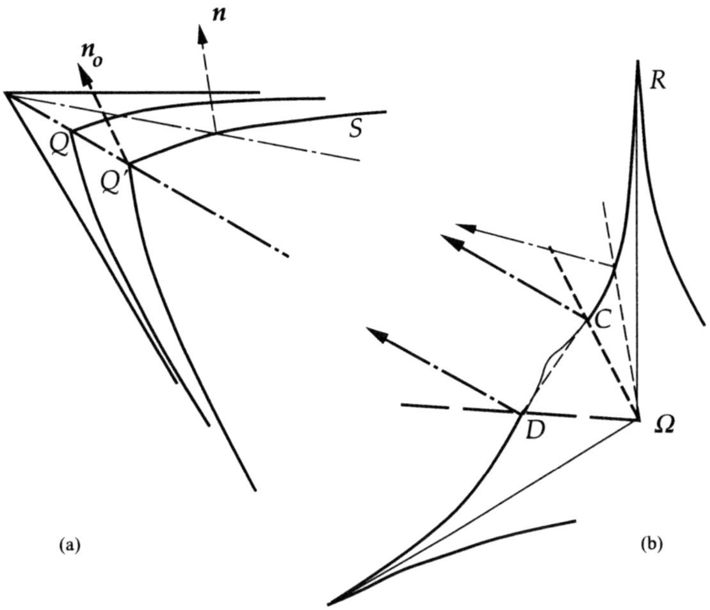

Fig. 4.6. Dissolution of a wedge-shaped crystal a) corresponding to the polar diagram of the reciprocal velocity shown in (b). Points C and D belonging to the latter correspond to the the crystal axis QQ'. The arc CD of the polar diagram has no physical meaning.

needs correcting, as well as Fig. 4.4. In fact, angular points are bound to appear in the course of dissolution, and formula (4.4) cannot apply. For instance, suppose that that sample possesses two axes of rectangular symmetry (Fig. 4.7); each straight line of slope $y' =$ constant will certainly eventually cut one of the symmetry axes in a point Q at a time $t(y')$. At this time, the crystal has an angular point in Q. Note, however, that if the surface tension is included, the angular points in general disappear (Yokoyama & Sekerka 1992).

3) The last example, borrowed from Ives (1961), concernes the case of a concave initial profile (Fig. 4.8). The polar plot of the reluctance $\mathscr{R}$ (being that of lithium fluoride) is shown as a dotted curve. The initial profile (bold face continuous line) has a cylindrical symmetry of axis O. In order to know the trajectory of all points whose tangent is parallel to the tangent in M to the initial surface, we draw the line OM cutting the reluctance plot $\mathscr{R}$ in P, and we find the normal $\mathbf{n}$ in P to $\mathscr{R}$. The sought-for trajectory is the straight line parallel in M to $\mathbf{n}$. It cuts in Q the trajectory $\mathscr{A}$ of the points whose tangent is at 45° with respect to the

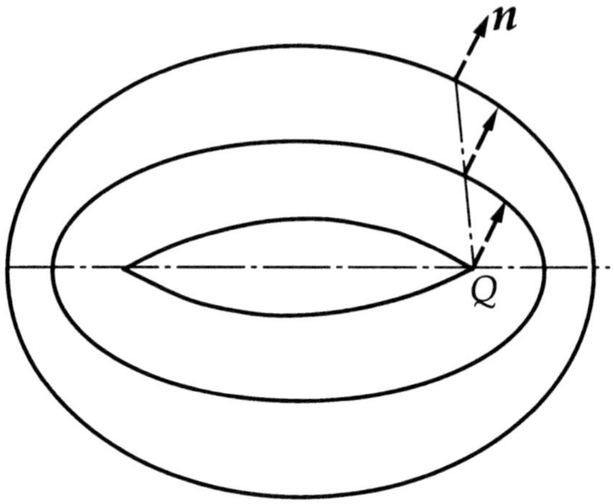

Fig. 4.7. Dissolution of an ellipsoid, and formation of an angular point.

surface. Thus, the tangent in Q to S takes different left and right values. $S(t)$ has therefore an angular point (thin continuous line).

One might wonder about the stability of Frank's construction face to shape fluctuations during time evolution. Fluctuations would certainly lead to a smoothing of the angular points, and thus to the appearence of a fast growing, 45°-tangent point. This would therefore lead to an instability, with the appearence of a thin finger. However, the geometric model is no longer physically valid in this case, because the solvent flow in the narrow channel which forms would become impossible.

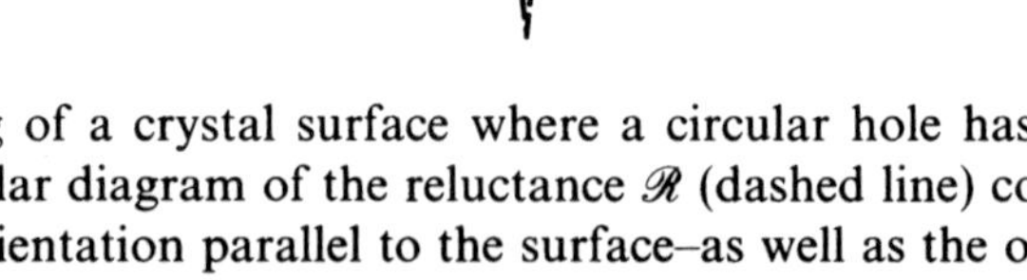

Fig. 4.8. Etching of a crystal surface where a circular hole has been dug (bold full line). The polar diagram of the reluctance $\mathcal{R}$ (dashed line) corresponds to the case when the orientation parallel to the surface–as well as the orthogonal one–is very hard to etch. According to Ives (1961) the case of lithium fluoride shows such a diagram. The trajectory of points on the crystal surface whose normal is parallel to OM, is a straight line cutting a symmetry axis in Q. This is therefore an angular point of the crystal surface S.

Note in Figs. 4.6 and 4.8 the very sharp maxima of the function $1/v(\mathbf{n})$, corresponding to the deep minima of $v(\mathbf{n})$. This explains the spectacular facets typical of growth shapes.

Fig. 4.8 shows facet formation during dissolution. In fact, the criterion for facet formation is that the surface, but not necessarily the crystal itself, should expand. Maybe in this case one should speak of pseudo-facets rather than of facets.

4.6 Experiments

Heyraud & Métois (1987) published several images of crystal growth shapes characterized by sharper edges than on the equilibrium shapes, and by better defined facets. Within the framework of Frank's model, this is explained by the form of the function $v(\mathbf{n})$ (Fig. 4.8).

On the other hand, dissolution shapes are, in general, rounded. This is due to the fact that the orientations which dominate dissolution are the fast ones, and that for such orientations the velocity is nearly orientation-independent. However, one may find surprising not seeing any angular points such as those predicted in section 4.5. We think that this is due to a failure of Frank's model. One should include diffusion on the surface (in vapour phase growth) and far from the surface (in liquid phase growth).

4.7 Roughening and faceting

Faceting of equilibrium shapes was associated to non-rough orientations. This is not the case for growth or dissolution. We will see in chapters 11, 12 and 13 that any growing surface is rough, in the sense of section 1.3.

4.8 Conclusion

This chapter was only meant as a first contact with surface dynamics. We will see in chapters 8, 12, 13 and 14 that it is often necessary to invoke the surface curvature beside the local orientation. In chapter 10 instabilities will enter the game and complicate the picture. We will nonetheless give an example, in chapter 6, where Frank's model can be applied (formulae (6.9) and (6.11)).

5
Crystal growth: the abc

Il y a longtemps que personne	Since long nobody ever thinks
ne songe plus à devancer	of overtaking experience,
l'expérience, ou à construire	nor of building the world
le monde de toutes pièces	on hasty hypotheses.
sur quelques hypothèses hâtives.	Of all this buildings ...
De toutes ces constructions ...	only ruins lasted to our days.
il ne reste plus aujourd'hui que des ruines.	

Henry Poincaré (*La valeur de la science*)

In the preceding chapter we got acquainted with crystal growth. The framework was that of a phenomenological and macroscopic model. In this chapter, we give a short and qualitative, but rather more realistic, introduction to the problem of far-from-equilibrium crystal growth. We will start treating what might be called *wetting* of a solid by another solid. Then, we will see that both in the case of partial (Stranski-Krastanov) and complete (Frank-Van der Merwe) wetting, the two solids can either be commensurate or incommensurate with each other. The essential role of steps and dislocations during growth, will be recalled. Finally, we will introduce the concept of supersaturation, and we will show how the supersaturation can make a smooth surface become rough. This short chapter gives only a sketch of a vast subject; a more complete picture can be found in textbooks such as Tiller's (1992).

5.1 The different types of heterogrowth

We shall use the term *heterogrowth* to designate the growth of a material on another material, for instance Pb on Ge (Fig 5.1c).

Crystals can be grown from a solution, from the melt, from the vapour and even under ultra-high vacuum (UHV), with techniques like molecular beam epitaxy (MBE), and atomic layer epitaxy (ALE).

The theory of growth from dense phases is extremely complex. 'A correct treatment requires solution of Navier-Stokes and the energy transport equations, and the mass transport equations, simultaneously' (Cadoret *et al.* 1977). For this reason, this book is mainly concerned with MBE and related growth methods. However, many concepts introduced here are also useful in other types of growth, especially from a vapour phase. A much more complete study will be found in books by Tiller (1992) and by Chernov (1984) or in review papers (Müller-Krumbhaar 1978, Cadoret 1980, Boistelle 1985, Mutaftschiev 1988).

It is becoming increasingly important, for instance in view of applications to micro- and optoelectronics, to be able to grow different solids one on another. One could for example make 'superlattices' by depositing successively a precise number of layers of CdTe and of ZnTe.

For various reasons, this requires a vapour phase growth technique, or, even better, an ultra-high vacuum procedure such as MBE. It is in fact very difficult to control precisely the number of deposited atomic layers in growth from a fluid phase. Sometimes, layer growth is thoroughly impossible. For example, if one deposits lead on graphite (Fig. 5.1a), one gets a bunch of droplets, very much like the oil droplets which form at the bottom of a frying pan (but much smaller, see problem 5.1). This is called the *Volmer-Weber* type of growth.

Conversely, rare gases adsorbed on graphite form monolayers, and eventually well-spread multilayers. Thus for xenon, at appropriate temperature, five successive layers can be seen as plateaux in the curve of Fig. 5.1b, representing the xenon equilibrium coverage as a function of the pressure or of the chemical potential. This type of layer-by-layer growth is called *Frank-Van der Merwe.*

One often observes an intermediate type of growth, called *Stranski-Krastanov.* The adsorbate grows layer-by-layer for a while, then it forms droplets (Fig. 5.1c), as in Volmer-Weber growth.

These three types of growth processes can be compared to wetting phenomena in liquids. The solid has however two additional features: elasticity and commensurability. Elasticity will be addressed in chapters 15 and 16, and commensurability in section 5.3.

Another type of growth is columnar growth. It is not really crystal growth, since the crystalline regions form fairly narrow columns. It is of interest, however, to keep in mind that crystal are not always easily grown. As a matter of fact, magnetic tapes for recording are often made

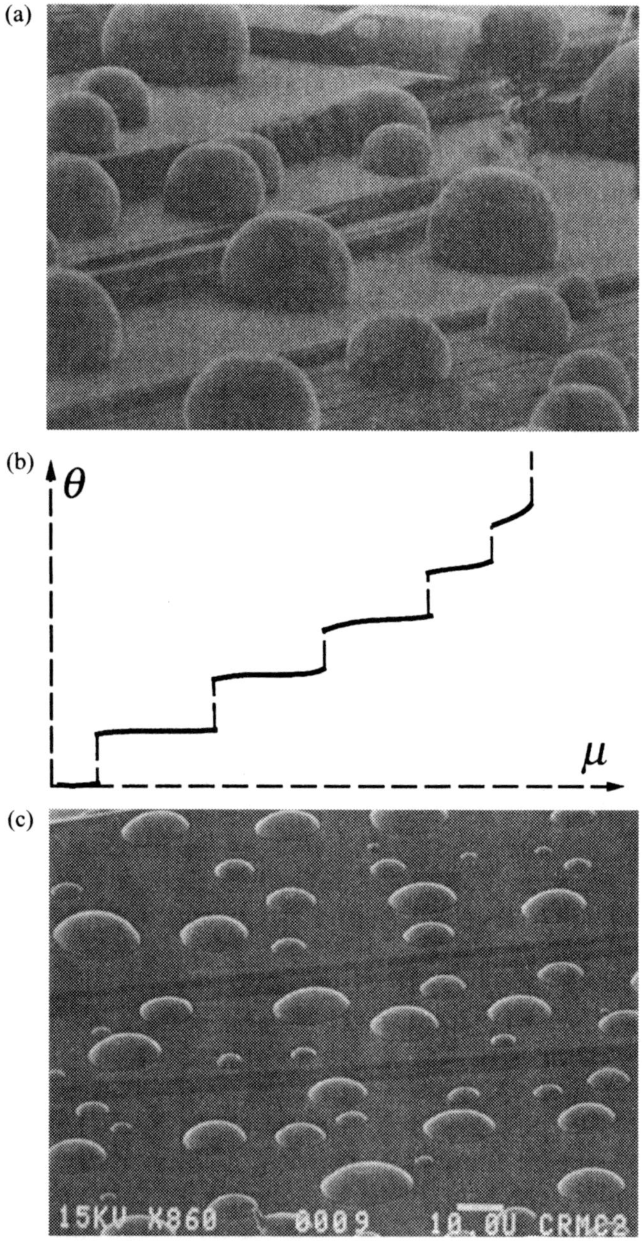

Fig. 5.1. a) Volmer-Weber type of growth (Pb on graphite). b) Wetting isotherms corresponding to a Frank-Van der Merwe type of growth (Xe on graphite), according to Thomy *et al.* (1981). On the abscissae the chemical potential μ, and the coverage θ is on the ordinates. c) Stranski-Krastanov type of growth (Pb on Ge(111)). The diameter of the bubbles in a) and c) is of the order of 5 μm. (Pictures a) and c) with permission by J.C. Heyraud & J.J. Métois.)

of columnar deposits of, e.g. Co-Ni alloys evaporated on polyesters (Bales & Zangwill 1990b, Buschow *et al.* 1993, Abelmann 1994).

5.2 Wetting

Nearly two centuries ago, the British physicist Young was able to show that a liquid drop in thermodynamical equilibrium on a flat substrate, forms with the latter an angle θ which only depends on the 'surface tensions' σ_{sa}, σ_{av}, and σ_{sv} of the three interfaces (substrate-adsorbate, adsorbate-vapour, and substrate-vapour, respectively). Fig. 5.1 shows that the same can be true for a solid adsorbate, too. Assuming that the surface tension is isotropic, as it is for a liquid, Young's relation reads:

$$\cos\theta = \frac{\sigma_{sv} - \sigma_{sa}}{\sigma_{av}} . \tag{5.1}$$

The rounded forms observed in Fig. 5.1a suggest that such an assumption is not too bad also in the case of lead on graphite.

Formula (5.1) loses its meaning when

$$\sigma_{sv} > \sigma_{av} + \sigma_{sa} . \tag{5.2}$$

In this case a Frank-Van der Merwe type of growth is expected. Qualitatively, one immediately sees that the substrate-adsorbate interface cannot be stable unless its free energy (per unit surface) σ_{sa} is not too large.

The transition from wetting to non-wetting, and related phase transitions, have been the subject of many investigations in the second half of the twentieth century (Cahn 1977, De Gennes 1985, Dietrich 1989)

5.3 Commensurate and incommensurate growth

A layer of atoms adsorbed on a foreign substrate at equilibrium is subjected to two competing effects: the adatom-adatom interaction (described by a Hamiltonian $\mathscr{H}_{aa}$) favours an interatomic distance, a, which is the natural distance in a free adatom layer; the adatom-substrate interaction $\mathscr{H}_{as}$, on the other hand, forces the adsorbate to the lattice parameter of the substrate, b. It is nearly obvious that, for a given *lattice misfit* $(a-b)/b$ the incommensurate structure ($a \neq b$) is stabilized by strong adatom-adatom couplings, compared to the adatom-substrate interaction; if the latter is the stronger, the commensurate ($a = b$) structure is stable. In this case, we say there is *epitaxy*. It is less obvious that the commensurate structure is stable even when $\mathscr{H}_{as}$ is weak, provided the misfit is small enough. In fact, the adsorbate-adsorbate coupling energy we lose by letting the interatomic distance vary from a to b is proportional to $(a-b)^2$, whilst the adsorbate-substrate energy gain is independent of $a - b$.

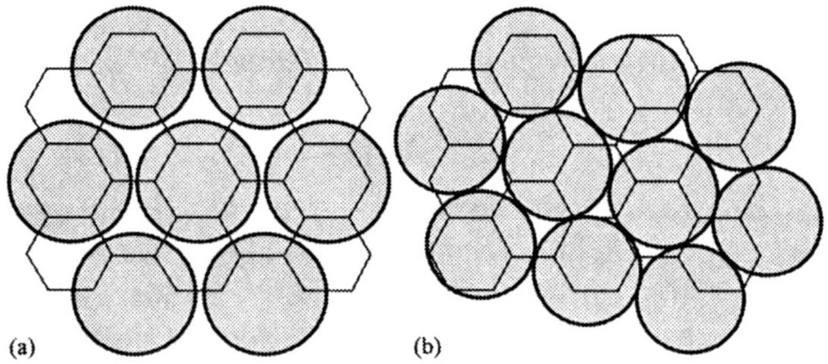

Fig. 5.2. Commensurate structure of a monolayer of krypton adsorbed on the cleavage face of graphite (a), and incommensurate structure of a monolayer of argon on graphite (b).

The rare gas monolayers adsorbed on graphite illustrate this phenomenon (Thomy *et al.* 1981). They are always incommensurate for neon and argon, whose lattice misfit with graphite is large. On the other hand, krypton is commensurate if its partial pressure P is not too high. Increasing P leads to a commensurate-incommensurate transition. In the case of xenon, which is normally incommensurate, the reverse happens: raising P makes a solid monolayer of xenon on graphite become commensurate, just before the second monolayer starts forming. Helium appears as quite a special case: the smallness of its Bohr radius is compensated by its large quantum zero-point motion, allowing a commensurate phase to exist.

The commensurate structure of krypton on graphite (Fig. 5.2) has a rhomb-shaped lattice cell, whose side is equal to $b_0\sqrt{3}$, where b_0 is the

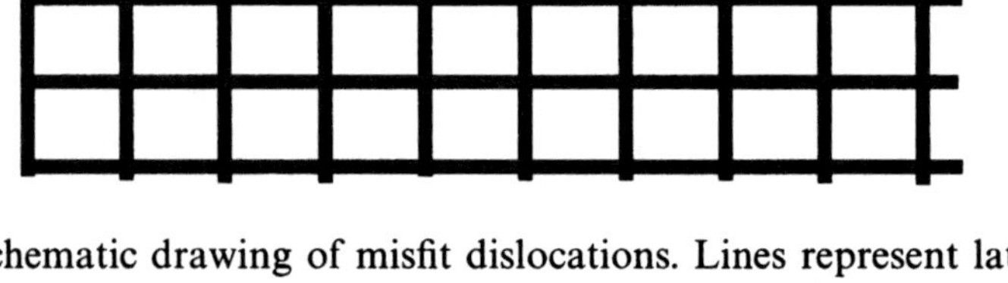

Fig. 5.3. Schematic drawing of misfit dislocations. Lines represent lattice planes. Thick lines are for the substrate lattice planes, and thin lines for the adsorbate ones.

parameter of the adsorption site lattice. This is the reason why this structure is called $\sqrt{3} \times \sqrt{3}$.

What happens when the first monolayer is commensurate and one increases the number h of layers? Clearly, when h becomes very large the structure becomes incommensurate–at equilibrium. In fact, the energy gain in letting the adsorbate take its natural interatomic distance a is proportional to the layer volume, and eventually overcomes the energy loss due to giving up commensuration, which is only proportional to the interface area. If the first layers stay nonetheless commensurate, the appearance of the incommensurate structure requires the introduction of *misfit dislocations* (see Fig. 5.3). The critical thickness for the appearance of misfit dislocation has been investigated in many works. They are reviewed for instance in Jesser & van der Merwe (1988).

5.4 Effects of the elasticity of the solid

Layer-by-layer growth is hindered by the elastic constraints at the solid-solid interface.

To see this, consider the case when the first few layers are commensurate. The following layers will in general force the appearance of misfit dislocations. This costs an energy proportional to the interface area L^2. We now take all the adsorbate above the dislocations, whose height is h, and we make a sphere out of it. The sphere radius is of order $(L^2h)^{\frac{1}{3}}$ and its surface is of order $(L^2h)^{\frac{2}{3}} = L^2(h/L)^{\frac{2}{3}}$. Thus, if L/h is large enough, it is advantageous to make a spherical adsorbate.

A deeper discussion of elasticity is deferred until chapter 15. A sample discussion can be found in Bruinsma & Zangwill (1986) and in the last chapter of Zangwill's textbook (1988). Let us only mention here that elasticity may force into different orientations two crystals in contact. This effect is known as orientational epitaxy and was predicted by Novaco & McTague (1977, 1979). It has been observed for monolayer films of rare gases physisorbed on solid substrates. The Novaco-McTague effect takes place in the incommensurate phases, when the adatom-adatom coupling is larger then the adsorbate-substrate interaction. This allows a harmonic approximation to be used (see problem 5.3). Near a commensurate-incommensurate transition, anharmonicities become important, and the treatment more complicate: see Villain (1978).

5.5 Nucleation, steps, dislocations, Frank-Read sources

Imagine growing, for instance by MBE, a crystal delimited by a high-symmetry plane, (111) or (001). Atoms land onto the surface from the beam, are adsorbed there and become adatoms. What do they do next?

We have taken care of choosing a substrate temperature where surface diffusion is effective. Thus, each single adatom wanders about until it finds a good reason for making a halt.

The good reason might be another adatom. They form then a pair, or *dimer*, which is likely to be stable–that is, not willing to break up into two

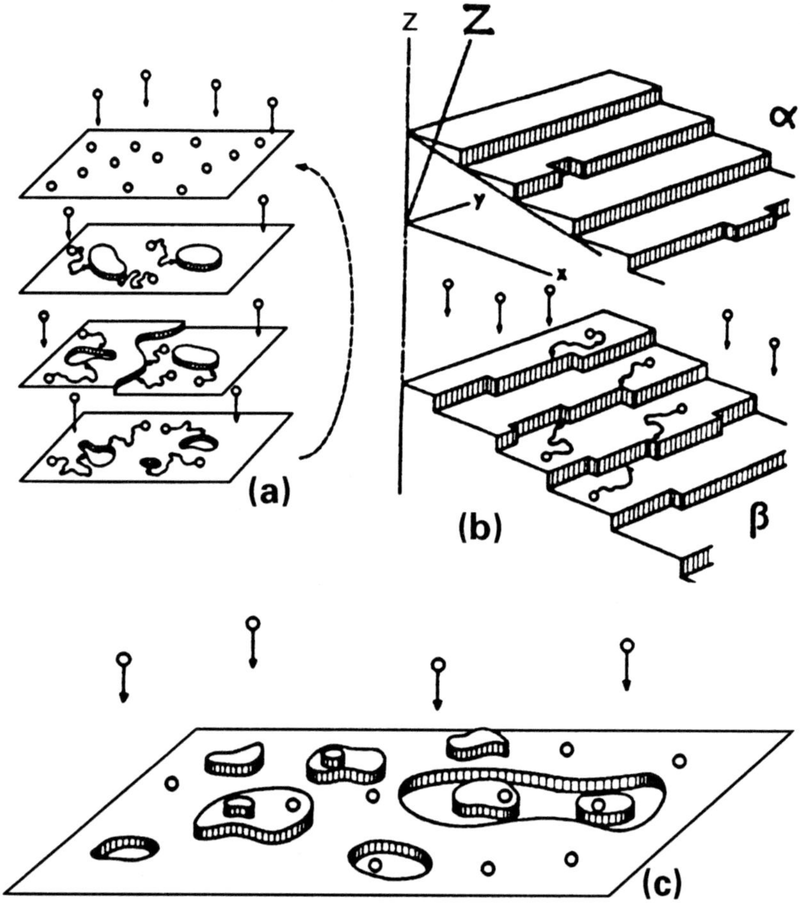

Fig. 5.4. Growth of a high-symmetry face during MBE is sketched in (a) and (c), while (b) describes deposition and growth of a vicinal (stepped) surface. In case (a), starting from a relatively smooth surface, one gets finite-size islands, or terraces, after whose coalescence finite-size holes are left. The surface then becomes again locally smooth, and the cycle repeats. This is the origin of diffraction oscillations. Case (b) is, in principle, the best technique for growing a good crystal. The steps are smoother at rest (α) than during growth (β). Case (c) may be representative of low-temperature growth, when surface diffusion is slow. The surface is rough due to, for instance, random fluctuations in the beam intensity.

adatoms again–at low enough temperature. Other adatoms will come to join the dimer, which constitutes thus the nucleus of a whole terrace. The same process is going on farther away, and terraces keep on forming on the surface. When terraces reach a given size, they will efficiently compete with each other for incoming adatoms, no new nucleation will take place, and terraces grow until they coalesce to yield a new complete layer.

If we now increase the substrate temperature, the dimers will no longer be stable enough to act as nuclei. It is likely that *trimers* will be, and so on. The whole problem will be addressed in detail in chapter 11.

At still higher temperature, critical nuclei are clusters of many atoms, and the distance between them increases. Eventually, growth no longer takes place through nucleation of terraces: rather, adatoms diffuse far enough to directly reach pre-existing steps. In fact, any real surface is atomically flat only up to a certain typical lengthscale, depending mostly on sample preparation condition. Steps are usually present due to an 'imperfect' cut (*miscut*) away from the high-symmetry direction. Their distance rarely exceeds a few microns, and is typically around 100 nm (Fig. 5.4).

Also, steps are created by screw dislocations cutting the surface. They are very rare in MBE deposited samples, more common in large crystals grown from the liquid phase. One often finds such dislocations grouped into pairs of opposite sign, known as *Frank-Read sources* (Fig. 5.5). A spiralling step originates from the end of both dislocations. This structure, predicted by Burton, Cabrera & Frank (1951), has been observed by means of various microscopy techniques, for example (Fig. 5.6) by electron microscopy following 'decoration' (Bethge 1990).

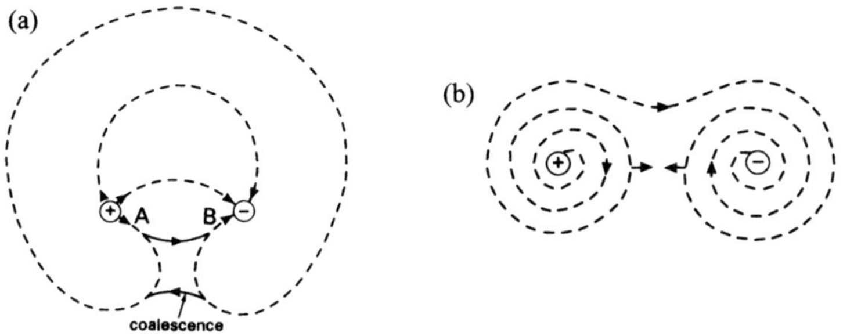

Fig. 5.5. Frank-Read sources at weak (a) and at strong (b) supersaturation (after Nozières 1991).

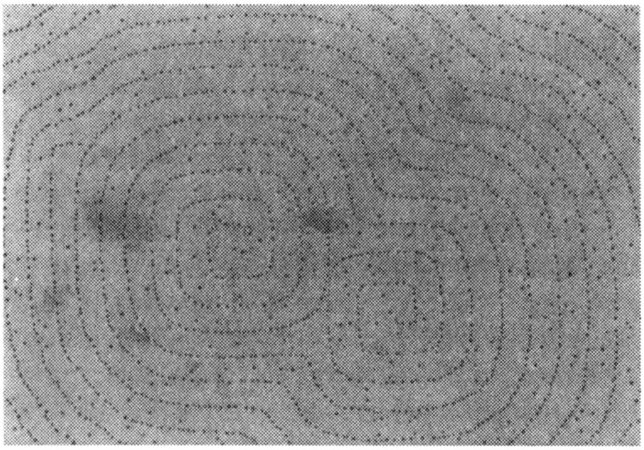

Fig. 5.6. Growth spirals formed by steps issuing from screw dislocations. The picture has been taken by electron microscopy following decoration of the surface with gold clusters with a radius of approximately 5 nm. Gold sticks to the steps and thus makes them visible (with permission by Heyraud & Métois, CRMC2, Marseille.)

5.6 Supersaturation

Liquid or vapour phase growth is the faster the larger the difference

$$\Delta\mu = \mu - \mu_{eq} \tag{5.3}$$

between the chemical potential μ of the fluid phase and the equilibrium one. This difference is called *supersaturation.* It can be related to the adatom density ρ_1, since the latter is related to the chemical potential by the grand-canonical distribution,

$$\rho_1 = \exp(-\beta W_a + \beta\mu)$$

where W_a is the energy of the adatoms. This formula is correct for an appropriate choice of the origin of μ. However, it is not necessary to specify this choice because one can also write

$$\rho_1 = \rho_1^0 \exp(\beta\Delta\mu) \tag{5.4}$$

where $\rho_1^0 = \exp(\beta\mu_{eq})$ is the equilibrium adatom density.

The excess chemical potential $\Delta\mu$ depends on the partial pressure (in the case of growth from the vapour), on the concentration (in the case of growth from a solution), etc. The concept of chemical potential can also be used to take into account the beam intensity in MBE (see chapter 6), or the curvature of the surface (chapter 8) or of steps (chapters 2 and 10).

One can also define a reduced supersaturation

$$\frac{\rho_1 - \rho_1^0}{\rho_1^0} = \beta\Delta\mu\,.$$

The relation of the excess chemical potential with the density ρ_v of the vapour in contact with the surface can be easily derived if the vapour is supposed to be an ideal gas–a good approximation even for water vapour–at pressure P. Then, the density in the vapour is $\rho_v = \rho_{sat}\exp(\beta\Delta\mu)$, and

$$\frac{\rho_v - \rho_{sat}}{\rho_{sat}} = \frac{P - P_{sat}}{P_{sat}} \approx \beta\Delta\mu = \frac{\rho_1 - \rho_1^0}{\rho_1^0} \tag{5.5}$$

where P_{sat} is the pressure of the saturated vapour (in equilibrium with the solid).

Note that in the language of MBE, where atoms reach the surface at a rate F under ultra-high vacuum, and leave the surface at a rate ρ_1^0/τ in the absence of the beam, the reduced supersaturation results from the equality $P/P_{sat} = F\tau/\rho_1^0$, so that

$$\beta\Delta\mu = \frac{\rho_1 - \rho_1^0}{\rho_1^0} \approx \frac{F\tau - \rho_1^0}{\rho_1^0}$$

and the adatom density is given by $\rho_1 \approx F\tau$.

5.7 Kinetic roughening during growth at small supersaturation

To an experimentalist, a surface looks rough if it is all bumps and holes, that is, if the free energy of such defects is not too large. Consider a bump of monatomic height–also called *terrace* or *island*–and radius R. If γ is the step line tension, that we assume isotropic for the sake of simplification, the free energy of an island is (Fig. 5.7):

$$\Delta F = 2\pi\gamma R - \pi\Delta\mu R^2\,. \tag{5.6}$$

This quantity vanishes at $R = 0$, is negative for large R and has thus a maximum of order $\gamma^2/\Delta\mu$ for R equal to

$$R_c = \frac{\gamma}{\Delta\mu}\,. \tag{5.7}$$

If we want large terraces to form easily, we need a *nucleation barrier* $\Delta F(R_c) = \pi\gamma^2/\Delta\mu$ not larger than the thermal energy $k_B T$. Elwenspoek & van der Eerden (1987) define a rough surface as a surface for which this

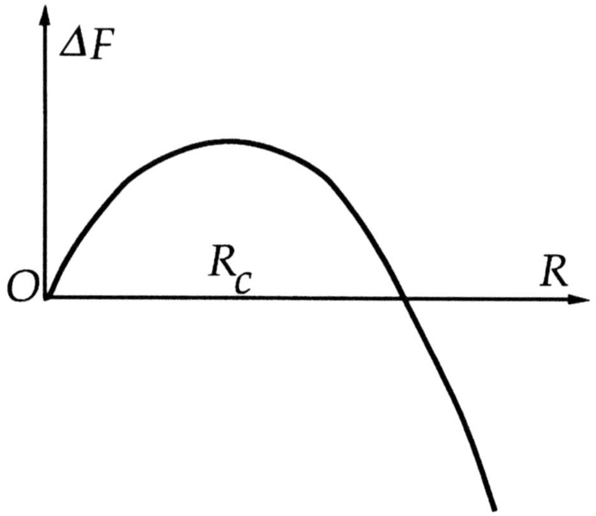

Fig. 5.7. Qualitative aspect of the nucleation barrier, function (5.6).

is true, that is, for which $\Delta\mu > \Delta\mu_c$, with

$$\Delta\mu_c = \frac{\pi\gamma^2}{k_B T} \,. \tag{5.8}$$

In principle, and to the theorist's eyes, in the infinite-volume ('thermodynamic') limit, a growing surface is always rough–thus, for any $\Delta\mu > 0$. We will see this in chapter 12. In practice, even a theorist cannot disregard steps and dislocations: it is their presence that sustains growth when $\Delta\mu < \Delta\mu_c$. Beyond the critical value the role of preexisting defects becomes negligible and nucleated growth dominates.

As a matter of fact, relation (5.7) defines a characteristic length for growth at very low supersaturation from a fluid phase. In particular, if growth takes place from screw dislocations (Fig. 5.6) the distance betweeen successive turns of a growth spiral turns out to be of order R_c. More precisely, it is equal to $4\pi R_c$ with a very good approximation (Burton, Cabrera & Frank 1951). This result, which has an appealing simplicity, is nevertheless not very easy to derive. It requires the approximate solution of a non-linear, second order differential equation.

Many compounds cannot grow at small supersaturation because they are unstable, for instance due to some mechanism such as those described in section 5.4. On the other hand, a too large supersaturation can lead to kinetic instabilities, such as we will investigate in chapter 10. The great advantage of MBE or related techniques is to allow building metastable structures and compounds.

5.8 Evaporation rate, saturating vapour pressure, cohesive energy, and sticking coefficient

Evaporation is often an important factor in crystal growth. The evaporation rate of an element per unit surface (or per site) and unit time can be related to its saturating vapour pressure through global balance at equilibrium. It is however unavoidable to make assumptions which are far from being general.

Let ρ_1 be the adatom concentration per surface site, and ρ_1^0 be its value at equilibrium. If each adatom has a probability $1/\tau_v$ per unit time to evaporate, the evaporation rate is $1/\tau_{ev} = \rho_1/\tau_v$. At equilibrium with the vapour, evaporation must be compensated by sticking. We will assume a sticking coefficient Q independent of energy, which is only true if Q is near unity. In fact, according to Zangwill (1988) Q is often of order unity, most of all for those elements whose vapour is monatomic, such as silicon and a number of metals. On the contrary, in problems of catalysis or molecular adsorption, Q may differ substantially from 1. Molecules are in that case physisorbed, that is, weakly bound, until the moment they dissociate.

Another possible cause for reducing the sticking coefficient is the existence of potential barriers (Iche & Nozières 1976). We will assume the vapour to be monatomic. If ρ_v is the number of vapour atoms per unit volume, the arrival rate of particles per adsorption site is

$$\frac{1}{\tau_{inc}} = \rho_v \int_0^\infty dv_z p(v_z) v_z$$

where $p(v_z)dv_z$ is the probability of the z-component of the velocity being found between v_z and $v_z + dv_z$. The particles hitting the surface with velocity v_z during the time interval δt are contained into a layer of thickness $v_z \delta t$. Note that we are assuming that ρ_v and $p(v_z)$ are the same everywhere in the vapour, even near the surface. This is in fact correct in classical mechanics. The particle velocity distribution function p is then given by Maxwell's law, and at equilibrium we find

$$\frac{1}{\tau_{inc}} = \rho_v \sqrt{\frac{m}{2\pi k_B T}} \int_0^\infty dv_z \exp\left(-\frac{mv_z^2}{2k_B T}\right) v_z = \rho_v \sqrt{\frac{k_B T}{2\pi m}}, \tag{5.9}$$

where m is the particle mass.

At equilibrium, ρ_v equals the saturating vapour density ρ_{sat}, so that the global evaporation rate reads:

$$\frac{1}{\tau_{ev}} = \frac{\rho_1^0}{\tau_v} = \frac{Q}{\tau_{inc}} = Q\rho_{sat}\sqrt{\frac{k_B T}{2\pi m}}. \tag{5.10}$$

The density $\rho_{sat} = N_v/V_v$ of the saturating vapour is obtained by minimizing with respect to the particle number N_v, at constant volume

V_v, the vapour free energy, which is (Landau & Lifshitz 1959a)

$$F(N_v) = F_0 + N_v W_{coh} - k_B T N_v \ln \left\{ \frac{eV_v}{N_v} \left(\frac{mk_B T}{2\pi\hbar^2} \right)^{\frac{3}{2}} \right\}$$

for a monatomic ideal gas. Here F_0 is a constant, and W_{coh} is the cohesive energy of the solid, that is, the energy needed to break up the solid into its constituent atoms at zero temperature, or in other words to transform the solid into an ideal gas. The cohesive energy of a number of elements is shown in Table 1.1, and in Fig. 1.11. All the preceding, as well as the following, formulae neglect the entropy of the solid. Their validity is thus, strictly speaking, limited to low temperature. However, experience shows that they provide order-of-magnitude estimates even very near the bulk melting point.

The minimization yields

$$\rho_{sat} = \frac{N_v}{V_v} = \left(\frac{mk_B T}{2\pi\hbar^2} \right)^{\frac{3}{2}} \exp(-\beta W_{coh}) \,. \tag{5.11}$$

Inserting (5.11) into (5.10) we finally get

$$\boxed{1/\tau_{ev} = Qm(k_B T)^2/(4\pi^2\hbar^3) \exp(-\beta W_{coh}) \,.} \tag{5.12}$$

Note that even when the sticking coefficient Q is 1, the pre-exponential factor has a non-trivial temperature dependence. The Arrhenius activation energy in (5.12) (at low temperature and without additional potential energy barriers) coincides with the cohesive energy. This simple result is maybe surprising, given the complicated character of the evaporation process!

The evaporation probability of a single adatom per unit time $1/\tau_v$ is given by (5.10). According to the Gibbs formula, at not too high temperature the adatom equilibrium density is

$$\rho_1^0 = \exp(-\beta W_a) \tag{5.13}$$

where W_a is the energy spent in transforming an atom bound to a kink (*kinkatom*) into a terrace adatom. In fact, removing an atom from a kink simply results in the motion of the kink position, which does not change the energy of the crystal (the number of dangling bonds per atom does not vary when the kink moves). Equations (5.10), (5.12) and (5.13) yield

$$\frac{1}{\tau_v} = Q\frac{\rho_{sat}}{\rho_1^0}\sqrt{\frac{k_B T}{2\pi m}} = Q\frac{m(k_B T)^2}{4\pi^2\hbar^3} \exp(-\beta W_v) \tag{5.14}$$

where

$$W_v = W_{coh} - W_a \,. \tag{5.15}$$

In the absence of a potential barrier to adatom desorption, W_v is the energy spent in transforming an adatom into a vapour atom. In fact, (5.15) can be straightforwardly obtained with the following argument: the energy W_{coh} needed to vaporize a bulk atom equals the energy needed to vaporize a kinkatom, since the crystal can be dissolved by removing atoms from kinks, and kink motion does not require any energy. On the contrary, removing an atom from the bulk leaves a vacancy, and this process is energetically expensive. Therefore, W_{coh} must equal the energy cost W_a of transforming a kinkatom into an adatom, plus the energy W_v needed for transferring the adatom into the vapour.

If Q is unity for all crystal faces, the evaporation velocity $1/\tau_{ev}$ given by (5.12) is independent of the surface orientation! The result agrees with the observations by Souchière & Vu Thien Binh (1986) for Si. However, we will see in next chapter that it needs not be so. In fact, it is possible (see Fig. 6.2) that the adatom density during evaporation, ρ_1 differs from its equilibrium value ρ_1^0.

Additional details on adsorbate-vapour equilibria can be found in section 5.1 of Tiller's textbook (1992). In particular, one might wonder how the foreign atom density ρ varies with the corresponding partial pressure P. The simplest formula is Langmuir's, $P = K(T)\rho/(1-\rho)$.

Formula (5.12) illustrates the importance of the cohesive energy. As of most bulk properties, this quantity is precisely known for all materials. It is shown in Fig. 1.11 for a few common elements. Its value is typically of a few electronvolts per atom.

5.9 Growth from the vapour

When a material grows from the vapour, its growth rate $\dot{z}$ is equal to the difference between the flux of atoms from the vapour (multiplied by the sticking coefficient Q) and the evaporation rate (5.12). The incoming flux from the vapour has been computed above and it is proportional to the vapour pressure P, and thus to the vapour density ρ_v. The growth rate is therefore

$$\dot{z} = Q(\rho_v - \rho_{sat})\sqrt{\frac{k_B T}{2\pi m}}$$

or, using (5.9) and (5.12):

$$\dot{z} = Q\left(\frac{P}{P_{sat}} - 1\right)\frac{m(k_B T)^2}{4\pi^2\hbar^3}\exp(-\beta W_{coh}) \,. \tag{5.16}$$

This formula coincides with equation (50) of Burton, Cabrera & Frank (1951), only if the T-dependent fraction in front of the exponential is replaced by a temperature-independent phonon frequency ν. In fact, Burton *et al.* arrived at this result by assuming that the evaporation rate per adatom $1/\tau_v$ is given by a purely Arrhenius law, $1/\tau_v = \nu \exp(-\beta W_v)$. Does this assumption lead to a reliable expression for the sticking coefficient Q? It is indeed reasonable to believe that the sticking coefficient has something to do with lattice vibrations, since they ensure energy conservation during adsorption. However, there is no reason to postulate an Arrhenius law with a temperature-independent pre-exponential term. The calculation of Q is surely a difficult problem (Iche & Nozières 1976). In any case, when comparing their formula with experiments, Burton *et al.* did obtain the reasonable result $\nu = 10^{13}$ s^{-1} from the rate of growth of liquid mercury at 0 °C, but in the case of iodine and phosphorus they found $\nu \approx 10^{15}$ to 10^{17} s^{-1}, which is too high for a phonon frequency, but agrees well with our formula (5.16) if $Q \approx 1$.

The preceding discussion may appear confusing to the reader. However, even though the pre-exponential factor is not well known, a simple conclusion can be drawn: *there is an activation energy for crystal growth from the vapour at high supersaturation, and this is the cohesive energy* W_{coh}. This simple result is somewhat spoiled in practice by the fact that the activation *free* energy is somewhat modified at the temperatures where growth and evaporation kinetics are observable. Moreover, the rate of growth is expected to vary *linearly* with the supersaturation (see equation (5.16)). On the other hand, for very low supersaturations, growth proceeds by attachment at spiralling steps issuing from dislocations, and the rate of growth is predicted to be proportional to the *square* of the supersaturation in this case (Burton, Cabrera & Frank 1951).

5.10 Segregation, interdiffusion, buffer layers

When a material contains impurities, these may be attracted by the surface (this is segregation), or on the contrary they may prefer to diffuse towards the bulk. To avoid such 'interdiffusion', the impurities and the bulk can be separated by 'buffer layers' of a third substance.

Fig. 5.8 shows the simulation of diffusion of a nickel atom on a gold substrate at room temperature. Nickel has a strong affinity with gold, nevertheless it diffuses on the surface for a few picoseconds before sinking underneath. Beside their aesthetic attractiveness, these unpublished images provided by T. Deutsch, F. Lançon & J. Eymery (Centre d'Études Nucléaires de Grenoble) have the merit of being derived from relatively sophisticated models, at least more sophisticated than the pair potentials

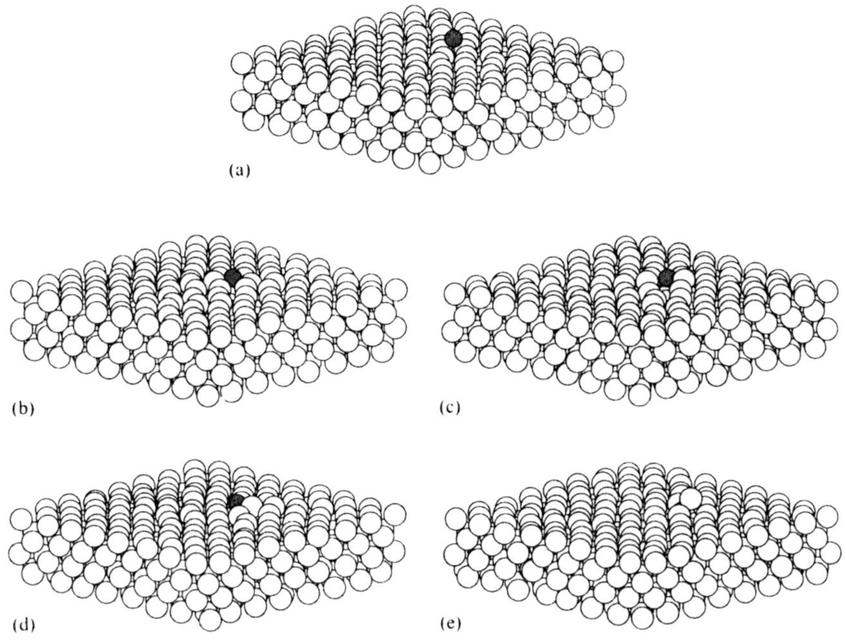

Fig. 5.8. Exchange process between an adatom and a surface atom. A nickel atom (black) has been deposited on the (001) face of Au, and a molecular dynamics simulation based on the embedded atom approximation has been run. Note the tumultuous appearance of the surface during the echange, and its relatively quiet aspect before and after it. Approximately 4 ps have elapsed between the first (a) and the last (e) image (with permission of T. Deutsch, F. Lançon & J. Eymery).

(used in appendix M) and even than the tight-binding approximation, rather succesful for transition metals... if appropriate phenomenological parameters are used. The technique employed by Deutsch *et al.* is a variant of the so-called *embedded atom* approximation. The simulations are performed by molecular dynamics: the equation of motions are explicitly solved. More microscopic than the Monte Carlo method, which consists of determining the system evolution by randomly sampling its phase space, molecular dynamics is extremely time consuming, and only allows considering very short times.

Problems

5.1. Compare the process of lead droplet formation of Fig. 5.1 with that of oil droplet in a frying pan. Why are the latter much bigger?

Solution – The droplets of Fig. 5.1 result from the diffusive motion of individual atoms, and therefore their size, without being atomic, is microscopic. On the contrary, the breaking up of a liquid film into droplets results from a collective and even macroscopic motion, and the surface tension forbids the formation of microscopic droplets. Generally speaking, the size of a liquid droplet is dictated by the competition between gravity and surface tension, and is given by formula (1.13), while in Fig. 5.1 the size is essentially determined by the chemical bonds of the material in the droplet.

5.2. Prove Young's relation.

Solution – One has to write the free energy per unit surface of the system at the line of contact as a function of θ, and minimize with respect to it. That is, the free energy variation corresponding to a small change in the contact line must vanish. The variation of the free energy per unit length is

$$\sigma_{sa}\delta x + \sigma_{av}\delta x \cos\theta - \sigma_{sv}\delta x.$$

Equating this to zero yields equation (5.1).

5.3. Investigate orientational ordering in harmonic approximation.

Solution – The major difficulty lies on the use of linear elasticity, that we will investigate at length later on in chapters 15 and 16; the calculations are quite straightforward. However, the problem is tricky.

Let the two-dimensional vectors $\mathbf{r}$ label the ideal adatom lattice, with reciprocal lattice vectors $\mathbf{g}$. Let $\mathbf{R} = \mathbf{r} + \mathbf{u}$ be the actual adatom positions, due to the (weak) adsorbate-substrate interaction potential $V(\mathbf{R})$. The substrate is characterized through its reciprocal lattice vectors $\mathbf{G}$. At $T = 0$, the Hamiltonian of the adsorbate, considered as an elastic medium, reads (see equation (16.21) below):

$$\mathscr{H} = \sum_r [\frac{1}{2}\lambda(\sum_\alpha \epsilon_{\alpha\alpha})^2 + \mu \sum_{\alpha\gamma}(\epsilon_{\alpha\gamma})^2] + \sum_r V(\mathbf{R})$$

where $\epsilon_{\alpha\gamma} = \frac{1}{2}\partial_\alpha(R_\gamma - r_\gamma) + \frac{1}{2}\partial_\gamma(R_\alpha - r_\alpha)$ is the strain tensor (equation (16.2)) and μ and λ the Lamé coefficients (see section 16.6).

The interaction potential can be expanded in a Taylor series:

$$V(\mathbf{R}) = V_0(\mathbf{r}) + \sum_\alpha \partial_\alpha V(\mathbf{r}) u_\alpha + \dots$$

and Fourier transformed

$$\sum_r V(\mathbf{R}) = \sum_G \tilde{V}_0(\mathbf{G})\delta(\mathbf{G} - \mathbf{g}) + i\sqrt{N} \sum_{G,q,\alpha} \tilde{V}_1(\mathbf{G}) G_\alpha u_q^\alpha \delta(\mathbf{G} + \mathbf{q} - \mathbf{g}) + \dots$$

For an incommensurate phase the first term on the right side, proportional to a Dirac delta which would enforce commensuration, vanishes. We retain only the linear term, and insert it into the Hamiltonian, where we Fourier transform the elastic part; we get thus

$$\mathscr{H} = \frac{1}{2} N \sum_{q,\alpha,\gamma} (K q_\alpha q_\gamma + \mu q \delta_{\alpha\gamma}) u_q^\alpha u_{-q}^\gamma + i\sqrt{N} \sum_{G,q,\alpha} \tilde{V}_1(\mathbf{G}) G_\alpha u_q^\alpha \delta(\mathbf{G} + \mathbf{q} - \mathbf{g})$$

where $K = \lambda + \mu$ is the compression modulus. Minimizing with respect to u_q^α, yields the energy

$$E = \frac{1}{2} \sum_{G,g} \delta(\mathbf{G} + \mathbf{q} - \mathbf{g}) |V_G|^2 \frac{1}{q^2} \left[\frac{1}{\mu} (\mathbf{G} \wedge \frac{\mathbf{q}}{q})^2 + \frac{1}{K + \mu} (\mathbf{G} \cdot \frac{\mathbf{q}}{q})^2 \right] .$$

Let us call θ the angle between $\mathbf{G}$ and $\mathbf{g}$, and let us use the relation $\mathbf{G} + \mathbf{q} = \mathbf{g}$; then

$$E = \frac{1}{2} \sum_{G,g} |V_G|^2 \frac{G^2}{(G^2 + g^2 - 2Gg\cos\theta)^2} [\frac{1}{\mu} g^2 \sin^2\theta + \frac{1}{K + \mu} (G - g\cos\theta)^2] .$$

The angle θ between the incommensurate lattices of the adsorbate and of the substrate is still free: therefore, the system can still gain energy by adjusting θ. It is easy to show that the minimum energy is attained at

$$\cos\theta = 1 + \frac{(1-\rho)^2(1-\eta\rho)}{\rho[2 + \eta(1+\rho^2)]} \qquad \eta \geq \frac{1}{\rho}$$

$$\theta = 0 \qquad \eta < \frac{1}{\rho}$$

where $\rho = g/G$ and $\eta = K/\mu$; the latter is related to the longitudinal and transversal sound velocities in the medium through $\eta = (c_L/c_T)^2 - 1$.

6
Growth and evaporation of a stepped surface

Denn die Zeit ist vorbei,
da die Sibyllen unter der Erde
weissagten.

J.W. Goethe

For the time has gone when
the Sibyls prophesized under the earth.

Although most experimental growth studies are made on (nominally) high-symmetry surfaces, where growth is dominated by nucleation of adatom clusters (discussed in chapter 11), it would seem that the best way to grow a good crystal is to use a substrate cut along a direction 'vicinal' to a high-symmetry orientation, i.e. a 'stepped' surface. In this way, by appropriately choosing the experimental conditions, the crystal may be grown through a regular flow of the steps, avoiding all possible problems related to random nucleation. The theory of the growth of a stepped surface by molecular beam epitaxy (MBE) was devised at a time when MBE was not known, by Burton, Cabrera & Frank in 1951. In their prophetic article, crystal growth was indeed described as a regular flow of the ledges (step flow). Modern, *in situ* experimentation is able to check the model of Burton, Cabrera & Frank (BCF model) by, for instance, reflection electron microscopy (REM).

Indeed, if the temperature is low enough, adatom diffusion slows down and additional terraces and steps can form between preexisting steps (this is just the nucleated growth mentioned before). Similarly, the BCF model can be used to describe step flow during evaporation, and at high enough T clusters of vacancies may appear between the steps. This is a limitation to the validity of the BCF theory. On the other hand, a tendency of the step to form close groups or 'bunches' (*step bunching*) is observed in some instances. Step bunching may be due to impurities, or (in the case of evaporation) to the fact that atoms are preferentially incorporated into the upper step of a terrace. This is the 'Schwoebel effect'.

6.1 General equations

In this chapter, the beginning of a famous article published by Burton, Cabrera & Frank (BCF) in 1951 will be reproduced and somewhat generalized.

Let us consider the growth by MBE of an infinite vicinal surface, formed by terraces of high-symmetry orientation separated by parallel, equidistant steps (Fig. 5.4b). This is, in principle, the best epitaxial mode because it avoids random nucleation of new terraces–a subject addressed in chapter 11–and thus should allow growing better crystals.

Let x be the direction perpendicular to the steps. Between two steps, there is a low density $\rho(x,t)$ of adatoms freshly deposited at a rate F. The density evolution is described by the following equation where $1/\tau_v$ designates the evaporation probability of an adatom per unit time (Fig. 6.1), and primes mean x-derivative, e.g. $\rho'' = \partial^2\rho/\partial x^2$:

$$\dot{\rho}(x,t) = D_s\rho''(x,t) - \frac{1}{\tau_v}\rho(x,t) + F \,. \tag{6.1}$$

The first term of the right-hand side of (6.1) corresponds to diffusion, the second term describes evaporation and the last term corresponds to the deposition of adatoms, either by a beam or from the vapour. The constant D_s is the microscopic surface diffusion constant of adatoms.

Equation (6.1) must be complemented by boundary conditions which are to be satisfied by $\rho(x,t)$ at the steps. It is often a good approximation to write that emission and capture of adatoms at the steps are so rapid

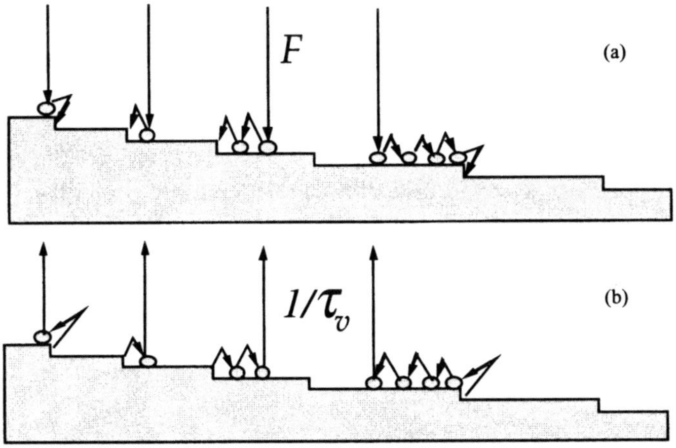

Fig. 6.1. One-dimensional picture of the growth (a) and evaporation (b) of a stepped surface.

that $\rho(x,t)$ has, near any step of position x_n, its equilibrium value ρ_0:

$$\rho(x_n) = \rho_0 \, . \tag{6.2}$$

This value, which depends on the temperature T, is that of the adatom density on the infinite surface at equilibrium with its saturating vapour.

Adatoms, or small adatom clusters, can sometimes be seen by STM or by field ion microscopy (FIM). Step motion (especially during evaporation) is most easily observed by reflection electron microscopy (REM) (Latyshev *et al.* 1989, 1990, 1993, Alfonso *et al.* 1993) or low-energy electron microscopy (LEEM) (Bauer 1992, Bartelt *et al.* 1994). In certain cases, one sees the steps all moving with a well-defined velocity v (*step flow*). The growth or evaporation rate of the crystal is obviously

$$\dot{z} = \frac{v}{\ell} \tag{6.3}$$

if ℓ is the (average) distance between steps. We shall now compute v from (6.1) and eliminate the adatom density $\rho(x,t)$ which is difficult to observe, and therefore not so interesting.

6.2 The quasi-static approximation

Adatom diffusion is generally much faster than step motion. Therefore, it is a good approximation to neglect $\dot{\rho}$ in (6.1). One can thus deduce the current of adatoms,

$$\mathbf{j} = -D_s \nabla \rho \tag{6.4}$$

and therefore the step velocity, since this velocity is proportional to the net flux of adatoms arriving at or leaving the step. Thus, we assume no step motion, and we deduce the step motion. This is just the opposite of a self-consistent treatment! In the case of equidistant steps, it is easy to avoid this quasi-static approximation (see problem 6.1), but in more complicated situations the latter is very useful. The quasi-static approximation may be compared to the Born-Oppenheimer approximation in chemistry, where the electronic motion is treated assuming immobile ions. Here, the adatoms play the part of the electrons, and the steps that of the ions.

For $\dot{\rho} = 0$, (6.1) is solved by

$$\rho(x) = F\tau_v + \lambda \exp(\kappa x) + \mu \exp(-\kappa x) \tag{6.5}$$

where κ (often called $1/x_s$ in the literature) is given by

$$\kappa^2 D_s \tau_v = 1 \, . \tag{6.6}$$

For a given terrace, the calculations may be simplified if the origin is chosen at the middle of the terrace. The values of λ and μ which verify

(6.2) are easily obtained, and (6.5) reads (Fig. 6.2):

$$\rho(x) = F\tau_v + \frac{\cosh(\kappa x)}{\cosh(\kappa\ell/2)}(\rho_0 - F\tau_v) \; . \tag{6.7}$$

This quantity has not yet been experimentally observed. However, the velocity of a step can be measured. Let for instance the left-hand step at $x = -\ell/2$, be considered. The current j from the right side is

$$j(-\ell/2) = -D_s\rho'(-\ell/2) \, . \tag{6.8}$$

Adding the contribution from the left-hand terrace, and writing that the sum of both current densities is v in our units (v/a^2 in normal units) one finds that the velocity of a step separating two terraces of respective widths ℓ and ℓ' is

$$\left[v/a^2 =\right] v = D_s(F\tau_v - \rho_0)\,\kappa[\tanh(\kappa\ell/2) + \tanh(\kappa\ell'/2)] \, . \tag{6.9}$$

This function is shown in Fig. 6.3 for $\ell = \ell'$. The form (6.9) has been observed experimentally by Keller (1975) in NaCl.

Generally, and certainly for silicon (Alfonso *et al.* 1993, Métois & Wolf 1993), $\kappa\ell \ll 1$, and (6.9) reads

$$v = D_s(F\tau_v - \rho_0)\,\kappa^2[\ell + \ell']/2 \, . \tag{6.10}$$

In particular, in the absence of a beam, and in vanishing vapour pressure ($F = 0$), the evaporation velocity of the surface is independent of ℓ:

$$\mid \dot{z} \mid = v(\ell)/\ell = \frac{\rho_0}{\tau_v} \, . \tag{6.11}$$

This equation can of course be immediately written without calculation, since ρ_0/τ_v is the macroscopic evaporation rate of the crystal. What is indeed remarkable is that (6.11) is not always true (if $\kappa\ell > 1$).

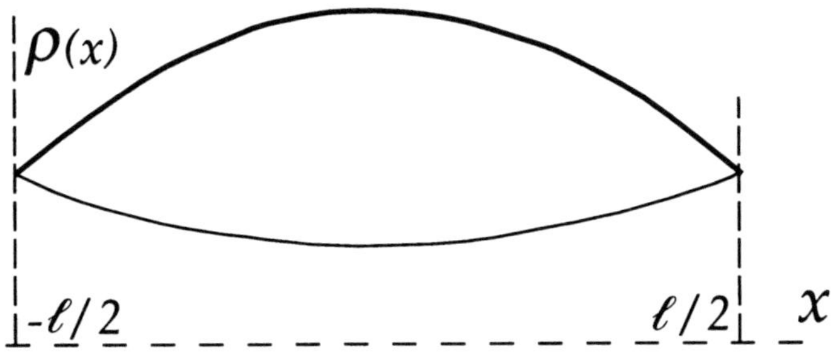

Fig. 6.2. Variation of the adatom density between two terraces. Upper curve: growth. Lower curve: evaporation.

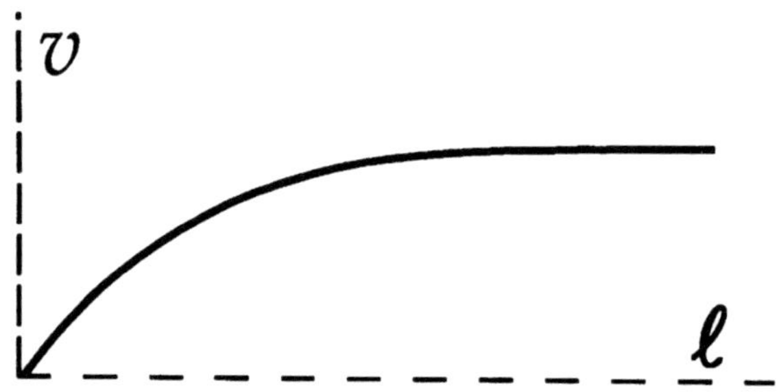

Fig. 6.3. Variation of the step velocity with the distance between steps.

As we said above, the saturation of (6.9) has been observed experimentally by Keller (1975) in NaCl. On the other hand, Métois has observed the opposite effect on Si(001): the evaporation velocity increases with ℓ! (Métois & Wolf 1993). This indicates a breakdown of the BCF theory, whose origin is not really clear. In fact, as we discuss in the next section for growth, and in section 6.6 for evaporation, one does expect the theory to break down for very large terraces. However, an increase of the *macroscopic* evaporation rate is very hard to understand.

6.3 The case without evaporation: validity of the BCF model

If the residence time τ_v is long, κ is small according to (6.6). If $\kappa\ell$ is also small (and on silicon, for instance, it is always small) one can replace (6.7) by its Taylor expansion

$$\rho(x) = \frac{\cosh(\kappa x)}{\cosh(\kappa\ell/2)}\rho_0 + F\tau_v \frac{\cosh(\kappa\ell/2) - \cosh(\kappa x)}{\cosh(\kappa\ell/2)}$$
$$\simeq \rho_0 + F\tau_v\kappa^2\left(\frac{\ell^2}{8} - \frac{x^2}{2}\right).$$

Thus, according to (6.6):

$$\rho(x) = \rho_0 + (F/D_s)\left(\frac{\ell^2}{8} - \frac{x^2}{2}\right). \tag{6.12}$$

We have argued that $\rho(x)$ is not easy to measure. On the other hand, it is obvious without calculation that the growth velocity is equal to F. Burton, Cabrera & Frank's theory has therefore no great predictive power in the absence of evaporation. However, from formula (6.12), one can deduce the conditions of validity of this theory.

The BCF theory, which assumes step flow, will cease to be valid if terraces form between preexisting steps. This will happen if the adatoms have no time to reach a step before meeting other adatoms and nucleating new terraces (Fig. 5.4a). This will in turn happen if the adatom density $\rho(x)$ becomes too large. It is seen from (6.12), for instance by making $x = 0$ (half-way between two steps), that the density ρ becomes large if $F\ell^2/D_s$ becomes large. For a given miscut $1/\ell$, the step flow regime is not stable at low temperature, when the diffusion constant D_s is small.

The problem can in fact be analysed in a more quantitative way. For the sake of simplicity, it will be assumed that two adatoms which meet each other form the nucleus of a new terrace. It is so at low temperature, because the breaking of a pair has a low probability there. The step flow regime will therefore be stable if and only if an adatom has a small probability of meeting another one. In other words, single adatoms have a finite lifetime τ before meeting a fellow adatom and forming a pair. This defines a diffusion length $\ell_s = \sqrt{D_s\tau}$. Step flow is stable if $\ell_s \gg \ell$. The probability of two adatoms meeting each other is of order $\nu(\ell_s)\bar{\rho}$, where $\nu(\ell_s)$ is the average number of distinct sites visited by a given adatom, whose diffusion length is ℓ_s, and $\bar{\rho} = \rho_0 + (F/D_s)\,\ell_s^2/12$ is the average value of (6.12) over the diffusion lengthscale. The equilibrium density ρ_0 is generally negligible.

One has now to calculate the diffusion length ℓ_s. From its definition it follows that in an area of order $\ell_s \times \ell_s$ only one pair formation (in order of magnitude) must take place. This condition reads

$$\nu(\ell_s)\bar{\rho}\ell_s^2 \approx \nu(\ell_s)F/D_s\ell_s^4 \approx 1 \,. \tag{6.13}$$

We need to know the average number $\nu(\ell_s)$ of distinct sites visited by each adatom. The diffusion motion will be identified with a random walk with hopping between neighbouring sites. The average lifetime τ of an adatom is equal to ℓ_s^2/D_s, and therefore each adatom makes an average number of hops approximately equal to ℓ_s^2. A rough evaluation of the average number of distinct sites visited is $\nu(\ell_s) = \ell_s^2/n_v$, where n_v is the average number of visits to a given site, e.g. the origin. Now, the probability that an adatom leaving the origin at time 0 is back at the origin at time t, is of order $1/(D_s t)$. Since each adatom spends on each site an average time of order $1/D_s$, one deduces

$$n_v \approx \int_{1/D_s}^{\tau} \frac{dt}{t} \approx \ln(D_s\tau) \approx \ln\left(\ell_s^2\right) \,.$$

The mathematically-minded reader will object that this argument has no reason to be correct. In fact, correlations inside a given random walk have been neglected: if an adatom has visited one site many times, it has a high probability of having visited the neighbouring sites many times. Surprisingly, the logarithmic behaviour *is* correct! (See Montroll & Weiss

1965, Shlesinger 1974, Zumofen & Blumen 1982, Henyey & Seshadri 1982, Kehr & Argyrakis 1986.) The diffusion length can now be found from (6.13) as

$$\left[\frac{\ell_s^6/a^2}{\ln(\ell_s^2/a^2)} = \right] \frac{\ell_s^6}{\ln(\ell_s^2)} \approx \frac{D_s}{F} . \tag{6.14}$$

It is usual to neglect the logarithmic correction and to write (Stoyanov & Kashchiev 1981, Venables *et al.* 1984)

$$\ell_s \approx (D_s/F)^{1/6} . \tag{6.15}$$

This formula also gives the typical size of nucleated terraces on a high-symmetry surface (for which $\ell = \infty$ in the language of this chapter), at least at low temperature. This problem will be addressed in chapter 11, where the generalization of formula (6.15) to more commonly encountered cases (e.g., when adatom pairs are not stable) will be presented.

6.4 The Schwoebel effect

In their original work, Burton *et al.* assumed, as we have done above, that steps release and capture atoms to and from the upper and the lower terrace with the same probability. There is no ground for this assumption as demonstrated experimentally for the first time by Ehrlich & Hudda (1966). This property has important consequences for crystal growth as pointed out by Schwoebel (1968) and as will be seen throughout this book. Fig. 6.4a shows the potential seen by an adatom according to Bourdin *et al.* (1988). The curve has a maximum which should be overcome by adatoms willing to go downstairs and prevents them to stick easily to the down step. One can easily understand this maximum ('Schwoebel barrier'), since an adatom willing to step down should go through an uncomfortable position (Fig. 6.4c) where it has few neighbours. This is the intuitive basis of the calculation of Bourdin *et al.* On the other hand, the energy barrier may be lower if the edge atom is pushed away by the incoming adatom (Fig. 6.4d) which takes its place. Numerical calculations within the *effective medium theory* (see for instance the review by Stoltze 1994) show that on the (100) face of metals such an 'exchange' mechanism has a lower energy than the over-edge 'jump' for Au, but the opposite is true for Cu, Ag, Ni, Pd and Pt. On the (111) face the jump mechanism seems to be always favoured. Both barriers are always higher than for diffusion on the flat surface. This Schwoebel barrier is presumably responsible for the stability of tips used, for instance, in scanning tunneling microscopy. The Schwoebel effect has been

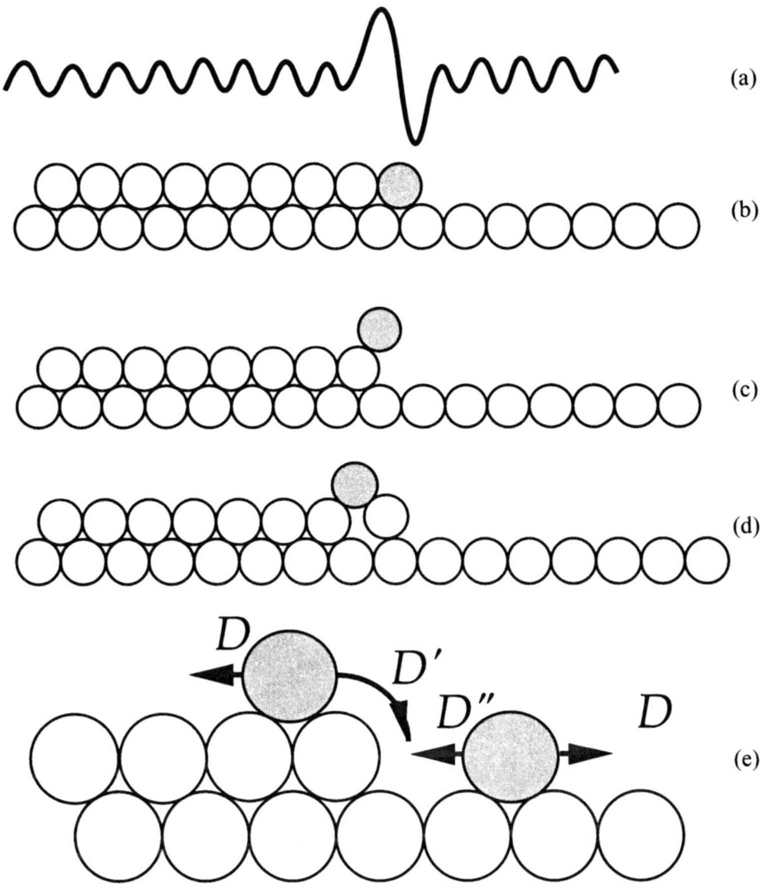

Fig. 6.4. a) Theoretical potential felt by an adatom (grey circle). b) Intuitive reason for the minimum minimorum. c), (d) Intuitive reason for the maximum maximorum. Scheme (c) corresponds to the mechanism considered by Bourdin *et al.* (1988), and scheme (d) to the exchange process suggested by Hansen *et al.* (1991). e) Interpretation of the parameters D, D' and D'' of section 6.4.

investigated by *ab initio* calculations in a single case, namely aluminium (Stumpf & Scheffler 1994). These calculations confirm the importance of the exchange mechanism, but predict that, at least for certain orientations of the steps, there is no Schwoebel barrier, in the sense that the potential is maximum at the center of the terrace rather than on the terrace edge.

It is easy to take into account an *infinite* Schwoebel barrier. Then, adatoms cannot step down at all. In that case, the adatom current on the upper side of each step should be zero. Therefore $\nabla\rho = 0$ there. It is now convenient to take the origin of each terrace on its right-hand edge, and the appropriate solution of (6.1) is then

$$\rho(x) = F\tau_v + \frac{\cosh(\kappa x)}{\cosh(\kappa\ell)}(\rho_0 - F\tau_v) \ . \tag{6.16}$$

Consider now the case of a partial Schwoebel effect (a finite Schwoebel barrier). Equation (6.1) is obviously valid with or without the Schwoebel effect. However, the boundary conditions at steps depend on the strength of the Schwoebel effect, because adatom sticking from the upper side of a step is difficult. Let $\rho(x_n - a)$ be the adatom density at the upper side of a step, at one atomic distance from the step edge. The quantity of adatoms which stick to this step per unit time and per unit length is of course proportional to $\rho(x_n - a)$. In the absence of a Schwoebel effect it would be $D\rho(x_n - a)/a$ (we suppress the subscript s in the following). Because of the Schwoebel effect, the proportionality factor becomes D'/a, smaller than D. The quantity of adatoms which stick per unit time and length is then $D'\rho(x_n - a)/a$. Because of the detailed balance principle, the number of atoms detaching from a step and going to the upper side (per unit time and length) should be equal to $D'\rho_0/a$. Therefore, the net current of atoms at the upper side of a step is $D'[\rho(x_n - a) - \rho_0]/a$. On the other hand, this current should be equal to $-D\mathrm{d}\rho/\mathrm{d}x$. Therefore the boundary condition at the upper side of a step is

$$D\frac{\mathrm{d}\rho}{\mathrm{d}x}(x_n) = D'[\rho_0 - \rho(x_n - a)]/a \approx D'[\rho_0 - \rho(x_n)]/a + D'\frac{\mathrm{d}\rho}{\mathrm{d}x}(x_n)$$

or

$$D\frac{\mathrm{d}\rho}{\mathrm{d}x}(x_n) = \Delta'[\rho_0 - \rho(x_n)] \tag{6.17a}$$

where

$$\Delta' = \frac{D}{a}\frac{D'}{D - D'} \ .$$

At the lower side of a step, a similar equation holds, with a different kinetic coefficient Δ'', namely

$$-D\frac{\mathrm{d}\rho}{\mathrm{d}x} = \Delta''(\rho_0 - \rho) \ . \tag{6.17b}$$

In the absence of the Schwoebel effect, $\Delta' = \Delta'' = \infty$, and one recovers (6.2). The resolution of equations (6.1) and (6.17) requires a lengthy, but straightforward calculation given in appendix F. This calculation yields the velocity v of a step separating an upper terrace of width ℓ' from a lower terrace of width ℓ'' in the form

$$v = v' + v'' \tag{6.18}$$

where (see appendix F)

$$v' = \kappa D\,(\rho_0 - F\tau_v) \times \\ \times \frac{2 - (1 - \kappa D/\Delta'')\exp(-\kappa\ell') - (1 + \kappa D/\Delta'')\exp(\kappa\ell')}{\mathscr{D}(\ell')} \tag{6.19a}$$

is the part of the velocity due to down-stepping atoms. The denominator $\mathscr{D}(\ell)$ is equal to

$$\mathscr{D}(\ell) = (1 + \kappa\frac{D}{\Delta''})(1 + \kappa\frac{D}{\Delta'})\exp(\kappa\ell) - (1 - \kappa\frac{D}{\Delta''})(1 - \kappa\frac{D}{\Delta'})\exp(-\kappa\ell)\,.$$

If $\Delta' = 0$, v' vanishes. The other contribution, due to atoms sticking to the upper step, is

$$v'' = \kappa D\,(\rho_0 - F\tau_v) \times \\ \times \frac{2 - (1 - \kappa D/\Delta')\exp(-\kappa\ell'') - (1 + \kappa D/\Delta')\exp(\kappa\ell'')}{\mathscr{D}(\ell'')}\,. \tag{6.19b}$$

Instead of the kinetic coefficients Δ' and Δ'', which have dimensions of velocities, one can use the lengths

$$d' = D/\Delta' \tag{6.20a}$$

$$d'' = D/\Delta''\,. \tag{6.20b}$$

6.5 Advacancies and evaporation

On a terrace, there are adatoms, but also surface vacancies or ‘advacancies’. They may become important at evaporation. Indeed, while the adatom density becomes lower at the centre of a terrace, the advacancy density increases. Thus, advacancies may cluster and form ‘macrovacancies’.

On the right-hand side of (6.1), it is therefore necessary to add a term which takes recombination of adatoms and advacancies into account. This term is proportional to the product of the adatom density $\rho(x)$ by the advacancy density $\sigma(x)$. The proportionality coefficient Δ should be of the same order as D. There is another term which corresponds to the reverse process (creation of adatom-advacancy pairs). From the detailed balance principle, this term is equal to $\Delta\rho_0\sigma_0$, where ρ_0 and σ_0 are the equilibrium concentrations. If Λ designates the advacancy diffusion constant, the evolution equations are therefore

$$\begin{cases} \dot{\rho} = D\rho'' - \rho/\tau_v + \Delta\rho_0\sigma_0 - \Delta\rho\sigma + F = 0 & (6.21) \\ \dot{\sigma} = \Lambda\sigma'' + \Delta\rho_0\sigma_0 - \Delta\rho\sigma = 0\,. & (6.22) \end{cases}$$

Unfortunately, these equations, in contrast with (6.1), are not linear. They can be only approximately solved (between two steps at $-\ell/2$ and $\ell/2$), either by expanding ρ and σ in series of powers of x, or by linearizing with respect to $\delta\rho = \rho - \rho_0$ and $\delta\sigma = \sigma - \sigma_0$, which are actually small in general (see appendix F). The final result will only be given, in ultra-high vacuum and no deposition ($F = 0$). In the case of silicon, evaporation is slow in comparison with diffusion, that is, adatoms have a high probability to diffuse from one step to the next one without evaporating, so that it is sufficient to work at order 0 in $1/\tau_v$. Finally, at the high temperatures where evaporation is important, the Schwoebel barrier can easily be overcome, so that it is reasonable to neglect it altogether. With these approximations and assumptions, the solution of (6.21) and (6.22) is

$$\begin{cases} \delta\rho(x) = -\rho_0 + \dfrac{\rho_0}{\cosh(\kappa\ell/2)}\cosh(\kappa x) & (6.23) \\ \delta\sigma(x) = \sigma_0 - \dfrac{\sigma_0}{\cosh(\kappa\ell/2)}\cosh(\kappa x) & (6.24) \end{cases}$$

where

$$\kappa^2 = \frac{1}{2D\tau_v} - \frac{1}{2D\tau_v}\frac{\Lambda\sigma_0 - D\rho_0}{\Lambda\sigma_0 + D\rho_0} = \frac{1}{\tau_v}\frac{\rho_0}{\Lambda\sigma_0 + D\rho_0}\,. \tag{6.25}$$

Formulae (6.23) and (6.7) turn out to be identical for $F = 0$. Equation (6.9), however, is modified, but in a very simple way. Assuming equidistant terraces, the new equation is

$$v = 2D\rho'(\ell/2) - 2\Lambda\sigma'(\ell/2) = (D\rho_0 + \Lambda\sigma_0)\tanh(\kappa\ell/2) \tag{6.26}$$

where the prime means derivative with respect to x. Thus, the quantity $D\rho_0$, which appeared in (6.9), is just replaced by the sum $D\rho_0 + \Lambda\sigma_0$. Very often, the microscopic surface diffusion constants D and Λ of adatoms and advacancies, the adatom and advacancy densities ρ_0 and σ_0, and the adatom evaporation rate $1/\tau_v$ *appear in macroscopic measurable quantities only through the two combinations* ρ_0/τ_v and $D\rho_0 + \Lambda\sigma_0$. Formula (6.25) illustrates this property. As will be seen in the next chapter, the term $\Lambda\sigma_0$, which corresponds to advacancies, is generally much smaller than the corresponding quantity for adatoms, $D\rho_0$.

6.6 Validity of the BCF model in the case of evaporation

In section 6.3, the stability of step flow against adatom cluster formation between steps during growth has been discussed. Similarly, during evaporation, one should discuss the stability of step flow against nucleation of advacancy clusters (or 'surface macrovacancies') between steps. These macrovacancies have been observed on Si(100) (Mundschau *et al.* 1989, Bauer 1992, Métois & Wolf 1993, Pimpinelli & Métois 1994).

Similarly to section 6.3, macrovacancies are expected to appear when the distance between steps exceeds some value ℓ_c. However, the calculation of ℓ_c is rather different from that of section 6.3, because an evaporating surface is generally closer to thermal equilibrium than a growing surface. In particular, small polyvacancies are certainly almost at equilibrium, constantly decaying and recombining.

Only the case $F = 0$ will be considered in the following.

Between two steps, the decrease of the adatom density $\rho(x)$ due to evaporation results in an increase of the advacancy concentration $\sigma(x)$, due to the creation of adatom-advacancy pairs. This effect is described by equations (6.21-22) and their approximate solutions (6.23-24).

In the step flow regime (i.e., when no macrovacancies form) one can expect to be near an equilibrium state characterized by the 'reactions'

n-vacancy + adatom $\rightleftharpoons$ $(n-1)$-vacancy

and

n-vacancy + advacancy $\rightleftharpoons$ $(n+1)$-vacancy .

The detailed theory, using the rate equations corresponding to small deviations from this equilibrium, has been given by Pimpinelli & Villain (1994). The following simplified argument is not too bad.

The density of surface macrovacancies of radius R may be expected to be proportional to $\exp[-\beta\mathscr{F}(R)]$, where the free energy $\mathscr{F}(R)$ has the form

$$\mathscr{F}(R) = 2\pi\gamma R + \pi\delta\mu R^2 \tag{6.27}$$

where γ is the line tension of a step (averaged on its orientation). The effective chemical potential $\delta\mu$ depends on the position x and can be calculated as follows. The adatom density $\rho(\mathbf{r}, t)$ at point $\mathbf{r}$ on the surface is related to the local chemical potential $\delta\mu(\mathbf{r}, t)$ (counted from the equilibrium chemical potential) by the relation

$$\rho(\mathbf{r}, t) = \rho_0 \exp\{\beta\delta\mu(\mathbf{r}, t)\} \ . \tag{6.28}$$

Similarly, the local vacancy concentration is

$$\sigma(\mathbf{r}, t) = \sigma_0 \exp\{-\beta\delta\mu(\mathbf{r}, t)\} \ . \tag{6.29}$$

Comparison with (6.23) or (6.24) yields

$$\beta\delta\mu(x) = \ln\frac{\cosh(\kappa x)}{\cosh(\kappa\ell/2)} \simeq \frac{\cosh(\kappa x)}{\cosh(\kappa\ell/2)} - 1 \ . \tag{6.30}$$

This formula holds if $\beta\delta\mu$ is not too large, which is anyway the condition for (6.23) and (6.24) to be correct.

The maximum of (6.27) corresponds to $R = R_c$, with

$$R_c = \frac{\gamma}{|\delta\mu|} \tag{6.31}$$

and the corresponding free energy barrier is

$$\Delta\mathscr{F} = \pi\gamma^2/|\delta\mu| \,. \tag{6.32}$$

An approximate condition for the stability of step flow is that no surface vacancy of radius of order R_c exists in the area $\ell \times \ell$. This condition may approximately be written as

$$\ell^2 \exp(-\beta\Delta\mathscr{F}) < 1 \tag{6.33}$$

i.e.

$$\ell^2 \exp\left(-\beta\pi\gamma^2/|\delta\mu|\right) < 1 \,.$$

Replacing $|\delta\mu|$ by its maximal value obtained by setting $x = 0$ in (6.30), the following condition is obtained for step flow stability:

$$\frac{8\beta^2\pi\gamma^2\tau_v\,(\Lambda\sigma_0 + D\rho_0)}{\rho_0\ell^2} > 2\ln\ell \,. \tag{6.34}$$

For example, if $\ell \approx 1000$ atomic distances (6.34) yields

$$\kappa\ell < \beta\gamma \,. \tag{6.35}$$

We see from equation (6.3) that Burton, Cabrera & Frank's theory predicts a slope-dependent (equivalently, an ℓ-dependent) sublimation rate as soon as the condition $\kappa\ell \approx 1$ is satisfied. Our result (6.35) implies that this condition might never be satisfied, if $\beta\gamma \ll 1$. In fact, κ increases with increasing temperature, so that the condition $\kappa\ell \approx \beta\gamma$ would be reached first, and macrovacancies will start nucleating before one attains the ℓ–dependent regime of (6.3). This is what is seen on Si(001) (Métois & Wolf 1993), whose step stiffness is indeed quite small at high temperature ($\approx 0.9T_M$) (Pimpinelli & Métois 1994, Bartelt *et al.* 1994).

The combination $\Lambda\sigma_0 + D\rho_0$, which appears in (6.34) is the same as that already encountered in (6.26), and will be found again in chapter 7.

6.7 Step bunching and macrosteps

In sections 6.3 and 6.6, the stability of step flow growth and evaporation with respect to adatom and advacancy clustering has been investigated.

Other instabilities are possible. One is step bunching, which results in the formation of macrosteps. Macrostep formation is often observed, although it is not always clear at which stage of the history of the crystal it has taken place. It will be argued below that step bunching may occur during growth from the liquid or from the vapour, because of impurity deposit, or during evaporation. Step bunching during MBE growth is very rarely observed.

a) Impurities (Van der Eerden & Müller-Krumbhaar 1986)

There are always a few impurities in the melt, the vapour, the solution or the beam from which a crystal is grown.

It will be assumed that these impurities hinder crystal growth. This is plausible if the crystal is grown from a melt, a vapour or a solution with a weak supersaturation. Indeed, the impurities can change the local chemical potential. In the case of MBE, the effect of impurities on the sticking coefficient is not expected to be so dramatic.

Moreover, it will be assumed that impurities diffuse more slowly than the adatoms of the growing crystal. Thus, their density is high near the upper step of a terrace (Fig. 6.5) and practically zero near the lower step. If the impurity diffusion constant is zero, their concentration is indeed proportional to the distance to the lower step edge.

We will now argue that in this model, step bunching occurs (Van der Eerden & Müller-Krumbhaar 1986). The general method to check the stability of a particular state (here the state with equidistant steps) is to see whether a small perturbation increases or decreases with time. In the

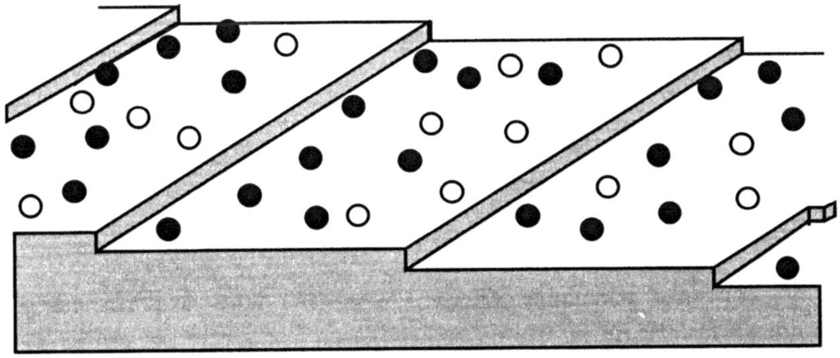

Fig. 6.5. Effect of slowly diffusing impurities (black circles) in growth from the vapour. Their concentration is higher ahead of steps and they can impede the adatoms (white circles) to reach the steps. If some terrace is broader than the other ones, there are more impurities ahead of the upper step, which thus proceeds more slowly, and the broad terrace becomes broader. Step bunches result.

former case, there is an instability. In the present instance, we assume that, at a particular time, because of fluctuations, a particular terrace happens to be broader than the other ones. This broader terrace receives more impurities; therefore the concentration of impurities near the upper edge of this terrace is larger than near the other steps. This step goes slower than the other ones, is caught up by them, and a bunch of steps eventually results.

b) Schwoebel effect

Consider an evaporating, stepped surface subject to a Schwoebel effect. We want to show that an instability takes place. To prove this, we assume again that one terrace, A, is broader than the other ones (Fig. 6.6). If the crystal were growing (Fig. 6.6a) the broad terrace A would gather more atoms, which would preferably go upward to the step edge B, and the width of the broad terrace A would therefore decrease: *the Schwoebel effect is stabilizing (with respect to step bunching) during growth.* During evaporation, on the contrary, the broad terrace A evaporates more atoms, as can be seen from (6.19) and problem 6.3 below. These atoms predominantly come from the upward step edge of B (because of the Schwoebel effect). Therefore, this

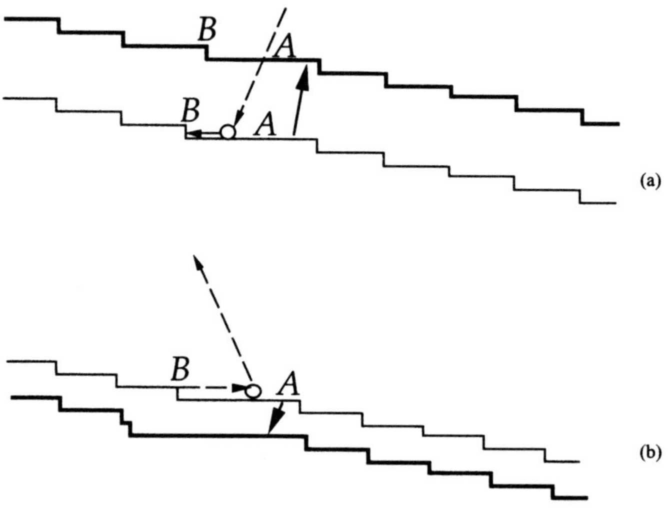

Fig. 6.6. Evolution of a broader terrace A when the Schwoebel effect is present. The thin line shows the initial profile and the thick line shows the profile at a later time. a) During growth, the terrace A gathers more atoms which preferably go to the upper ledge B. This ledge moves forth faster, so that the terrace A becomes narrower and the regular profile is stabilized. b) During evaporation, the terrace A evaporates more atoms which mostly come from the upper ledge B. This ledge recedes faster and catches up the next uppermost step, thus forming a bunch of two steps.

step recedes faster than the other ones and the broad terrace becomes even broader. *The Schwoebel effect is destabilizing during evaporation.*

One may wonder whether the instability of a regular array of steps leads to a stable, less regular array. The step just above the broad terrace is caught up by the upper step. However, if the broader terrace is initially just a little broader, its width will then stop to increase. But this step pairing is not the final stage, because all steps will pair (Kandel & Weeks 1992). When all steps are paired, the system of double steps is again unstable as the system of single steps was. Then the pairs form pairs, i.e. step quadruplets. Then the quadruplets form pairs of quadruplets, which are octuplets, etc. The step bunching process is not limited and does not lead to a steady state. However, the formation of large bunches may require a long time.

c) Effect of a d.c. electric current

In experiments on silicon evaporation, the crystal is often heated by an electric d.c. current. Latyshev *et al.* (1989, 1990, 1993) have observed step bunching for a well-defined sign of the electric field (with respect to the slope of the surface). Stoyanov (1991) has proposed an explanation of this phenomenon, assuming that the electric field exerts a bias on adatom diffusion. This could be due to adatoms being charged, or

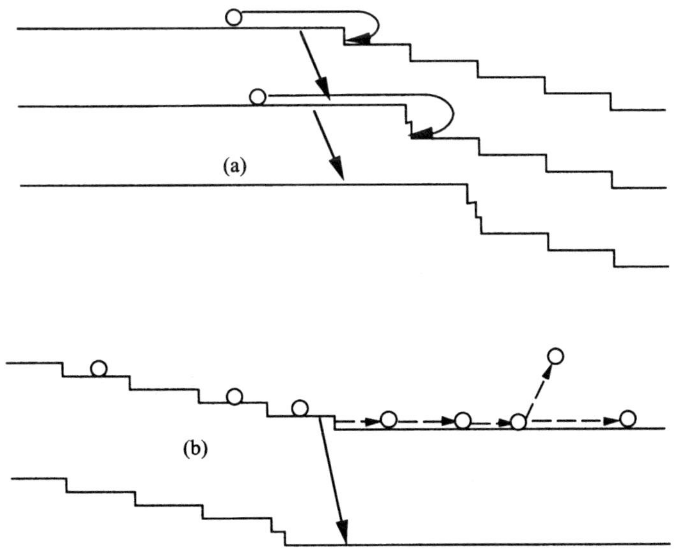

Fig. 6.7. Growth (a) and evaporation (b) of a train of steps limited on one side by an infinite terrace. In both cases, a step bunch appears at the edge of the infinite terrace.

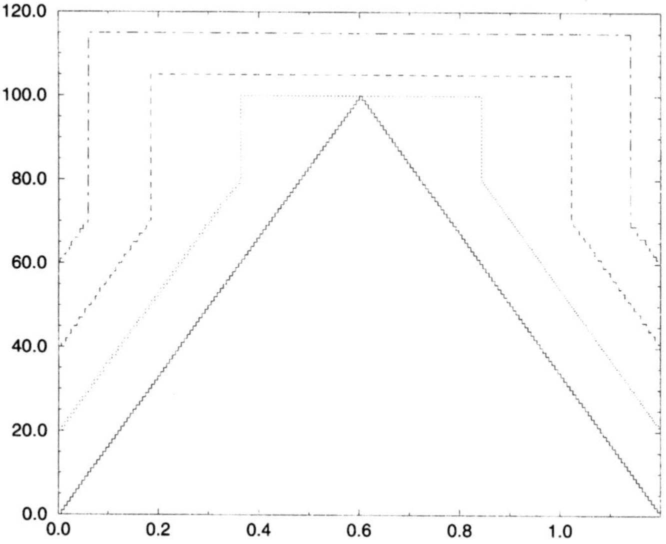

Fig. 6.8. Transient shapes, in MBE growth, of a periodic profile which has initially a sawtooth shape. Dotted, dashed and full outer line correspond to increasing times. Only one period has been represented (from Elkinani & Villain 1994). There is no Schwoebel effect.

to a momentum exchange between adatoms and electrons (Lodder 1989). Normally, an adatom has no charge, but an electric dipole moment. *A priori*, one might argue that an electric field does not apply any force on an electric dipole, or that the detailed balance principle forbids the hopping probability to be the same in all directions since the hop does not modify the energy. However, since there is an electric current, the system is not at equilibrium and the detailed balance principle is not necessarily obeyed. Thus, we assume, following Stoyanov, that the hopping probability of an adatom against the field and in the field direction is different. This is not quite enough. We have to assume that steps are not perfect sinks, or sources, for adatoms. That is, atoms are captured and emitted by steps at a finite rate. In that case, there is a field direction which pushes adatoms downward while after field reversal they are pushed upward. When they are pushed downward, they accumulate near the descending step edge, since the step cannot equilibrate instantaneously their density. Thus, the adatoms diffuse *backwards* to the ascending step. This is analogous to the Schwoebel effect, when the adatoms preferentially go to the upward step of each terrace. Then step bunching is expected during evaporation. Reversing the current is, on the contrary, stabilizing. Advacancies should also be affected by the electric current. The complex behaviour of adatoms and advacancies in the presence of a diffusion bias has been investigated by Misbah, Pierre-Louis & Pimpinelli (1995).

d) Effect of the initial profile

For certain irregular shapes of the initial profile, step bunching may occur during MBE growth, even without Schwoebel effect. Consider a convex sample, one half of which is initially limited by a semi-infinite terrace of high symmetry, while the other half is limited by a regular stepped surface (Fig. 6.7a). The edge of the semi-infinite terrace will gather a large number of adatoms and will rapidly catch up the lower step, thus forming a pair of steps which will catch up the next step, etc. The process becomes more complicated when new terraces nucleate on the semi-infinite terrace. A similar step bunching phenomenon occurs during evaporation of the

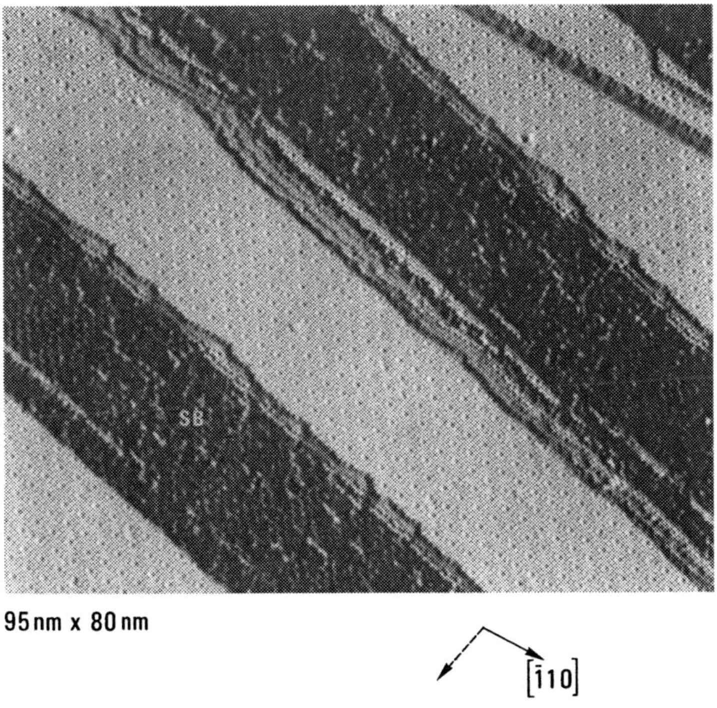

Fig. 6.9. STM micrograph of step bunching on a thermodynamically unstable vicinal of the (111) face of silicon. The size of the region shown is 95 nm × 80 nm. The steps are parallel to the direction [$\bar{1}$10]. The average slope of the surface is 7.3° and the slope of the bunches is about 15°. (Courtesy of Margrit Hanbücken, CRMC2 - CNRS Marseille, Bernd Röttger, Ruhr-Universität Bochum and Henning Neddermeyer, Martin-Luther-Universität Halle-Wittenberg.)

concave profile shown by Fig. 6.7b. Finally, Fig. 6.8 shows what happens in the growth of a periodic, sawtooth profile (Elkinani & Villain 1994, Politi & Villain 1996). Note that the final result is a perfectly flat surface! A systematic study of the effect of substrate irregularities has been published by Colas *et al.* (1989).

e) Thermodynamic faceting

Another reason for step bunching may of course be the thermodynamical instability of the stepped surface. For instance, a (111) vicinal surface of silicon, misoriented by 4°, is in equilibrium at high temperature but facets at moderate temperature (~ 850 °C) (Williams *et al.* 1993, Williams 1994) (Fig. 6.9).

Another example, shown in Fig. 6.10, is the (433) face of gold. Faceting can be deduced from the Wulff construction and the double tangent construction of chapter 3, if the orientation-dependent surface tension is known (Mullins 1963). The microscopic reason of the instability may be an attraction between steps, due for instance to the tendency to form $(11\bar{1})$ facets. In the case of vicinals of Si(111), these facets do not appear, and the formation of large (111) terraces may be related to the competing

Fig. 6.10. STM micrograph of step bunching on the thermodynamically unstable (433) face of Au. The area shown is 70 nm × 70 nm wide. The monoatomic steps are 0.235 nm high. (Courtesy of Sylvie Rousset, Universités Paris 6-7.)

surface tensions of the 7×7 reconstruction, and of the 'disordered'-1×1 surface phase. The competition results in a non-convex surface free energy, whence the faceting, as explained in chapter 3.

Problems

6.1. Rewrite the Burton, Cabrera & Frank theory for equidistant steps without making the quasi-static approximation, neglecting the Schwoebel effect.

Solution – Between two steps, equation (6.1) holds. Let $\kappa = 1/\sqrt{D_s \tau_v}$, $f = F\tau_v$, $\tilde{x} = \kappa x$, and $\tilde{t} = t/\tau_v$. Let $u(\tilde{x}, \tilde{t}) = \rho(\tilde{x}, \tilde{t}) - f$. Then equation (6.1) becomes

$$u_{\tilde{t}} = u_{\tilde{x}\tilde{x}} - u \,. \tag{1}$$

We look for a propagating solution, with velocity $v = \tilde{v}/(\kappa \tau_v)$: $u(\tilde{x}, \tilde{t}) = \varphi(\tilde{x} - \tilde{v}\tilde{t})$. Define now $\xi = \tilde{x} - \tilde{v}\tilde{t}$; (1) reads

$$\varphi''(\xi) + \tilde{v}\varphi'(\xi) - \varphi(\xi) = 0 \,.$$

The solution has the form

$$\varphi(\xi) = \varphi_1 \exp(\alpha_1 \xi) + \varphi_2 \exp(\alpha_2 \xi)$$

where the α_is are solutions of $\alpha^2 + \tilde{v}\alpha - 1 = 0$, i.e. $\alpha_{1,2} = [-\tilde{v} \pm \sqrt{\tilde{v}^2 + 4}]/2$, and the constants φ_1 and φ_2 are determined by boundary conditions at steps. In the absence of the Schwoebel effect, (6.2) yields

$$\varphi(\pm\tilde{\ell}/2) = \rho_0 - f \equiv \varphi_0$$

or

$$\begin{aligned}\varphi_0 &= \varphi_1 \exp(\alpha_1 \tilde{\ell}/2) + \varphi_2 \exp(\alpha_2 \tilde{\ell}/2) \\ &= \varphi_1 \exp(-\alpha_1 \tilde{\ell}/2) + \varphi_2 \exp(-\alpha_2 \tilde{\ell}/2) \,.\end{aligned}$$

It follows

$$\varphi(\xi) = \frac{(f - \rho_0)}{\sinh\left(\frac{\tilde{\ell}}{2}\sqrt{\tilde{v}^2 + 4}\right)} \left[\sinh\left(\frac{\alpha_2 \tilde{\ell}}{2}\right) e^{\alpha_1 \xi} - \sinh\left(\frac{\alpha_1 \tilde{\ell}}{2}\right) e^{\alpha_2 \xi}\right]$$

where we used $\alpha_1 - \alpha_2 = \sqrt{\tilde{v}^2 + 4}$.

The velocity must be found self-consistently from matter conservation at the step:

$$\tilde{v} = \varphi'(\xi)|_{-\tilde{\ell}/2} - \varphi'(\xi)|_{\tilde{\ell}/2} \,.$$

The result reads

$$\tilde{v} = (f - \rho_0)\sqrt{\tilde{v}^2 + 4}\times$$
$$\times\left[\tanh(\frac{\tilde{\ell}}{4}\sqrt{\tilde{v}^2+4}) - \frac{\sinh^2(\frac{\tilde{\ell}\tilde{v}}{4})}{\sinh(\frac{\tilde{\ell}}{4}\sqrt{\tilde{v}^2+4})\cosh(\frac{\tilde{\ell}}{4}\sqrt{\tilde{v}^2+4})}\right] . \quad (2)$$

Recalling that $\tilde{v} = \kappa\tau_v v$ it is now an easy matter to verify that equation (6.9), giving the step velocity in the quasi-static approximation, is recovered when $\tilde{v} \ll 1$. This means

$$v \ll \sqrt{D_s/\tau_v} .$$

The step flow velocity must be negligible with respect to the 'adatom velocity' $\sqrt{D_s/\tau_v}$ for the approximation to be valid.

In the absence of evaporation, $\tau_v \to \infty$, (2) reduces to $v = F\ell$–an equality which is a special case of (6.10) for $\ell = \ell'$ in this limit–to first order in $v\ell/D_s$. Thus, the condition of validity of the quasi-static approximation is

$$F/D_s \ll 1$$

for pure growth. This condition is always satisfied in MBE.

6.2. Derive the step flow velocity in the absence of evaporation.

Solution

$$\rho(\xi) = \rho_0 + \frac{F\ell}{2v}\coth\left(v\ell/2D\right) - \frac{F\xi}{v} - \frac{F\ell}{2v\sinh\left(v\ell/2D\right)}\exp\left(-v\xi/D\right)$$

and

$$v = D\rho'|_{-\ell/2} - D\rho'|_{\ell/2} = F\ell .$$

6.3. Write (6.19) for $D = D''$, ℓ much larger than the atomic distance, and $\kappa\ell \ll 1$.

Solution – Using the lengths $d' = D/\Delta'$ and $d'' = D/\Delta''$, equation (6.19a) reads

$$v' = \kappa D\left(\rho_0 - F\tau_v\right)\times$$
$$\times\frac{2-\left(1-\kappa d''\right)\exp\left(-\kappa\ell'\right)-\left(1+\kappa d''\right)\exp\left(\kappa\ell'\right)}{(1+\kappa d'')(1+\kappa d')\exp(\kappa\ell')-(1-\kappa d'')(1-\kappa d')\exp(-\kappa\ell')} .$$

The assumption $D = D''$ implies that Δ'' is infinite, and d'' vanishes. Expanding to first order in $\kappa\ell$ we get

$$v' = (\rho_0 - F\tau_v)\frac{2-\left(1-\kappa\ell'\right)-\left(1+\kappa\ell'\right)+(\kappa\ell')^2}{(1+\kappa d')(1+\kappa\ell')-(1-\kappa d')(1-\kappa\ell')}\kappa D$$

and

$$v' = D\left(F\tau_v - \rho_0\right)\frac{(\kappa\ell')^2}{2(d'+\ell')} .$$

In the case of growth, the product $\rho_0\kappa^2$ may be neglected, but not $D\tau_v\kappa^2$ which is equal to 1 according to (6.6). One thus finds

$$v' = \frac{F\ell'^2}{2(d'+\ell')} .$$

Equation (6.19b) reads:

$$v'' = \kappa D\,(\rho_0 - F\tau_v)\times$$
$$\times\frac{2-(1-\kappa d')\exp(-\kappa\ell'')-(1+\kappa d')\exp(\kappa\ell'')}{(1+\kappa d'')(1+\kappa d')\exp(\kappa\ell'')-(1-\kappa d'')(1-\kappa d')\exp(-\kappa\ell'')} .$$

Making again $d''=0$, and expanding the exponentials, yields

$$v'' = D\,(F\tau_v - \rho_0)\,\kappa^2\ell''\,\frac{2d'+\ell''}{2(d'+\ell'')} .$$

The step velocity v is then in pure growth $(1/\tau_v = \kappa = 0)$

$$v = v' + v'' = F\left[\frac{\ell'^2}{2(d'+\ell'')} + \frac{2d'\ell''+\ell''^2}{2(d'+\ell'')}\right] .$$

Note that, if $\ell' = \ell'' = \ell$, v becomes

$$v = \frac{F\ell}{2(d'+\ell)}\left(\ell + 2d' + \ell\right) = F\ell$$

which expresses the fact that the step velocity is equal to the flux of atoms on each terrace.

6.4. Write and solve the equations of evolution of a stepped surface during growth with desorption, when all terrace widths are much smaller than $x_s = 1/\kappa$. Discuss the stabilizing or destabilizing action of the Schwoebel effect.

Solution – We only discuss the case of a total Schwoebel effect. During growth with evaporation, the positions x_n of the steps obey the equations

$$\dot{x}_n = (F - \rho_0/\tau_v)\,(x_{n+1} - x_n) .$$

Defining $\ell_n = x_{n+1} - x_n$, and letting $F - \rho_0/\tau_v = K$ one obtains

$$\dot{\ell}_n = K\,(\ell_{n+1} - \ell_n).$$

The Fourier transform λ_k of ℓ_n satisfies

$$\dot{\lambda}_k = K\left(e^{ik} - 1\right)\lambda_k ,$$

or

$$\dot{\lambda}_k = K\,(\cos k - 1 + i\sin k)\,\lambda_k .$$

The regular array of steps is stable if all eigenvalues $K(\cos k - 1 + i\sin k)$ have a non-positive real part. Therefore it is stable if $K > 0$, (growth) and unstable if $K < 0$ (evaporation). This result has been obtained in section 6.7.

6.5. Discuss the evolution of a stepped profile with an initially random distribution of step widths during MBE growth with Schwoebel effect. Show that the width of the distribution decreases approximately as the inverse fourth root of the deposited coverage.

Solution – See Gossmann *et al.* (1990a,b).

6.6. Discuss the diffusion and incorporation of impurities during the MBE growth of a stepped surface. Consider the following cases. a) Impurities are incorporated with probability 1 when they meet a step. b) Impurities can jump over a step (case of 'surfactants'). What are the modifications with respect to section 6.1?

Solution – See Andrieu *et al.* (1989).

7
Diffusion

Une même équation, celle
de Laplace, se rencontre dans la
théorie de l'attraction newtonienne,
dans celle du mouvement
des liquides, dans celle du potentiel
électrique, dans celle du magnétisme,
dans celle de la propagation
de la chaleur et bien d'autres encore.

Henri Poincaré, *(La valeur de la Science)*

The same equation–the Laplace
equation, appears in the
theory of gravitation, in that of
motion in liquids, in that of the
electric potential, in that of
magnetism, in that of heat
propagation as well as
in many other theories.

In the bulk or on a surface, the theory of diffusion is full of unexpected traps. The most obvious one is that the diffusion constant which appears in Fick's law is not the same which describes the random walk of an isolated atom. The latter is called tracer diffusion constant. This means that you need a tracer, e.g. a radioactive tracer, to measure it.

In the case of diffusion on a clean surface (self-diffusion) there is no equivalent of Fick's law. One can however define the response to a chemical potential gradient, which turns out to be the quantity measured by many macroscopic techniques... This introduces a 'surface mass diffusion constant' $\tilde{D}_s$ which, on a high-symmetry or a vicinal surface, is a linear combination of the diffusion constants of adatoms and advacancies.

At ordinary temperatures, the surfaces of usual metals do not evolve spontaneously because $\tilde{D}_s$ is very low. On the contrary, diffusion is already quite important on these surfaces in MBE growth, because the adatom diffusion constant is much higher.

7.1 Mass diffusion and tracer diffusion

In this introductory section, certain general notions concerning diffusion in a d-dimensional space will be recalled. Most of them apply to surface diffusion if d is set equal to 2. More details will be found in textbooks (Adda & Philibert 1966, Philibert 1990) and review articles (Bonzel 1975, 1990, Gomer 1990).

It is convenient to distinguish two diffusion processes, to which correspond two diffusion constants (or 'coefficients').

The first process is collective mass diffusion: if the density $\rho(\mathbf{r}, t)$ of some material in a homogeneous medium is not uniform, the motion of atoms tends to equalize it. The process is described by Fick's law, in which appears what we will call the *Fick diffusion constant* $\tilde{D}$:

$$\dot{\rho}(\mathbf{r}, t) = \tilde{D}\nabla^2 \rho(\mathbf{r}, t) . \tag{7.1}$$

The second process is the individual random walk (Fig. 7.1) of a particular atom. A usual and often good approximation in (and on) a solid at low enough temperature is to assume that the atom moves by uncorrelated jumps. If the atom jumps by a vector $\mathbf{a}_n$ at its n-th step, the mean square path after m steps is

$$\left\langle \left| \sum_{n=1}^{m} \mathbf{a}_n \right|^2 \right\rangle = \sum_{n,n'} \langle \mathbf{a}_n \cdot \mathbf{a}_{n'} \rangle = \sum_{n} \langle \mathbf{a}_n \cdot \mathbf{a}_n \rangle = n \left\langle a_1^2 \right\rangle . \tag{7.2}$$

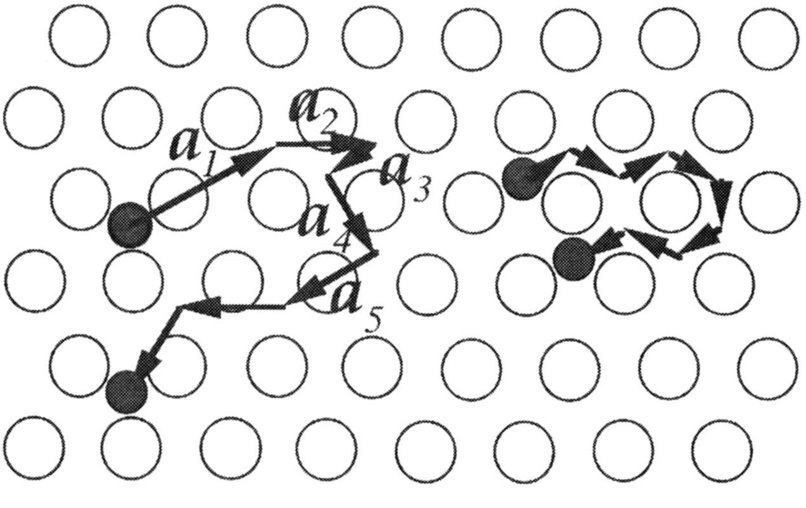

Fig. 7.1. Random walk of an interstitial impurity in a crystal. The left hand side represents a general random walk. It is often assumed that all jumps are of a single atomic distance as shown on the right hand side.

The second factor is independent of time and the first one is proportional to the time t. If the position of the atom of interest at time t is $\mathbf{r}(t)$, one can write

$$\left\langle \left[\mathbf{r}(t'+t)-\mathbf{r}(t')\right]^2\right\rangle = 2dD^*t \tag{7.3}$$

where D^* is called *tracer diffusion constant*, because this kind of diffusion is only observable if a few atoms are marked. This is generally achieved by observing the diffusion of a radioactive isotope or tracer.

One can be interested in the diffusion of a material in another one (heterodiffusion) or in the diffusion in a pure material (self-diffusion). When one deals with self-diffusion in a solid, one has generally in mind tracer diffusion.

There is a special, important case where $\tilde{D}$ is equal to D^*. This happens for dilute objects (let us call them 'impurities') which do not interact. The local impurity concentration can be written as $\rho(\mathbf{r},t) = \sum_j p_j(\mathbf{r},t)$, where $p_j(\mathbf{r},t)$ is the probability that the j-th impurity is at $\mathbf{r}$ at time t. This probability, after all, is a density, and therefore should satisfy (7.1). The sum of these probabilities also satisfies (7.1) with the same diffusion constant $\tilde{D}$. On the other hand, solving (7.1) with the initial condition $\rho(\mathbf{r},0) = \delta(0)$, one finds

$$\rho_j(\mathbf{r},t) = (4\tilde{D}t)^{-d/2}\exp\left(-r^2/4\tilde{D}t\right)$$

Hence

$$\left\langle \left[\mathbf{r}(t'+t)-\mathbf{r}(t')\right]^2\right\rangle = \int \mathrm{d}^d r\rho(\mathbf{r},t)r^2 = 2d\tilde{D}t\,.$$

Comparison with (7.3) yields $\tilde{D} = D^*$.

It is not quite obvious that this argument is incorrect when the concentration is not weak. This is the issue of problem 7.2. It is illuminating to consider the following examples, which show that D^* and $\tilde{D}$ are generally different.

1) Consider a lattice gas in which the only possible process is the exchange of 2 atoms. In this perhaps academic case, $\tilde{D} = 0$ and $D^* \neq 0$.
2) In a fluid at its critical point, $\tilde{D} = 0$ because a density fluctuation of weak amplitude and long wavelength feels no restoring force. On the other hand, D^* has no reason to vanish.

Other examples will follow in the next sections. The effect of interactions between particles on the difference between $\tilde{D}$ and D^* has been investigated by Ferrando & Scalas (1993) and in textbooks (Philibert 1990).

7.2 Conservation law and current density

It follows from (7.1) that the total number of particles, $\int \mathrm{d}^d r \rho(\mathbf{r}, t)$, integrated over the whole sample, is conserved, i.e. independent of time. If the integration is on a finite volume V, the time derivative is equal to the flux of the current density $\mathbf{j}$ through its boundary surface S, namely

$$\int_V \mathrm{d}^d r \dot{\rho}(\mathbf{r}, t) = -\int_S \mathrm{d}^{d-1} r \ \mathbf{n} \cdot \mathbf{j}(\mathbf{r}, t) = -\int \mathrm{d}^d r \ \mathrm{div}\mathbf{j}(\mathbf{r}, t) .$$

Since this should be true for any volume, one deduces the continuity (or conservation) equation

$$\dot{\rho}(\mathbf{r}, t) = -\mathrm{div}\mathbf{j}(\mathbf{r}, t) . \tag{7.4}$$

Comparison with (7.1) yields

$$\mathbf{j}(\mathbf{r}, t) = -\tilde{D} \nabla \rho(\mathbf{r}, t) . \tag{7.5}$$

In the case of a surface, generally considered in the following chapters, $\mathbf{j}$ is the surface current density, defined by the fact that the net flux through a line element on the surface in time δt is $\mathbf{n} \cdot \mathbf{j} \mathrm{d}\ell \delta t$, where $\mathrm{d}\ell$ is the length of the line element and $\mathbf{n}$ the unit normal vector.

7.3 Vacancies and interstitial defects in a bulk solid

Diffusion in solids takes place mainly through the motion of defects, especially vacancies and interstitial atoms (Fig. 7.2). It is possible to define diffusion constants of these defects. This is not quite obvious, since they have a finite lifetime. However, the defect densities satisfy equations analogous to (7.1) for times shorter than the lifetime, and the mean-square displacements satisfy equations analogous to (7.3) for times shorter than the lifetime.

Since the defect concentration is usually low in a solid, the Fick and tracer diffusion constants of a given defect are the same. Let D_{int} and D_{vac} be the diffusion constants of interstitials and vacancies.

If there are very few vacancies, or if they move very slowly, then diffusion is mainly due to interstitials. The density of the crystal (assumed to be pure) is in appropriate units $\rho(\mathbf{r}, t) = 1 + \rho_{\mathrm{int}}(\mathbf{r}, t)$. Inserting this relation into (7.1), one finds $\dot{\rho}_{\mathrm{int}} = \tilde{D} \nabla^2 \rho_{\mathrm{int}}$. This relation implies that ρ_{int} satisfies a Fick equation with a Fick diffusion constant $D_{\mathrm{int}} = \tilde{D}$. Thus, the Fick diffusion constant $\tilde{D}$ is equal to the diffusion constant D_{int} of interstitials.

However, as mentioned in section 7.1, when dealing with self-diffusion in solids, we are generally more interested in the tracer diffusion constant D^*. It is not equal to D_{int}, but to D_{int} times the average interstitial concentration $\bar{\rho}_{\mathrm{int}}$. Before proving this, note that formula (7.3), which

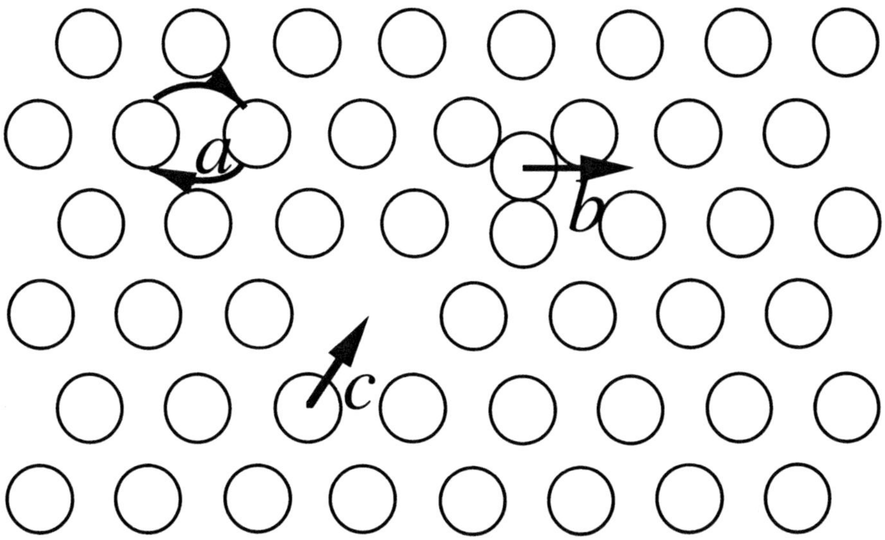

Fig. 7.2. Three self-diffusion processes in a crystal, represented as two-dimensional for graphical convenience. a) Exchange of atoms at normal sites. b) Diffusion of interstitial defects. c) Vacancy diffusion. Process (a) only contributes to tracer diffusion, not to mass diffusion.

defines D^*, is only valid for times so long that each atom has accomplished several jumps as an interstitial. This implies that any given interstitial atom exchanges rapidly its position with an ordinary atom.

It will now be proved that $D^* = D_{\text{int}}\bar{\rho}_{\text{int}}$ for interstitial-dominated self-diffusion. Indeed, the average displacement (7.3) is equal to the average displacement $\left\langle \delta r_{\text{int}}^2(t) \right\rangle$ of an interstitial, multiplied by the number of interstitials $N\bar{\rho}_{\text{int}}$, divided by the number of atoms N. Since $\left\langle \delta r_{\text{int}}^2(t) \right\rangle = 2dD_{\text{int}}t$, it follows that, if diffusion is due to interstitials,

$$D^* = D_{\text{int}}\bar{\rho}_{\text{int}} \, . \tag{7.6a}$$

Since $\tilde{D} = D_{\text{int}}$, this is a new example where $D^* \neq \tilde{D}$. Formula (7.6a) neglects correlations between jumps, which may be appreciable in certain cases (Adda & Philibert 1966). Note that there is no need to take vacancy-interstitial recombination into account in this argument.

It is remarkable that no relation analogous to (7.6a) exists when diffusion is dominated by vacancies. In fact, the hypothetical relation

$$D^* = D_{\text{vac}}\bar{\rho}_{\text{vac}} \, (?) \tag{7.6b}$$

is at best an approximation which is only acceptable if vacancies jump by a single atomic distance. Even in this case, correlation effects are important. These issues are discussed in problems 7.3 and 7.4.

On the other hand, for vacancy-dominated diffusion, the Fick diffusion coefficient defined by (7.1) is equal to D_{vac}.

Similar concepts will be found in the case of atoms diffusing in the two-dimensional space constituted by a crystal surface.

7.4 Surface diffusion

A typical case is that of heterodiffusion of impurities on a planar surface. Then, equations (7.1) and (7.3) hold, with $d = 2$ and $\nabla^2 = \partial^2_{xx} + \partial^2_{yy}$. Of course, they hold only for sufficiently short times, when the impurities do not penetrate in the bulk (see problem 7.5). One can define a surface Fick diffusion constant $\tilde{D}_s$ and a surface tracer diffusion constant D^*_s in the same way as in the bulk.

In the case of self-diffusion on a clean planar surface, one can define a tracer self-diffusion constant through (7.3). On the other hand, it is not easy to write any equation analogous to (7.1) on the surface. In the next section, a 'mass diffusion constant' will be defined, which describes the response to an external driving force. It coincides with the combination $D_s\rho_0 + \Lambda_s\sigma_0$ that we met in section 6.6, where, however, the indices s have been omitted, as they will generally be in the following chapters, when volume diffusion is ignored.

It is worth mentioning that the tracer diffusion constant is in principle not equal to this combination, although it is expected to be generally not very different. Indeed, in the absence of adatoms and advacancies, exchange of atoms can contribute to tracer diffusion. In addition, vacancies (and in particular advacancies) are not related to tracer diffusion in a simple way, as seen in section 7.2 and in problems 7.3 and 7.4.

It is also straightforward to define the surface current density of adatoms and advacancies on a planar surface. Their difference is the surface mass current, which is defined by a formula analogous to (7.4). Such a formula is only valid in the absence of deposition and evaporation, but it can be easily generalized (see for instance (6.1)).

The case of a curved surface requires some comments. Sometimes, the surface can be described as an array of planar terraces separated by steps (Tian & Rahman 1993, Merikoski & Ala-Nissila 1995) as in chapter 6. Otherwise, one can use the intrinsic surface mass current density $\mathbf{j}(x, y)$, which depends on the intrinsic curvature of the surface. Its employ requires some virtuosity in differential geometry, and we will never want to do that. We shall rather use the projection of the intrinsic current on a fixed plane, chosen to be the xy plane, and generally chosen parallel to a high-symmetry orientation. The drawback of this approach is that the diffusion coefficient defined by (7.11) becomes a (2×2) matrix which depends on the local orientation of the surface. However, this complication can be ignored if the surface does not depart very much from parallelism with the xy plane. We will assume this condition to be fulfilled generally in this book.

7.5 Diffusion under the effect of a chemical potential gradient

The density ρ which appears in equation (7.1) is generally not accessible to observation. It is, however, related to the local chemical potential $\mu(\mathbf{r}, t)$. The chemical potential is often related to easily observable quantities. For instance, as seen in chapter 2, the chemical potential of atoms on a clean surface is related to the surface shape by the Gibbs-Thomson relation. For that reason, the chemical potential will be often used in the next chapters.

Small variations $\delta\rho(\mathbf{r}, t)$ of the particle density are related to the corresponding variations $\delta\mu(\mathbf{r}, t)$ of the chemical potential by the relation

$$\delta\rho(\mathbf{r}, t) = \left(\frac{\partial\rho}{\partial\mu}\right)_{T,V} \delta\mu(\mathbf{r}, t) = -\frac{1}{V}\left(\frac{\partial^2\Omega}{\partial\mu^2}\right)_{T,V} \delta\mu(\mathbf{r}, t) \tag{7.7a}$$

where V is the volume (or, in two dimensions, the area) and $\Omega = U - TS - \mu N$ the grand potential. The temperature is assumed uniform.

In the case of dilute, non-interacting particles (such as adatoms at low temperature) one has (see section 5.6) the more explicit formula

$$\delta\rho = \rho_0 \frac{\delta\mu}{k_B T} \tag{7.7b}$$

where $\delta\rho$ is the deviation from the equilibrium density ρ_0.

If Fick's law (7.1) is applicable, its combination with (7.7) yields, for small variations $\delta\mu$:

$$\frac{\partial\rho}{\partial t} = \alpha\nabla^2\mu(\mathbf{r}, t) \tag{7.8}$$

where, in the case of a low concentration, (7.7b) yields

$$\alpha = D\rho_0/(k_B T) \,. \tag{7.9}$$

This relation is Einstein's formula, which is generally written as a relation between the diffusion constant and the mobility, which is α/ρ_0:

$$\alpha/\rho_0 = D/(k_B T) \,. \tag{7.10}$$

The general considerations above hold for volume or surface diffusion. They will now be applied to the special case of mass diffusion at the surface of a clean crystal. This mass diffusion is due to the motion of adatoms and advacancies under the influence of a chemical potential. It is a special case of a general external potential, which has the special feature that the chemical potential of advacancies is equal to minus the chemical potential of adatoms. Indeed the chemical potential μ corresponds to a term $-\mu N$ in the grand potential, where N is the number of particles; and the variation of N resulting from a variation δN_{adat} of the number of adatoms and a variation δN_{advac} of the number of advacancies is $\delta N_{\text{adat}} - \delta N_{\text{vac}}$. On the other hand, the total current density $\mathbf{j}_s$ on the surface is, for a similar reason, the difference $\mathbf{j}_s^{\text{adat}} - \mathbf{j}_s^{\text{advac}}$ of the current densities of adatoms and advacancies. Relations (7.8) and (7.4) yield $\mathbf{j}_s^{\text{adat}} = -\beta\rho D_s \nabla\mu$ and $\mathbf{j}_s^{\text{advac}} = \beta\sigma\Lambda_s\nabla\mu$, where, as in chapter 6, D_s and Λ_s are the adatom and advacancy diffusion constants, and ρ, σ are the adatom and advacancy densities, respectively. It follows

$$\mathbf{j}_s = \mathbf{j}_s^{\text{adat}} - \mathbf{j}_s^{\text{advac}} = -\beta\rho D_s\nabla\mu - \beta\rho\Lambda_s\nabla\mu\,.$$

It is of interest to define a 'surface mass diffusion constant' $\tilde{D}_s$ by

$$\mathbf{j}_s = -\beta\tilde{D}_s\nabla\mu\,. \qquad (7.11)$$

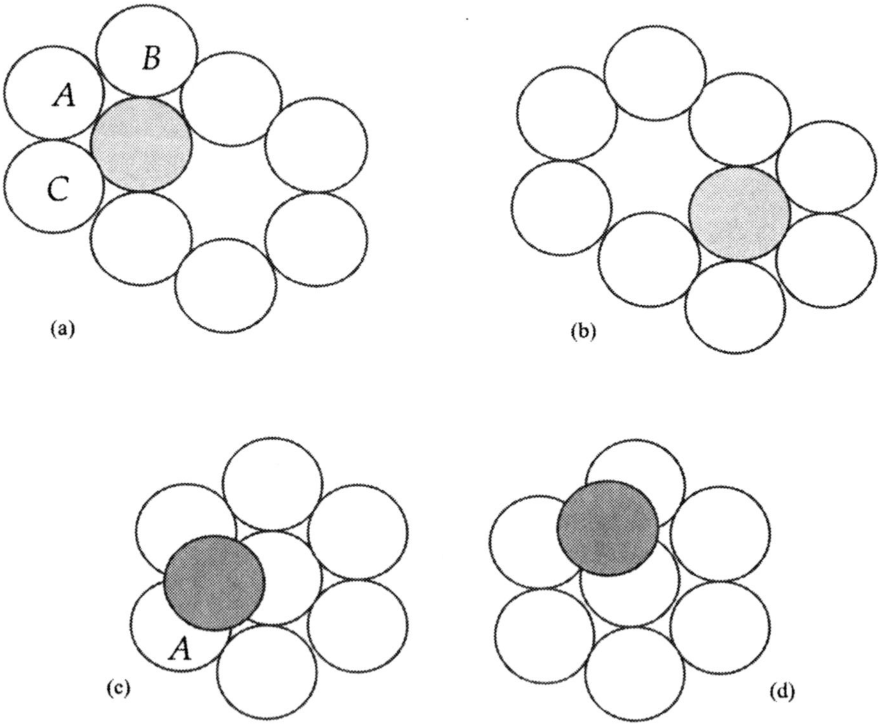

Fig. 7.3. a,b) Diffusion of an advacancy. The jump from (a) to (b) requires breaking 4 chemical bonds (with atoms A, B, C and with one atom of the underlying layer). c,d) Diffusion of an adatom. The jump from (c) to (d) requires breaking a single bond (with atom A).

Surprisingly, this quantity is easy to measure (see section 8.3). Comparing the last two equations yields

$$\tilde{D}_s = D_s\rho + \Lambda_s\sigma \ . \tag{7.12}$$

This is the combination encountered in section 6.6. It will appear again in section 8.3. Relation (7.12) makes sense only on a high-symmetry surface, while the definition (7.11) applies to mass transport on any clean surface.

The diffusion constant of advacancies, Λ_s may be expected to be smaller than that of adatoms, D_s (Fig. 7.3) so that the first term of (7.12) dominates the second one. Since ρ is usually small at moderate temperatures, $\tilde{D}_s$ is appreciably smaller than D_s. This property is illustrated by the surface dynamics in MBE and in the absence of growth. At the same temperature, the latter (which depends on D_s) is faster than the former, which depends on $\tilde{D}_s$. For instance, for the (001) surface of copper, Ernst *et al.* (1992b, 1993) have observed appreciable dynamics at 160 K on the growing surface while Lapujoulade (1990) and Fabre *et al.* (1986) report that the surface is already frozen at about 400 K in the absence of growth.

The microscopic diffusion constant D_s of adatoms can sometimes be measured by field ion microscopy (Ehrlich 1977). Mo and Lagally have used a more macroscopic (even though it involves using an STM) method which will be described in chapter 11.

7.6 Diffusion of radioactive tracers

The standard way to measure the tracer diffusion constant is to use radioactive tracers. This does not mean (Bonzel 1990) that radioactive tracers always provide an access to the tracer diffusion constant D^*. This is only true, as follows from the general discussion of section 7.1, if the concentration of radioactive tracers is kept very low. Experimentalists are not always able of doing that. For instance, Cousty *et al.* (1981) describe their measurement of the surface diffusion constant as follows: 'We deposit one monolayer of tracer on one half of the surface and we measure how it spreads.'

On the other hand, the measured quantity is not equal to (7.12) either, although (7.12) is presumably a good approximation in many cases (see problem 7.4).

7.7 Activation energy

At low temperature, the motion of an adatom on a surface is hindered by the 'corrugation' of the substrate, i.e. the potential modulation due to the atomic structure (see Fig. 6.4a). The minima of the potential will be called *easy sites*. In order to go from an easy site to a neighbouring one, the adatom must jump over an energy (or more correctly free energy) barrier W_{sd}. At equilibrium, the probability for this atom to reach this energy is, according to the Gibbs-Boltzmann formula of statistical mechanics, proportional to $\exp(-\beta W_{sd})$, where $\beta = 1/k_B T$. It is therefore natural to expect that the adatom diffusion constant is given by an expression of the form

$$D_s = D_s^0 \exp(-\beta W_{sd}) \tag{7.13}$$

where the 'prefactor' D_s^0 is presumably of the order of a typical atomic frequency, 10^{13} s^{-1}, say, and depends but weakly on temperature. Experiments are often interpreted assuming no temperature dependence at all.

This works fairly well, except that the prefactor is sometimes much larger than the Debye frequency (as seen later in this section) and therefore unphysical. The theoretical justification of (7.13) is actually extremely difficult. The *transition state theory* (Glasstone *et al.* 1941, Vineyard 1957) is an attempt to give a basis to the above argument. The problem is that, if the probability for the diffusing particle to reach the potential barrier is correctly given by (7.13), it is difficult to predict what this particle will do later. If it stops at the next energy minimum, (7.13) is all right. However, it can also come back, or propagate further if it is unable to release its energy (Iche & Nozières 1976).

As a matter of fact, most of the available experimental data (which are anyway not tremendously many) yield the mass diffusion constant $\tilde{D}_s$ defined by (7.12). The adatom and advacancy concentrations are given by the following Gibbs formulae where the (free) energies W_a and W_{adv} appear:

$$\rho = \exp(-\beta W_a)\,, \tag{7.14a}$$

and

$$\sigma = \exp(-\beta W_{adv})\,. \tag{7.14b}$$

Typical values of W_a and W_{adv} for metals are listed in Table 1.1.

Relation (7.12) now yields

$$\tilde{D}_s = D_s \exp(-\beta W_a) + \Lambda_s \exp(-\beta W_{adv})\,. \tag{7.15}$$

The second term is expected (Bonzel 1990) to be the smaller, at least on (110) and (111) faces of metals. Indeed, the jump of a vacancy is in reality the jump of an atom, which requires, for instance (Fig. 7.3), to break 3 atomic bonds on a compact face, instead of one in the case of adatom jumps. Therefore, (7.13) and (7.15) yield

$$\tilde{D}_s = D_s^0 \exp[-\beta(W_{sd} + W_a)]\,.$$

In this formula, an effective activation energy $W_{sd} + W_a$ appears, which is higher than the activation energy W_{sd} for adatom jumps. On a non-compact face of a typical metal like Cu or Ni, $W_{sd} + W_a$ is of order 1 eV and W_{sd} is of order 0.4 eV (Liu *et al.* 1991, Kuipers *et al.* 1993, Kuipers 1994). On the (111) face the barrier to adatom jumps is of the order of 0.1 eV according to numerical calculations (Stoltze 1994).

Fig. 7.4 shows that the Arrhenius law represents rather well the behaviour of the mass diffusion constant $\tilde{D}_s$ as a function of T (Freyer 1985). However, the prefactor is anomalously large at the high temperatures represented in the figure. As we said before, a value of D_s^0 of the order of a typical Debye frequency would be expected, namely ≈ 0.01 cm^2/s in CGS units. Indeed, this is the right order of magnitude for fcc metals, but only at low enough temperature, $T/T_M < 0.77$ (Bonzel 1975), if T_M is the bulk melting temperature. In this range, the activation energy is of order $6.5T_M$. For $T/T_M > 0.77$, the activation energy on a (110) face is about $15T_M$ and the prefactor is about 10 000 times as high. This

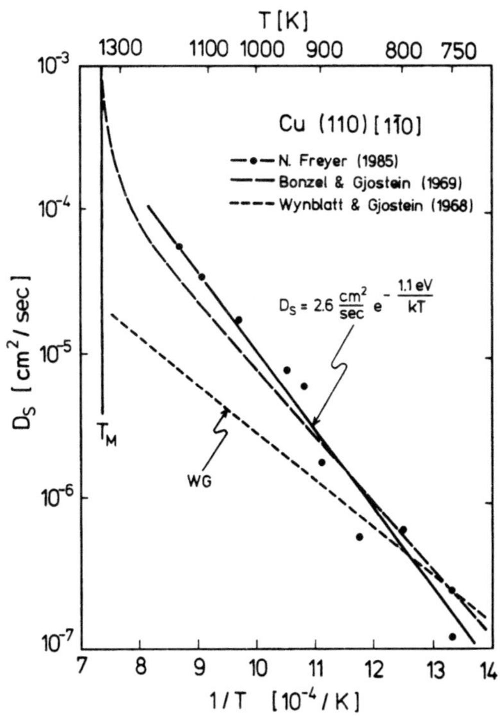

Fig. 7.4. Surface diffusion constant on the (110) and (111) faces of Cu (Courtesy of H. Bonzel). Continuous lines are guides to the eye. The dotted line is a theoretical calculation by Wynblatt & Gjostein (1968) based on pair interactions.

suggests that, near T_M, adatom dynamics becomes more collective and more complicated than a random walk on a lattice.

This change of behaviour is presumably related to the roughening of the (110) face (Kern 1987). Bonzel's measurements were done on a (110) face rather than on a (001) or (111) because, with the technique employed, the latter have a peculiar behaviour that will be analysed in section 8.6.

The calculation of the diffusion constant, or at least of the activation barrier W_{sd}, is only possible at lower temperatures (say below $0.77T_M$). It requires an approximation of the chemical bond which should be both simple enough and reliable. A good and reasonably simple approximation is the effective medium approximation (Jacobsen *et al.* 1987, Jacobsen 1988, Stoltze 1994). It is then necessary to find the 'diffusion path' (in the configuration space) which yields the lowest energy barrier. The naive expectation would be that a diffusing adatom just creeps (or 'bridges') on the surface, trying to disturb the underlying atoms as little as possible. This guess has been tested by Hansen *et al.* (1991) in the case of copper. They found that the bridging way was indeed the most favourable on the compact, (111) face, and also on the (110) face. However, the diffusion

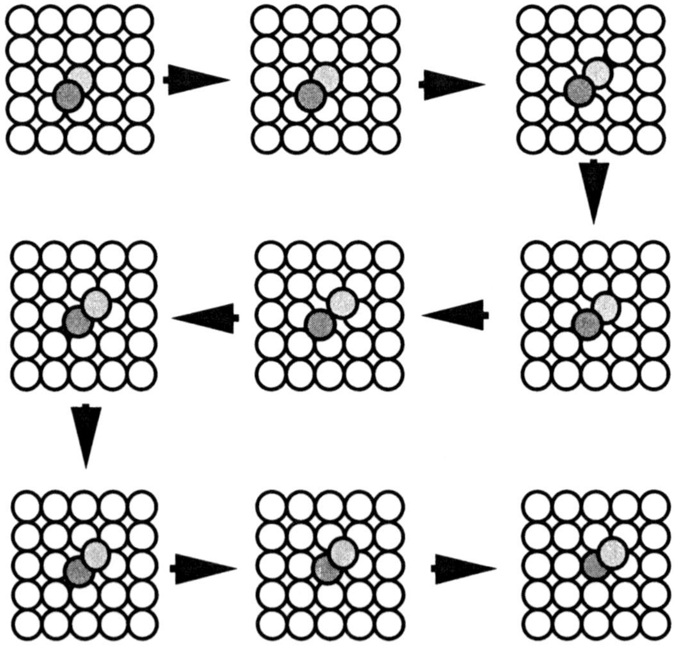

Fig. 7.5. The exchange mechanism for the (001) face of Cu. It turns out to be the most favourable 'diffusion path' for this surface (Hansen *et al.* 1991).

path for the (100) face turns out to be through the exchange of the adatom with a surface atom which then becomes an adatom (Fig. 7.5).

More recently, activation energies of metals have been determined by *ab initio* methods (Boisvert *et al.* 1995). The agreement with the effective medium approximation is good for (100) faces, while *ab initio* methods predict a much higher activation energy on (111) faces, typically twice as large as the effective medium approximation. Comparison with experiment is not always possible, but seems to be in favour of *ab initio* methods. Note that the same *ab initio* methods yield a cohesive energy which is typically 30% higher than the experimental value (Boisvert *et al.* 1995). Of course, partly due to the ever-increasing computing power, the situation will evolve considerably in the near future.

7.8 Surface melting

As noted in the previous section, the properties of surfaces at high temperatures (above $3T_M/4$) look different, more collective than at lower temperature. Atomic features become less apparent; for instance, at $T \approx 0.9T_M$ steps are no longer visible by electron microscopy. The solid surface begins to look like a liquid surface (Pavlovska & Nenow 1977, Bienfait 1987, Bienfait *et al.* 1988). The expression 'surface melting' has a more precise meaning than a progressive increase of the atomic thermal motion near the surface. A picture of the surface very near T_M is given by Fig. 7.6: there is a liquid film between the solid and the vapour, and the thickness goes to infinity when T approaches T_M (Frenken *et al.* 1986, 1988, Pluis *et al.* 1987, Van der Veen *et al.* 1990). Thus, melting of a solid with a surface becomes a continuous transition! This phenomenon is quite analogous to wetting, and may be regarded as the wetting of the solid by its own liquid.

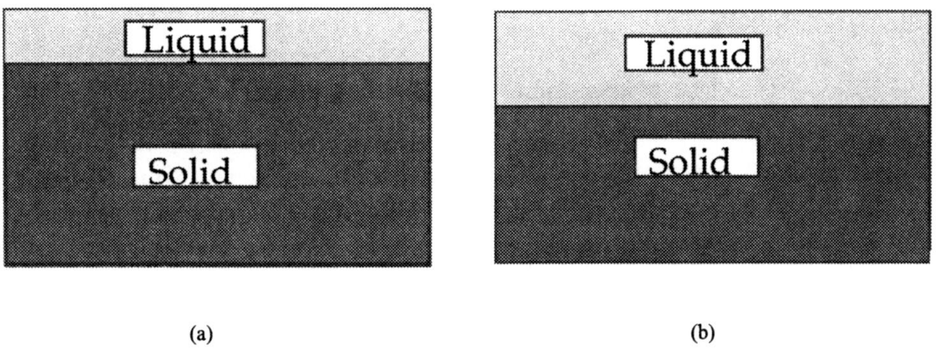

Fig. 7.6. A schematic representation of surface melting. When T approaches the bulk melting temperature T_M, a liquid layer covers the solid (a), and its thickness goes to infinity when T goes to T_M (b).

A mean field analysis shows that surface melting (as above defined) may occur or not, according to the particular properties of the material.

Surface melting implies that the steps which are visible on surfaces at lower temperatures (*cf.* Fig. 1.2) are no longer observed by electron microscopy. However, steps may disappear before surface melting occurs, for instance because of thermal meandering.

Surface melting affects the surface tension and therefore has *a priori* unexpected consequences on the crystal shape (Nozières 1989).

7.9 Calculation of the diffusion constant

The diffusion constant can be computed numerically using the so-called *molecular dynamics.* The method consists of solving the classical equations of motion, which read, if $\mathbf{R}_i$ is the position of the i-th atom, m_i its mass and $\mathbf{f}_i$ is the force acting on it:

$$m_i \frac{\mathrm{d}^2}{\mathrm{d}t^2} \mathbf{R}_i(t) = \mathbf{f}_i \ .$$

The force $\mathbf{f}_i$ is readily obtained as a function of the $\mathbf{R}_i$s if the potential energy $V(\mathbf{R}_1, \mathbf{R}_2, \mathbf{R}_3 \ldots)$ is known. The function V is an empirical potential. In the case of solid 'rare gases' and to some extent in the case of metals, it may be approximated by a sum of pair potentials. A more ambitious alternative is to determine the electronic wave function for each set of nuclear coordinates $(\mathbf{R}_1, \mathbf{R}_2, \mathbf{R}_3 \ldots)$, and to deduce the force.

Present-day computers do not allow to perform molecular dynamics simulations for times longer than a few nanoseconds. Therefore, it may be difficult to determine the diffusion constant since, as explained in section 7.7, it may be unclear whether an atom which has jumped from a lattice site to another will stop there or propagate further. Some of the solutions which can be proposed will be found in the article by Voter (1989) and references therein.

7.10 Diffusion of big adsorbed clusters

Big atom clusters adsorbed on surfaces are expected to diffuse very slowly. If the cluster is made of the same material as the substrate, cluster diffusion is obtained *via* the diffusion of atoms along the cluster perifery, or the detachment of individual atoms which then diffuse on the substrate and come back to another point on the cluster.

Let us briefly discuss the diffusion of individual atoms along the edge of a cluster of atomic thickness (2-dimensional cluster) of radius R. If D_2 is the diffusion constant along the edge, each atom needs a time of order R^2/D_2 to reach the opposite side of the cluster. In this time, if all the atoms of a particular side (e.g. the left hand side) move together to the other end, the net displacement of the cluster would be about 1 atomic distance. However, the atoms of the edge, whose number is of order R, actually move in random directions, so that the net displacement of the cluster must be divided by approximately $\sqrt{R}$. It should be related to the cluster diffusion constant $D(R)$ by the relation $1/\sqrt{R} \simeq D(R)R^2/D_2$. Therefore $D(R) \sim D_2/R^{5/2}$, which decreases rapidly with increasing R. However, fast diffusion of big clusters is occasionally (and mysteriously) observed (Masson *et al.* 1971, Bardotti *et al.* 1995, Steyer *et al.* 1992, Van Siclen 1995).

Problems

7.1. It has been mentioned in section 7.3 that equations (7.1) and (7.3) are not valid for times longer than the lifetime. Write equations which may be acceptable for longer times (Hint: look at section 6.5).

7.2. Spot the flaw in the following argument:

The density $\rho(\mathbf{r}, t)$ is the sum of the individual probabilities $P_i(\mathbf{r}, t)$ of the various atoms, labelled by i, to be at point $\mathbf{r}$. Each $P_i(\mathbf{r}, t)$ satisfies the equation $\dot{P}_i(\mathbf{r}, t) = D^*\nabla^2 P_i(\mathbf{r}, t)$. When adding these equations, one obtains $\dot{\rho}(\mathbf{r}, t) = D^*\nabla^2\rho(\mathbf{r}, t)$. Therefore $\tilde{D} = D^*$, which has been proven to be wrong.

Solution – The probabilities are not independent. It is not correct to write both $\dot{P}_i(\mathbf{r}, t) = D^*\nabla^2 P_i(\mathbf{r}, t)$ and $\dot{P}_j(\mathbf{r}, t) = D^*\nabla^2 P_j(\mathbf{r}, t)$ if $i \neq j$. The derivative $\dot{P}_i(\mathbf{r}, t)$ depends on all the other $P_i(\mathbf{r}, t)$ except if they are averaged out.

7.3. Show that successive jumps of a given atom in vacancy-dominated tracer diffusion are correlated. The vacancies will be assumed to jump by one atomic distance in an uncorrelated way. Derive an approximate expression of D^* as a function of D_{vac}, which takes correlations into account.

Solution – If a vacancy has jumped by a vector $\mathbf{a}_n$, there is a finite probability that the next jump is backward, i.e. $\mathbf{a}_{n+1} = -\mathbf{a}_n$. This probability is $1/\zeta$, if ζ is the lattice coordination (6 in a simple cubic lattice, etc.). Therefore, if the m-th jump of a particular atom is by the vector $\mathbf{b}_m$ there is a probability or order $1/\zeta$ that the next jump is by $-\mathbf{b}_m$. This is only approximate, because additional correlations between $\mathbf{b}_m$ and $\mathbf{b}_{m+1}$ result from closed loops accomplished by the vacancy in its motion. In this approximation, (7.2) should be corrected as follows:

$$\left\langle \left| \sum_{n=1}^{m} \mathbf{a}_n \right|^2 \right\rangle = \sum_{n,n'} \langle \mathbf{a}_n \cdot \mathbf{a}_{n'} \rangle \simeq \sum_{n} \langle \mathbf{a}_n \cdot \mathbf{a}_n \rangle + \sum_{n} \langle \mathbf{a}_n \cdot \mathbf{a}_{n+1} \rangle$$

$$\simeq \sum_{n} \langle \mathbf{a}_n \cdot \mathbf{a}_n \rangle - \frac{1}{\zeta} \sum_{n} \langle \mathbf{a}_n \cdot \mathbf{a}_n \rangle \ .$$

It follows that $D^* \simeq (1 - \frac{1}{\zeta})D_{\mathrm{vac}}$.

7.4. Compute the tracer diffusion constant, assuming that it results from uncorrelated adatom jumps, and from uncorrelated vacancy jumps of fixed length a in any direction. Correlations of the adatom motion will be neglected. Introduce the ratio $p = a/b$, where b is the atomic distance.

Solution – The quantity D^* of interest is defined by (7.3). In time t, a tracer runs a total length $\mathbf{r}'(t)$ because of vacancy motion (advacancy in the case of surface diffusion) and a total length $\mathbf{r}''(t)$ because of interstitial (adatom) motion. It is easy to show that

$$\langle r^2(t) \rangle = \langle r'^2(t) \rangle + \langle r''^2(t) \rangle \ .$$

As seen in section 7.3, the mean square path $\langle r''^2(t) \rangle$ due to interstitial motion is equal to the number $N\rho_{\mathrm{int}}$ of interstitials, multiplied the mean square displacement of an interstitial defect, which is $2dD_{\mathrm{int}}t$, and divided by the number N of atoms:

$$\langle r''^2(t) \rangle = 2d\rho_{\mathrm{int}} D_{\mathrm{int}} t \ .$$

The mean square path $\langle r'^2(t) \rangle$ due to vacancy motion is equal to the number $N\rho_{\mathrm{vac}}$ of vacancies, multiplied by the average number of displacements of a vacancy, which is $2dD_{\mathrm{vac}}t/a^2$, and by the probability that the tracer participates in a particular move, which is p/N. Therefore

$$\langle r'^2(t) \rangle = 2d\rho_{\mathrm{vac}} D_{\mathrm{vac}} t/p \ .$$

These equations yield

$$D^* = \frac{1}{p} \rho_{\mathrm{vac}} D_{\mathrm{vac}} + \rho_{\mathrm{int}} D_{\mathrm{int}} \ .$$

In the case of surface diffusion, one has

$$D_s^* = \frac{1}{p} \Lambda_s \sigma + D_s \rho$$

which is equal to $\tilde{D}_s$ if $p = 1$. However, even for $p = 1$, this equality is only approximate because correlations have been neglected.

7.5. Write diffusion equations for impurities on a solid limited by a surface, assuming that surface atoms can penetrate into the bulk.

Solution – Let $D_{\rm v}$ be the diffusion constant in the bulk, D_s the diffusion constant on the surface, $D'_{\rm v}$ the diffusion constant from the surface into the bulk. Let $\rho_n(x, y, t)$ designate the impurity concentration in the n-th layer. Defining $\nabla^2 = \partial^2/\partial x^2 + \partial^2/\partial y^2$, the diffusion equations are

$$\begin{aligned}\dot{\rho}_n(x, y, t) = {} & D_{\rm v}\nabla^2\rho_n(x, y, t) \\ & +D_{\rm v}\left[\rho_{n+1}(x, y, t) + \rho_{n-1}(x, y, t) - 2\rho_n(x, y, t)\right]\end{aligned}$$

for $n > 0$, and

$$\dot{\rho}_0(x, y, t) = D_s\nabla^2\rho_0(x, y, t) + D'_{\rm v}\left[\rho_1(x, y, t) - \rho_0(x, y, t)\right]$$

at $n = 0$.

Introducing the two-dimensional Fourier transforms $\tilde{\rho}_n(k_x, k_y, t)$, these equations read for $n > 0$:

$$\begin{aligned}\dot{\tilde{\rho}}_n(k_x, k_y, t) = {} & -D_{\rm v}k^2\tilde{\rho}_n(k_x, k_y, t) \\ & +D_{\rm v}\left[\tilde{\rho}_{n+1}(k_x, k_y, t) + \tilde{\rho}_{n-1}(k_x, k_y, t) - 2\tilde{\rho}_n(k_x, k_y, t)\right]\end{aligned}$$

and for n=0:

$$\dot{\tilde{\rho}}_0(k_x, k_y, t) = -D_sk^2\tilde{\rho}_0(k_x, k_y, t) + D'_{\rm v}\left[\tilde{\rho}_1(k_x, k_y, t) - \tilde{\rho}_0(k_x, k_y, t)\right] .$$

Let now $\tilde{\rho}_n(k_x, k_y, t) = K(k_x, k_y)R_n(t)$, and $k^2 = k_x^2 + k_y^2$. The preceding equations become, respectively:

$$\dot{R}_n(t) = -D_{\rm v}k^2R_n(t) + D_{\rm v}\left[R_{n+1}(t) + R_{n-1}(t) - 2R_n(t)\right] ,$$

$$\dot{R}_0(t) = -D_sk^2R_0(t) + D'_{\rm v}\left[R_1(t) - R_0(t)\right] .$$

Let us forget for the moment the term proportional to k^2. We can cast both equations into one, if we note that they describe a one-dimensional random walk between layers, which cannot cross the boundary. In other words, the layer $n = 0$ acts as a mirror, reflecting back the approaching walker. The classical method for solving this kind of problems is the *method of images* (Sommerfeld 1964): one solves the random walk problem without the mirror–that is, the boundary–adding in its place an initial condition $R_n(t = 0) = \delta_{n,0} + \delta_{n,-1}$. Therefore, the net particle current crossing the boundary at $n = 0$ is always nil. The equations read now:

$$\dot{R}_n(t) = D_{\rm v}\left[R_{n+1}(t) + R_{n-1}(t) - 2R_n(t)\right] ,$$

$$R_n(0) = \delta_{n,0} + \delta_{n,-1} .$$

These are easily solved using once more the Fourier transforms $\tilde{R}(q, t)$:

$$\dot{\tilde{R}}(q, t) = -2D_{\rm v}(1 - \cos q)\tilde{R}(q, t) ,$$

$$\tilde{R}(q, 0) = 1 = {\rm e}^{iq} ,$$

and by employing known properties of the modified Bessel function of imaginary argument:

$$I_n(z) = \frac{1}{\pi} \int_0^{\pi} e^{z\cos q} \cos(nq)dq \, ,$$

$$I_{n+1}(z) + I_{n-1}(z) = 2\frac{d}{dz} I_n(z) \, .$$

It is now easy to show that the solution reads

$$\tilde{\rho}_n(k_x, k_y, t) = K(k_x, k_y) e^{-[D(n)k^2 + 2D'(n)]t} \times \left[I_n(2D'(n)t) + I_{n+1}(2D'(n)t) \right]$$

where we defined $D(n) = D'(n) = D_v$ for $n \neq 0$, and $D(0) = D_s$, $D'(0) = D'_v$.

7.6. Calculate the chemical potential as a function of the density for a non-interacting lattice gas, using the formulae $\mu = (\partial F / \partial N)_{T,V}$ and $F = U - TS$. Rederive the result by using the grand-canonical distribution.

Solution – In the absence of interactions, $U = 0$ and $F = -TS = -k_B T \ln g$. For N particles on a lattice of $\mathcal{N}$ sites,

$$g = \frac{\mathcal{N}!}{N!(\mathcal{N} - N)!} \, .$$

The Stirling formula yields

$$\ln g = \mathcal{N} \ln \mathcal{N} - N \ln N - (\mathcal{N} - N) \ln(\mathcal{N} - N)$$

$$= -N \ln \left(\frac{N}{\mathcal{N}} \right) - (\mathcal{N} - N) \ln \left(\frac{\mathcal{N} - N}{\mathcal{N}} \right) = -\mathcal{N} \left[\rho \ln \rho + (1 - \rho) \ln (1 - \rho) \right] \, .$$

Hence

$$F = k_B T \mathcal{N} \left[\rho \ln \rho + (1 - \rho) \ln (1 - \rho) \right]$$

and

$$\mu = \frac{1}{\mathcal{N}} \left(\frac{\partial F}{\partial \rho} \right)_{T,\mathcal{N}} = k_B T \left[\ln \rho - \ln (1 - \rho) \right] = k_B T \ln \frac{\rho}{1 - \rho}$$

or

$$\frac{\rho}{1 - \rho} = \exp(\beta \mu)$$

and

$$\rho = \frac{\exp(\beta \mu)}{1 + \exp(\beta \mu)} \, .$$

These formulae may be obtained much more easily by the grand-canonical distribution. Indeed the probability of having n_R atoms at site R is given by

Const $\times \prod_R \exp(\beta\mu_R n_R)$, where μ_R is the local chemical potential at site R. Since n_R can be either 0 or 1, it follows

$$\rho(\mathbf{R}) =< n_R >= \frac{\exp(\beta\mu_R)}{1+\exp(\beta\mu_R)} \ .$$

8
Thermal smoothing of a surface

I was perpetually asking
not *for mathematical equations,*
but for physical circumstances
of what they were trying to work out.

R. Feynman (*Surely you are joking, Mr. Feynman*)

It is fortunate that things do not reach their equilibrium shape studied in chapter 3. It would be too bad, if Donatello's or Rodin's statues were to behave as hedge-hogs!

However, they would do so if they were heated at high enough temperatures. Annealing is indeed one of the methods used in laboratories in order to get smooth crystal surfaces.

During the annealing process, atoms diffuse until they have enough neighbours, i.e. enough satisfied chemical bonds, to be happy. Rather than considering the detailed atomic structure, it is generally preferable to introduce the chemical potential. Atoms go to the places where the chemical potential is low. Since the chemical potential is related to the shape through the Gibbs-Thomson formula, surfaces become smooth.

At very long distances, surface diffusion is less efficient than evaporation followed by condensation. In this chapter, we address the case where there is no net evaporation. This situation is easily obtained by putting the sample into a closed container.

The distances on which thermal equilibrium can be reached are:

- A few microns at temperatures of order $2T_M/3$.
- A few nanometres at temperatures of order $T_M/3$.

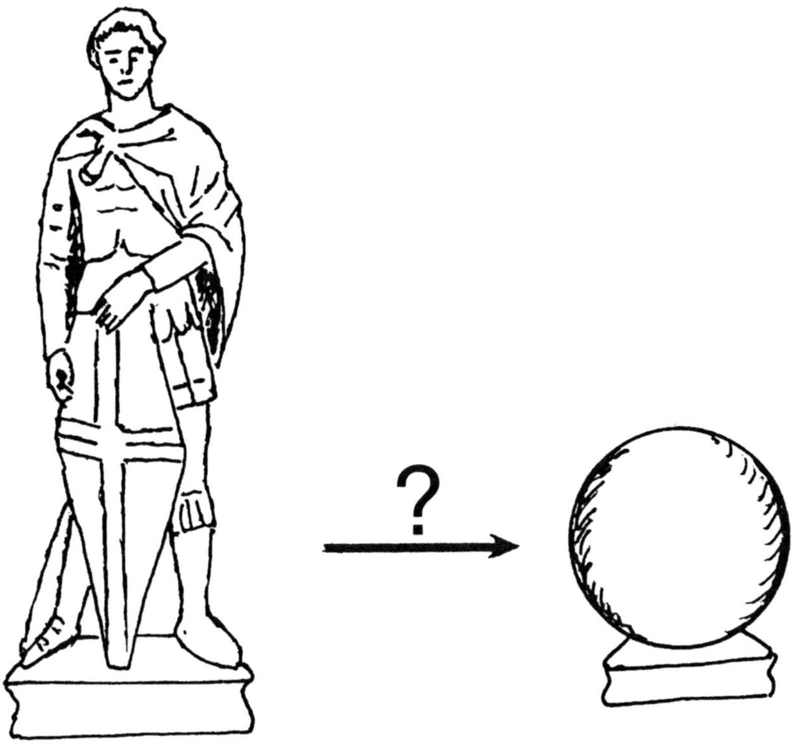

Fig. 8.0. Ancient works of art do not possess their equilibrium shape, which is not always the case with modern ones!

8.1 General features

In this chapter, we examine the possibility of reducing the roughness of a surface by annealing, i.e. by moderate but long heating. Surface annealing is often used after MBE growth or after cleaning by an ion beam. In addition, it is a method of measurement of the diffusion constant (Bonzel 1975). It will be assumed that there is globally no growth and no evaporation, or in other words, that the chemical potential inside the solid and inside the vapour is the same. This situation is easily obtained by putting the crystal into a closed container.

However, if the surface has not its equilibrium shape, the chemical potential is not uniform on the surface as seen in chapter 2. The variations of the chemical potential imply, according to (7.11) a flux of adatoms and advacancies from places where the chemical potential is high to where it is low. For instance, if there are islands and lakes (Fig. 8.1), matter will flow from islands to lakes. This flow generally takes place through surface diffusion and the current density is then given by (7.5). However, surface

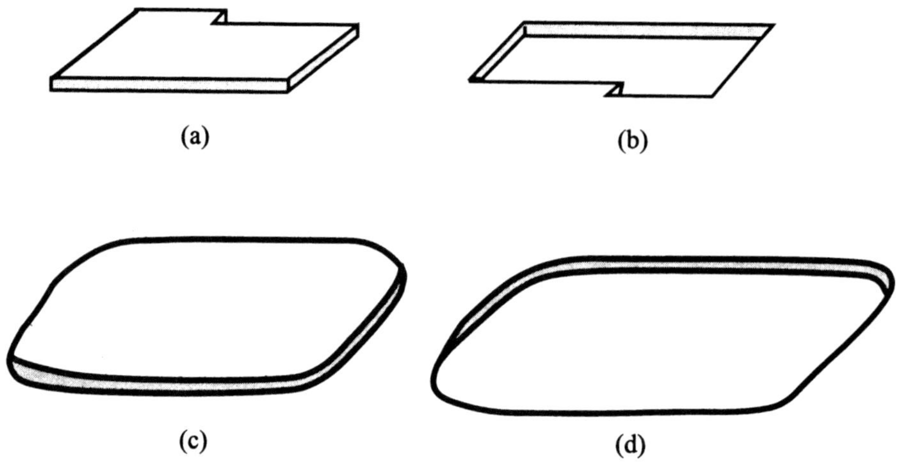

Fig. 8.1. The density of adatoms (and their chemical potential) is higher near a terrace (a or c) and weaker near a hole (b or d). Pictures (c) and (d) correspond to a continuous description, applicable to large terraces and holes.

transport is not always the most relevant process as will be seen in the next section.

8.2 The three ways to transport matter

Just as human beings do, atoms prefer to travel airborne or by surface transportation (Fig. 8.2). Pursuing this analogy, it may be said that going airborne is faster on long distances, but taking off requires some time (to jump over high energy barriers) so that surface transportation, i.e. surface diffusion, is more appropriate for short trips. These facts will be summarized by precise formulae in the next section.

Atoms also have the possibility, as Londoners or New Yorkers do, to travel underground (Fig. 8.2). This process, i.e. volume diffusion, may sometimes be effective at intermediate distances, but will generally be ignored in this book. It is just briefly addressed in section 8.4.

The characteristic length r_1 beyond which air transportation dominates surface transportation may be roughly evaluated as follows, on a high-symmetry surface. This length is the distance on which atoms can diffuse without evaporating. According to diffusion theory (see formula 7.3) the time t needed to diffuse on a distance r is of order $r^2/4D_s$. The evaporation probability during this time is $t/\tau_v \approx r^2/(4D_s\tau_v)$, where $1/\tau_v$ is the probability per unit time that a given adatom evaporates. Surface transport dominates if this quantity is much smaller than 1. It follows

$$r_1 \approx \sqrt{D_s\tau_v} \,. \qquad (8.1)$$

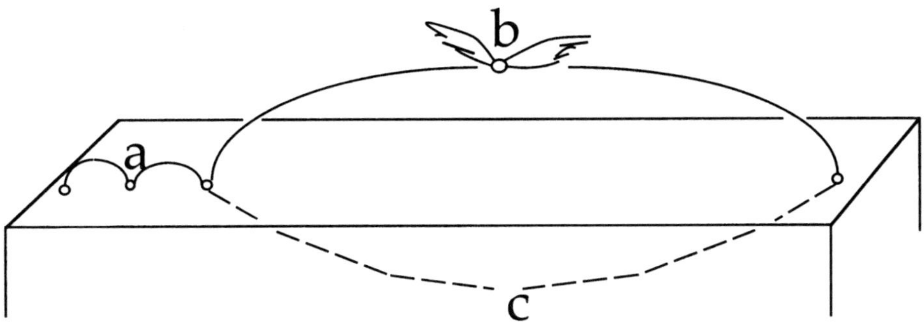

Fig. 8.2. Various ways of matter transport which contribute to smoothen a surface. Surface diffusion (a) is generally the dominant process. Transport through the vapour (b) becomes important at high temperatures and long distances. Diffusion through the bulk (c) will generally be neglected here.

This length coincides with the characteristic length $x_s = 1/\kappa$ encountered in chapter 6 (equation 6.6). At low temperature, τ_v is huge, and r_1 too, although D_s is rather small. Actually, evaporation is negligible for elements in most cases. It is not so for binary systems. In GaAs, for instance, in usual MBE conditions, arsenic evaporates.

8.3 Smoothing of a surface above its roughening transition

In this section we consider a crystal surface which is approximately planar, but for a small initial perturbation which slowly decays in time. We wish to study this decay. If the average orientation, supposed parallel to the xy plane, is above its roughening temperature, this decay turns out to be described by linear equations which may be solved by a Fourier transformation. Accepting this assumption, which will be checked self-consistently, it is enough to consider a single Fourier transform. Therefore, we assume a sinusoidal initial profile of the form (Fig. 8.3):

$$z(x, t = 0) = h(0) \cos \frac{\pi x}{2L} . \tag{8.2a}$$

It will be found that the equation of motion preserves the sinusoidal form and has the solution

$$z(x, t) = h(t) \cos \frac{\pi x}{2L} . \tag{8.2b}$$

The corresponding chemical potential is given, as seen in chapter 2, by

$$\mu = \mu_0 - \frac{\partial}{\partial x}\left(\frac{\partial \varphi}{\partial z_x}\right) - \frac{\partial}{\partial y}\left(\frac{\partial \varphi}{\partial z_y}\right) , \tag{8.3}$$

where φ is the projected free energy density per unit area of the xy plane, and $z_\alpha = \partial z/\partial x_\alpha$. For a weak modulation amplitude h, φ may be

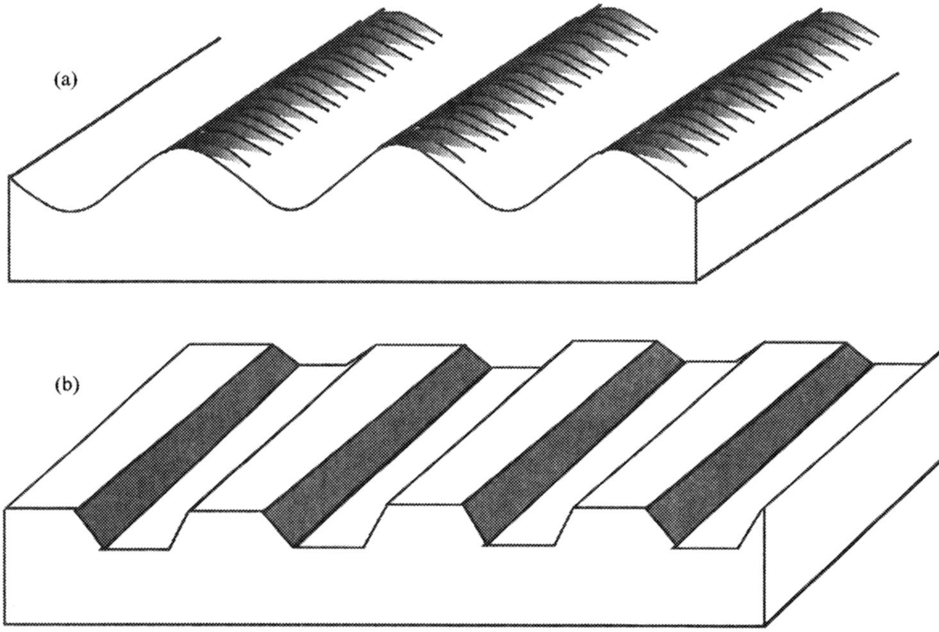

Fig. 8.3. A method, used in particular by Bonzel *et al.* (1984a,b) to measure the surface diffusion constant of a material, consists of making sinusoidal grooves (a) and measure the time they need to smoothen. When the average orientation of the surface is below its roughening transition, the formation of pseudo-facets (b) is observed.

approximated by a quadratic function which will be assumed isotropic for the sake of simplicity:

$$\varphi(z_x, z_y) \simeq \frac{\tilde{\sigma}}{2}\left(z_x^2 + z_y^2\right) . \tag{8.4}$$

It would be straightforward to take the anisotropy of the surface stiffness $\tilde{\sigma}$ into account. Relations (8.3) and (8.4) yield

$$\mu = \mu_0 - \tilde{\sigma}\left(z_{xx} + z_{yy}\right) . \tag{8.5}$$

The profile (8.3) may represent a single Fourier component of a complex accidental modulation. It may also represent an artificial, macroscopic profile cut in a surface in order to make an appropriate experiment (Bonzel 1975).

Evaporation-condensation kinetics

At high enough temperatures or long enough distances, evaporation dominates surface diffusion according to (8.1). It will be assumed that transport in the vapour is so fast, with respect to evaporation and adsorption (condensation), that the chemical potential μ_0 in the vapour is uniform. The chemical potential μ at each point of the surface is given by (8.5). Evaporation dominates adsorption at all places where $\mu > \mu_0$, while the reverse is true where $\mu < \mu_0$. According to the axioms of irreversible thermodynamics, the derivative $\dot{z}$ of the surface height with respect to time will be assumed to be a linear function of the chemical potential difference $(\mu - \mu_0)$, provided that this difference is small:

$$\dot{z} = -\text{Const} \times (\mu - \mu_0) \ . \tag{8.6}$$

Insertion of (8.5) into (8.6) yields

$$\dot{z} = \nu \left(z_{xx} + z_{yy} \right) \ , \tag{8.7}$$

where $\nu > 0$ is a constant. This equation of motion is linear. A consequence of this is that an initially sinusoidal profile keeps its sinusoidal shape. If the initial profile is given by (8.2a), the solution of (8.7) is given by (8.2b). More precisely, as first derived by Mullins (1957, 1959)

$$z(x, y, t) = z(x, y, 0) \exp\left(-t/\tau(L)\right) \tag{8.8}$$

where

$$\tau(L) = \text{Const} \times L^2 \ . \tag{8.9}$$

At usual temperatures, the vapour at equilibrium with the solid has generally an extremely low density, so that the coefficient ν in (8.7) is very small. Surface kinetics are therefore often dominated by surface diffusion, to which we turn now.

Surface diffusion kinetics

We now address the case where evolution of the surface shape is due to adatom or advacancy diffusion on the surface. It follows from (8.1) that this hypothesis is correct (Mullins 1957, 1959) for nickel at temperatures between 1000 K and 1000 °C with a surface modulation of wavelength $L \approx 10$ μm.

The first thing to write is a conservation equation which expresses the absence of evaporation:

$$\dot{z} = -\nabla \cdot \mathbf{j}_s \ . \tag{8.10}$$

The surface current density $\mathbf{j}_s$ is given by (7.11):

$$\mathbf{j}_s = -\beta \tilde{D}_s \nabla \mu \tag{8.11}$$

where $\nabla = (\partial_x, \partial_y)$. Inserting (8.5) into (8.11) and (8.10), one finds the equation of motion

$$\dot{z} = -K \nabla^2 \left(\nabla^2 z \right) \ , \tag{8.12a}$$

where $K = \beta\tilde{D}_s\tilde{\sigma} > 0$. This equation of motion is still linear, so that an initially sinusoidal profile again remains sinusoidal. This feature has been experimentally observed on the (110) face of nickel (Bonzel *et al.* 1984a,b). If z does not depend on y, (8.12a) reads

$$\dot{z} = -K\frac{\partial^4 z}{\partial x^4} . \tag{8.12b}$$

If the initial shape is given by (8.2), the solution of (8.12) has again the form (8.8), but now (8.9) should be replaced by

$$\tau(L) = \frac{16}{\pi^4}KL^4 . \tag{8.13}$$

The experimental determination of the constant yields the surface mass diffusion coefficient $\tilde{D}_s = k_B TK/\tilde{\sigma}$ if the surface stiffness $\tilde{\sigma}$ is known.

If both evaporation and surface diffusion are taken into account, the general equation of motion obtained combining (8.7) with (8.12a) is

$$\dot{z} = v\nabla^2 z - K\nabla^2\left(\nabla^2 z\right) . \tag{8.14}$$

The Fourier transform of wavevector q, z_q, has a relaxation rate

$$1/\tau_q = vq^2 + Kq^4 .$$

As said above, v is generally small excepted at very high temperatures, and the second term generally dominates. At high temperature, the first term is important for small wavevectors, i.e. at long distances.

8.4 Thermal smoothing due to diffusion in the bulk solid

The previous section has been devoted to the cases when mass transport is due to surface diffusion or to evaporation+condensation. We now investigate the case when transport through the bulk dominates these two processes. In practice, it is generally not so. However, the case of volume diffusion is of interest, not only for the sake of completeness, but also because the equations to be solved are of a very common type in surface physics. In the solid, the equation is the diffusion equation (7.1). As before, the profile is assumed independent of y, so that this equation reads

$$\dot{\rho} = -\tilde{D}\left(\frac{\partial^2\rho}{\partial x^2} + \frac{\partial^2\rho}{\partial z^2}\right) .$$

Note that, in this section, the matter density ρ and the current density j are three-dimensional densities, and $\tilde{D}$ is the bulk diffusion constant.

As in chapter 6, the quasi-static approximation may be used. The above equation reduces to

$$\frac{\partial^2\rho}{\partial x^2} + \frac{\partial^2\rho}{\partial z^2} = 0 .$$

The boundary condition at the surface is obtained by equating the expression (8.5) of the chemical potential μ with the expression (7.7a) which states that its deviation $\delta\mu$ from equilibrium is proportional to the deviation $\delta\rho$ of the density in the bulk, with a coefficient $1/C$. Anticipating a solution of the form (8.2b), (8.5) reads, if $k = 2\pi/L$:

$$\delta\mu = -\tilde{\sigma}k^2 h(t)\cos(kx)$$

so that the boundary condition for the density $\rho(x,z)$ is

$$\rho(x,z) = \rho_0 - C\tilde{\sigma}k^2 h(t)\cos(kx)\,.$$

The solution of $\rho_{xx} + \rho_{zz} = 0$ with this boundary condition (and of course $\rho = \rho_0$ for $z = -\infty$) is, as easily checked:

$$\rho(x,z) = \rho_0 - C\tilde{\sigma}k^2 h(t)\cos(kx)\exp(kz)\,.$$

The surface current has been computed in the preceding section. In addition, there is a bulk current given by (7.5). Near the surface, and for a weak modulation h, the bulk current is nearly normal to the surface and approximately given by

$$j(x,z) \simeq j(x,0) \simeq -\tilde{D}\frac{\partial\rho}{\partial z} = C\tilde{D}\tilde{\sigma}k^3 h(t)\cos(kx)\,.$$

Neglecting the surface contribution, this may be identified with $-\dot{z}$. The equation $\dot{h} = -C\tilde{D}\tilde{\sigma}k^3 h$ follows, and now (8.9) is to be replaced by (Mullins 1957, 1959)

$$\tau(L) = \text{Const} \times L^3\,. \tag{8.15}$$

From now on, only surface diffusion will be considered, and the index s will be omitted in j_s, D_s, Λ_s, $\tilde{D}_s$, etc.

8.5 Smoothing below the roughening transition

In the previous section, the smoothing of a surface above its roughening transition temperature T_R has been treated; the problem has been solved by straightforward linear equations. The case $T < T_R$ is much more complicated. As seen in chapter 1, it corresponds to high-symmetry surface at ordinary temperature. We shall first assume (Fig. 8.4) that terraces or 'islands' and holes or 'lakes' of atomic height have formed, and that their distance is large enough to allow the interaction between steps to be neglected. The structure of Fig. 8.4 is that of a growing surface. If growth is stopped, the surface smoothens. This smoothing process will also occur, in principle, after quenching a surface from above T_R to below T_R.

Indeed, the free energy associated to the surface in Fig. 8.4 is the integral of the line tension of steps on all the steps. We are considering

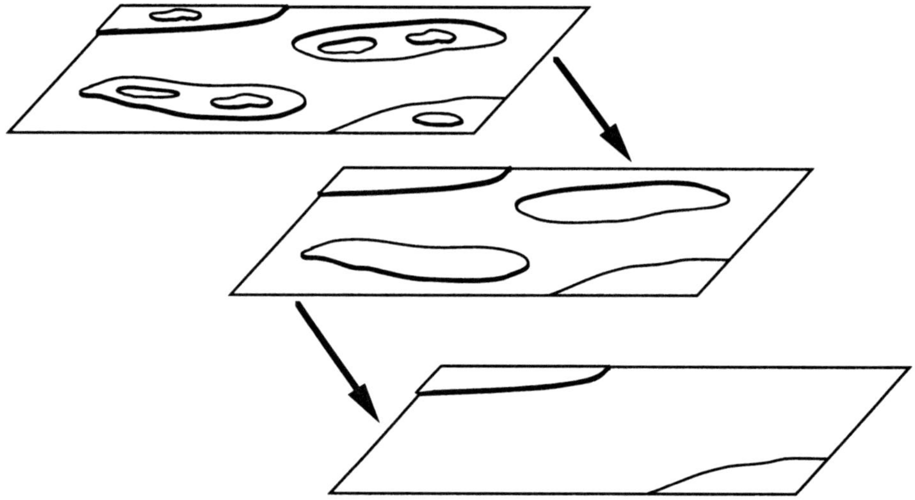

Fig. 8.4. Smoothing of a rough surface. The average terrace size increases with increasing time t as shown by the arrows.

low temperatures where, at equilibrium, there are practically no islands nor lakes. How will they disappear once formed?

The answer was given by Ostwald in 1908 in the case where only islands (no lakes) are present. Big islands grow at the expense of small ones. This evolution indeed reduces the step length, and therefore the free energy. To check this, we can consider a system of two circular islands of radius R and R' respectively. Their evolution should keep the total mass, and therefore the sum $R^2 + R'^2$, constant. Thus, small variations $\mathrm{d}R$ and $\mathrm{d}R'$ should satisfy $R\mathrm{d}R + R'\mathrm{d}R' = 0$. The corresponding variation of the total step length is $2\pi(\mathrm{d}R+\mathrm{d}R') = 2\pi(1-R/R')\mathrm{d}R$. This variation should be negative (since the energy must decrease) and therefore $\mathrm{d}R$ should be positive if $R > R'$, and vice versa. This proves (in this simple case) the statement that the big island grows at the expense of the small one. Ostwald's theory was improved by Lifshitz & Slyozov (1961) and extended by Chakraverty (1967) and Villain (1986). The most recent theoretical developments have been reviewed by Bray (1994). The agreement with experiment is often only qualitative (Ernst *et al.* 1992a, Sholl & Skodje 1995).

Ostwald's argument was actually related to a physically different problem: that of a supersaturated solid solution A_xB_{1-x}. For large x, the constituent B forms islands, which are of course three-dimensional. The above argument suggests that big islands grow at the expense of smaller ones.

The same argument also applies to the evolution of the Stranski-Krastanov or Volmer-Weber droplets of Fig. 5.1. The case of the clean surface of Fig. 8.4 is slightly different. Indeed, if the size distribution of terraces and lakes are the same, terraces and lakes can both shrink. Mass conservation does not require the increase of bigger terrace sizes. However, small terraces and lakes will disappear first for two reasons: because the quantity of atoms to be evacuated is smaller; and because the excess chemical potential is higher, in absolute value, for small defects. This is a consequence of the Gibbs-Thomson formula seen in chapter 2.

Qualitatively, it will be seen that, after a time t, only islands and lakes of size larger than some value $R_c(t)$ survive.

One might use a procedure similar to the method used in chapter 6 for straight steps. First deduce the adatom and advacancy densities $\rho(\mathbf{r})$ and $\sigma(\mathbf{r})$ from the diffusion equation (7.1) by neglecting the motion of steps (quasi-static approximation). Then the current $\mathbf{j}$ reaching the steps is deduced from the equation $\mathbf{j}(\mathbf{r}) = -D\nabla\rho(\mathbf{r}) + \Lambda\nabla\sigma(\mathbf{r})$, and the decay rate of terraces and lakes follows. This non-self-consistent procedure is justified because diffusion is much faster than step motion.

Here, an equivalent method will be used: the surface current of matter $\mathbf{j}(\mathbf{r})$ will be deduced from (7.11). The chemical potential $\mu(\mathbf{r})$ is in principle given by Laplace's equation $\nabla^2\mu = 0$ between terraces, with boundary conditions at steps given by the Gibbs-Thomson relation. The latter will just be written $\delta\mu = \gamma/R$ near a terrace of radius R, and $\delta\mu = -\gamma/R$ near a lake of radius R, where $\delta\mu$ is the chemical potential counted from its equilibrium value.

Solving the Laplace equation for all possible geometries and averaging over them would be a very hard task indeed. We shall be contented with a very simplified calculation. The terraces which decay first are the smaller ones as argued above. If a small terrace of radius R is surrounded by bigger ones (or by lakes) at distances of order r, the chemical potential near this terrace is γ/R; the chemical potential gradient is of order $\gamma/(Rr)$, the current density from this terrace is of order $\beta\gamma\tilde{D}/(Rr)$ according to (7.11) and the total number of atoms $-\mathrm{d}N/\mathrm{d}t$ emitted by the terrace per unit time is or order $\beta\gamma\tilde{D}/r$. With our usual units, N may be identified with πR^2. It will be assumed that all lengths are of the same order of magnitude, $r \approx R$. The result is that R decays according to the equation

$$\frac{\mathrm{d}}{\mathrm{d}t}\left(R^3\right) \approx \beta\gamma\tilde{D}$$

where $\tilde{D}$ is given by (7.12).

Integration of the previous equation yields

$$R^3(t) - R^3(0) \approx \beta\tilde{D}\gamma t = (D\rho + \Lambda\sigma)\,\beta\gamma t\,. \tag{8.16}$$

The exponent $1/3$ in $R(t) \sim t^{1/3}$ turns out to be the same as obtained for three-dimensional binary alloys by Lifshitz & Slyozov. In the case of the three-dimensional clusters on a surface (Fig. 5.1), Chakraverti (1967) has obtained an exponent 1/4 instead of 1/3.

8.6 Smoothing of a macroscopic profile below T_R

We now come back to the case addressed in section 8.3 and represented by Fig. 8.3, but it will now be assumed that the average orientation $z =$ Const is below its roughening transition temperature T_R. Experimentally (Bonzel *et al.* 1984a,b), the profile is seen to take a trapezoidal shape (Fig. 8.3b). The bottom and the top of the profile look like facets.

One might *a priori* think that this quasi-faceting is similar to the formation of facets at equilibrium. However, this dynamical faceting is not really expected, since it has been seen in chapter 4 that dissolution or evaporation shapes (which might be expected at the top of the profile) should be dominated by orientations with fast kinetics, which are generally not faceted. On the other hand, one cannot have too much confidence in the model of chapter 4. As will be seen, the dynamical faceting observed by Bonzel *et al.* is still subject to controversy.

Let us first address an easier problem, the case of a 'bi-directional' profile (Fig. 8.5). Such a profile is not easy to work out in practice, and has never been experimentally studied for this reason.

The basic point is that the surface is formed by terraces separated by steps, so that the evolution of the surface may be regarded as a motion of steps: where steps retreat (or advance, respectively), the solid evaporates

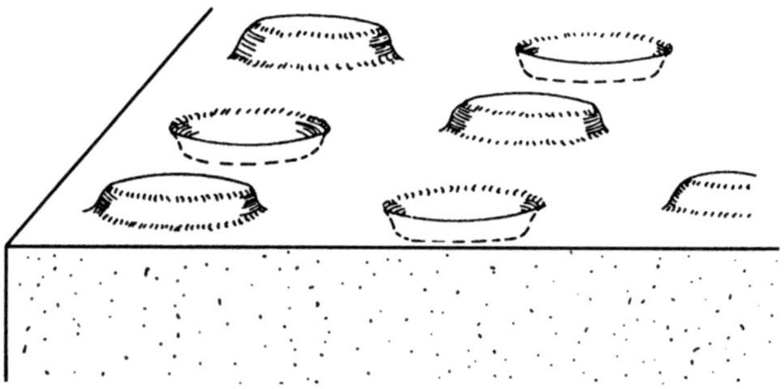

Fig. 8.5. Thermal smoothing of a surface in which a bi-directional profile has been cut. The initial shape may be for instance $z = \text{Const} \times \cos(Qx)\cos(Qy)$. Below the roughening transition, the transient shape before total smoothing should involve facets on three levels. The experiment has never been done because it is difficult to cut such shapes.

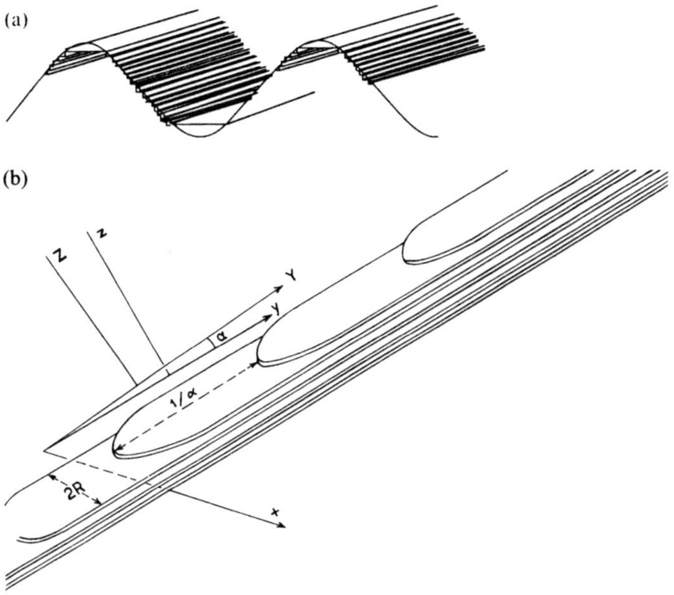

Fig. 8.6. a) If the grooves of Fig. 8.3a are perfectly parallel to a high-symmetry orientation, the transient shape (before total smoothing) should have sharp singularities at minima and maxima (formula 8.21). b) On the other hand, a miscut (but not too miscut) surface evolves from the sinusoidal shape of Fig. 8.3a to the trapezoidal shape of Fig. 8.3b. This is because the steps have, near the top, a high curvature which generates a high chemical potential and a fast recoil. The high chemical potential is analogous to the high electric potential near a tip in electrostatics.

(or grows, respectively). As seen from the previous section, small terraces and lakes decay faster than bigger ones. These small terraces are precisely those which are at the top and at the bottom of the profile, and one can therefore expect the top and the bottom to flatten. In addition, faceting also occurs at half-height. This is due to the fact that the chemical potential has an abrupt change of sign at this place.

Qualitatively, this result is due to the existence in the initial profile of steps with a high curvature. The argument does not apply to the profile of Fig. 8.3b or 8.6a, when all steps are straight. It does apply, however, if the profile is slightly miscut (Fig. 8.6b), as it is always in real experiments (Surnev *et al.* 1996). In that case, the qualitative experimental results can be reproduced, as well in the case of diffusion kinetics (Villain & Lançon 1989, Bonzel & Mullins 1996) as in the case of evaporation (Lançon & Villain 1990). If the miscut angle is too large, however, one recovers the linear problem of section 8.3, and the profile remains sinusoidal.

8.7 Grooves parallel to a high-symmetry orientation

This problem gives an opportunity to perform exciting exercises of non-linear mechanics (Rettori & Villain 1988, Lançon & Villain 1990, Ozdemir & Zangwill 1990). It is still subject to controversy (Spohn 1993, Selke & Bicker 1993, Selke & Duxbury 1994). The steps are parallel straight lines (Fig. 8.6a) which run parallel to the planes $z = \text{Const}$. The latter are below their roughening transition. Until now, such an ideal situation has never been realized, and the surfaces investigated so far were always slightly miscut.

An appropriate form of the continuous approximation will be used. It is not obvious that this is correct, but the same results can be obtained if steps are taken explicitly into account (Rettori & Villain 1988, Ozdemir & Zangwill 1990).

The first thing to do is to calculate the chemical potential which corresponds to a given surface shape. It is given by (2.9). Taking the y axis parallel to the steps and using the notation $\partial z/\partial x = z'$, (2.9) reads

$$\mu = \mu_0 - \frac{\partial}{\partial x}\left(\frac{\partial \varphi}{\partial z'}\right) . \tag{8.17}$$

The free energy density $\varphi(z')$ contains a term proportional to the density of steps, which is $|z'|$. This term has been discussed in section 2.3. There is an additional term due to interactions between steps. As seen in section 2.9, the interaction free energy per unit length between parallel steps at distance ℓ is proportional to ℓ^{-2} for all accepted interaction mechanisms. Therefore, the interaction free energy per surface area is proportional to ℓ^{-3} or $|z'|^3$. The projected free energy density is now

$$\varphi(z') = | z' | \left(\gamma_0 + \frac{1}{3}\gamma_1 (z')^2 + \ldots\right) . \tag{8.18}$$

The chemical potential is therefore, according to (8.17), and if $z' \neq 0$:

$$\mu = \mu_0 - 2\gamma_1 \mid z' \mid z'' . \tag{8.19}$$

Evaporation-condensation dynamics

It will be assumed that adatoms evaporate from steps, after diffusing on the surface on a length $1/\kappa$ given by (6.6), much shorter than the distance between steps. The evaporation rate is proportional to the product of the step density $|z'|$ by the excess chemical potential $|\mu - \mu_0|$ given by (8.19). Therefore

$$\dot{z} = \text{Const} \times (z')^2 z'' . \tag{8.20}$$

Note that this formula does not coincide with the equation (8.7), only valid above T_R.

It is reasonable to make the ansatz that the function $\delta z(\delta x, t) = z(L + \delta x, t) - z(L, t)$ behaves near its maxima (i.e for $\delta x \simeq 0$) as

$$\delta z = |\ \delta x\ |^b \ . \tag{8.21}$$

The ansatz (8.21), combined with the condition that $\dot{z}$ be finite and strictly negative for $x = L$, yields $b = 4/3$. The resulting profile is sharper than a parabola, at least after some time. However, for short times, a profile which is initially parabolic at its top and at its bottom keeps its parabolic shape. Indeed, if one assumes $z \simeq h_0 - \frac{1}{2}A\delta x^2$, insertion into (8.20) yields $\dot{A}/A^3 = \text{Const}$. It follows $A(t) \sim 1/\sqrt{t_c - t}$, where the integration constant t_c depends on the initial conditions. The form (8.21), with $b = 4/3$, is probably correct for $t > t_c$.

Surface diffusion dynamics

Using (8.10), (8.11) and (8.19), one obtains

$$\dot{z} = -\text{Const} \times \frac{\partial^2}{\partial x^2} (|\ z'\ |\ z'') \ . \tag{8.22}$$

Inserting (8.21) into (8.19) and assuming $\delta\mu$ strictly positive and finite for $x = L$, one finds $b = 3/2$. Again, the profile is sharper than parabolic at its tops and bottoms. A non-vanishing value of $\dot{z}$ can only be obtained from (8.22) if higher order terms are introduced, $\dot{z} = \lambda|\delta x|^b + v|\delta x|^{b'}$. It follows that $b' = 7/2$ in this case.

Thus, in all cases, the profile is found to sharpen at its top and bottom. This result is not confirmed by Monte Carlo simulations (Selke & Bicker 1993) according to which the profile remains essentially smooth. These simulations are done on very small samples where fluctuations (neglected in this section) are important. The disagreement cannot be solved by experiment, because grooves perfectly parallel to a high-symmetry direction cannot be prepared. On the other hand, a theory by Spohn (1993) predicts faceting even for the 'well-cut' profile of Fig. 8.6a, in disagreement with the arguments given here. The final solution of this problem is still an open question.

Problem

8.1. Show that $(\partial\rho/\partial\mu)_{T,V}$, and therefore $(\partial\mu/\partial\rho)_{T,V}$, is always positive.

9
Silicon and other semiconducting materials

He told us he had been thirty years
employing his thoughts for
the improvement of human life.

Jonathan Swift (*Gulliver's travels*)

Semiconductors and in particular silicon have completely changed our life. To the point that we could hardly imagine our cities without kids bobbing their heads to the deafening music of a portable laserdisc player, and grown-ups entertaining their neighbours with private-made-public phone calls on their cellular telephones–and even often the other way round!

But it is in the field of computers where the Silicon Revolution really struck hard. Crystalline samples can be now manufactured in such a way–thanks to silicon-based computers, as a matter of fact–that their surface has a high degree of perfection. Moreover, surface observation with atomic resolution in tunneling microscopy is even easier with semiconductors than with most metals. All these practical advantages are however compensated by several complications, whose main origin is the fourfold coordination of silicon atoms. Because of this, the silicon crystal lattice has two atoms per unit cell. This, in turn, originates surface structures which may be very complicated, such as the 7×7 reconstruction of the (111) face of silicon. Or also the alternation of inequivalent terraces on the vicinals of its (001) face.

9.1 The crystal structure

Silicon and germanium share their crystal lattice structure with diamond (Fig. 9.1a). This structure, which is due to the hybridization of s and p

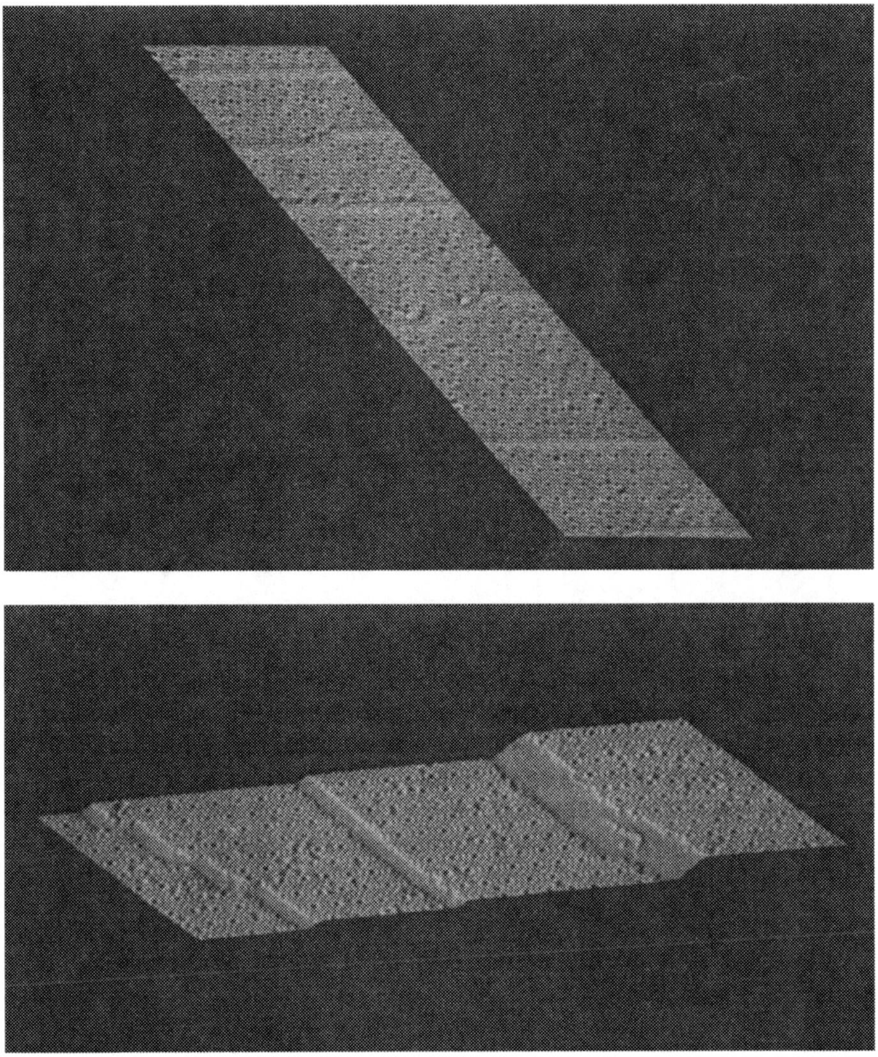

Fig. 9.0. STM pictures of the (111) face of silicon (courtesy of E. Williams).

electrons, stems from the fact that each atom 'likes' to have four neighbours arranged in a regular tetrahedron. The structure of many binary III-V semiconductors such as GaAs, and of several II-VI compounds, such as CdTe or HgTe is easily found by partitioning the diamond lattice in two identical cubic sublattices: one is occupied by Ga, the other one by As (Fig. 9.1b): it is the zincblende structure.

Cleaving a highly covalent body in two parts to create a surface, one breaks bonds which are now 'dangling'. They recombine to lower their

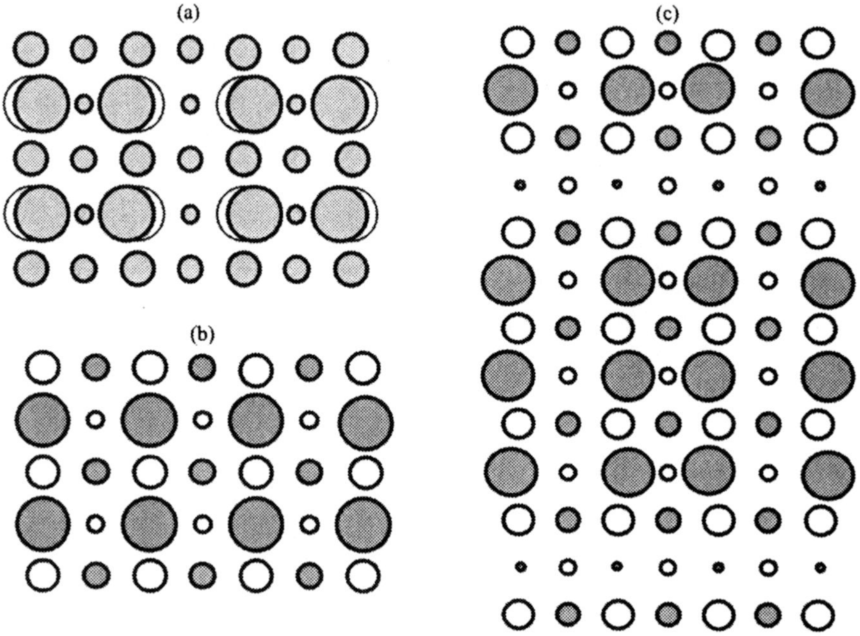

Fig. 9.1. a) Projection of the Si (or diamond) structure on the (001) face. The decreasing size of the closed circles represents the increasing distance of the atoms from the surface. The big open circles show atomic positions at the surface in the absence of reconstruction. b) Same projection as in a) for a compound semiconductor, e.g. GaAs (in the non-realistic absence of reconstruction). c) GaAs (001) face reconstructed 2×4. Note the missing Ga rows (Ga atoms shown as grey circles).

energy resulting in a *surface reconstruction*, a typical phenomenon for all semiconductor surfaces.

9.2 The (001) face of semiconductors

The (001) and (111) surface orientations are especially used for growth of microelectronic components. We will begin our discussion with the first one.

The diamond lattice has two, not one, atoms per unit cell. Thus, it is not a Bravais lattice. A lattice plane parallel to a quaternary axis has a rectangular symmetry, not a square one. This implies that the preferential step orientation (that is, the one which minimizes the step free energy) is unique, and not doubly degenerate. Of course, the cubic symmetry implies that this orientation turns by 90° from one (001) plane to the following.

In practice, (001) surfaces are most often used in MBE growth, and some steps are always present. In the case of [110] steps, STM images show alternating straight and meandering steps (see Fig. 1.1). The alternation of two types of steps is clearly due to the non-Bravais nature of silicon: two successive terraces are inequivalent, as well as two successive steps. The step line tension is thus alternately high and low. The sticking coefficient of adatoms to the step edge also alternates between strong and weak. Both effects contribute to the different look of the two step types.

9.3 Surface reconstruction

We mentioned in section 9.1 that silicon surfaces are reconstructed. This means that the surface possesses a lower symmetry than a planar section of the bulk. What reconstruction, i.e. what symmetry, is actually realized, depends on the considered orientation, on the preparation conditions,

Fig. 9.2. STM picture of a (001) vicinal of silicon. Note the alternation of dimer rows at 90° from one terrace to the following one. The wiggled appearance of some row is due to dimer buckling. The surface is cut at an angle $\theta = 2.7°$ in the [010] direction. Size is 30 nm × 30 nm. (Courtesy of Margrit Hanbücken, CRMC2 - CNRS Marseille, Bernd Röttger, Ruhr-Universität Bochum and Henning Neddermeyer, Martin-Luther-Universität Halle-Wittenberg.)

on the temperature, etc. For usual conditions for silicon growth, the (001) surface structure is a 2×1 reconstruction, where surface atoms form pairs or *dimers* (cf. Fig. 9.1a). The energy gain is evident: the number of unpaired electrons per atom is halved with respect to the non-reconstructed arrangement. The dimerization takes place in such a way as to minimize the disturbancy to the tetrahedral bond arrangement.

It is the dimer row structure which gives their striped appearence to the STM micrographs in Figs. 1.1 and 1.3. The dark spots one sees on the terraces are surface vacancies, probably produced during sample preparation. One may also see bright spots: they are adatom dimers. Isolated adatoms are not visible at room temperature (the microscope operating temperature).

The dimer rows, as one clearly sees in Fig. 9.2, turn by 90° from one layer to the following, and this determines the step properties: a step running parallel to the dimer rows of the upper terrace has a high stiffness, while a step orthogonal to the dimer rows of the upper terrace has a much lower stiffness. Also, adatom surface diffusion takes place more easily along than perpendicular to the dimer rows, which renders adatom diffusion very anisotropic on Si(001) (Roland & Gilmer 1992a,b).

9.4 Anisotropy of surface diffusion and sticking at steps

We mentioned that reconstruction may result in anisotropic adatom diffusion and sticking at kink sites on steps. It is the case for the 2×1 reconstruction of a (001) face.

During MBE growth, under appropriate conditions (see chapter 11), the forming layers consist of monoatomically high islands. Such islands, observed on silicon with a tunneling microscope by Mo *et al.* (1992), have an elongated shape (Fig. 1.3) in the direction of the dimer axis (perpendicular to the dimer rows of the substrate). Such an anisotropic shape is not due to the anisotropy of adatom diffusion, but rather to the anisotropy of sticking. In fact, if an adatom hits a dimer row along a side, it is more likely to bounce off than to stick–while sticking at a dimer row end is easy. It is quite surprising that in such conditions silicon epitaxy may efficiently take place, as it does at least at high enough temperature. At lower temperatures, Fig. 1.3 suggests that things may not be so easy.

Even at reasonably high temperatures, growth on high-symmetry surfaces may in fact suffer from some inconveniencies. This is what we are going to see.

9.5 Crystalline growth vs. amorphous growth

Unlike metals, semiconductors easily give rise to amorphous structures. Intuitively, one understands that a four-fold coordinated material results more easily in a glass state than an element which prefers more compact structures. One expects that, at equal deposition fluxes, MBE growth results in a good crystal at high temperature, where surface diffusion is fast, and in an amorphous structure at low temperature. However, the result depends also crucially on the surface orientation and on the number of deposited layers. During molecular beam homoepitaxy of silicon on a (001) substrate, one obtains at fixed temperature a good crystal only up to a critical thickness, beyond which one gets amorphous silicon. This critical thickness is reproducible, and it depends as an activated exponential on temperature (Eaglesham *et al.* 1990, 1991). The phenomenon has been ascribed, on the basis of numerical simulations (Eaglesham & Gilmer 1992), to a surface instability due to the Schwoebel effect (see section 6.4). Subsequent study has focussed on the role of impurities, and mostly the ubiquitous hydrogen, on the crystal-amorphous transition. The main conclusion (Eaglesham *et al.* 1993b, Adams *et al.* 1993) is that hydrogen affects strongly the critical epitaxial thickness, possibly by slowing down Si adatom surface diffusion and step crossing. We will come back to the problem of surface-active impurities in section 11.12.

9.6 The binary compounds *AB*

Many III-V compounds, such as GaAs, or II-VI, such as CdTe, share the crystal structure of Fig. 9.1c. Also in this case, the most used substrate orientation for MBE deposition is (001). One may then ask whether–at equilibrium–the surface layer is composed of *A* or *B* atoms, or maybe both alternately. In fact, it depends on the chemical potential. During growth, an *A*-layer could conceivably alternate with a *B*-layer. But sending both types of atoms at the same time looks dangerous for a neat growth. In MBE, for instance, *A* atoms would be likely to land on *A* atoms, whilst one would prefer them to land on *B* atoms. Are not inclusions or other defects likely to occur in such conditions? A solution is to send one type of atoms at a time, in what is called Atomic Layer Epitaxy (ALE) (Veron *et al.* 1996).

The case of GaAs is especially simple, because of the nice property of arsenic of evaporating quickly if not in contact with gallium. It is thus possible to deposit both species simultaneously, just taking care to send a larger quantity of arsenic. Fig. 9.3 gives an idea of the resulting structure.

GaAs growth has been studied in detail by many groups. We briefly mention the experimental works by Arthur (1974), Foxon & Joyce (1975), Shitara *et al.* (1992), and the theoretical works by Madhukar (1983), Madhukar & Ghaisas (1988) and Kaxiras *et al.* (1989). It is remarkable that several physical properties of this system can be described by an

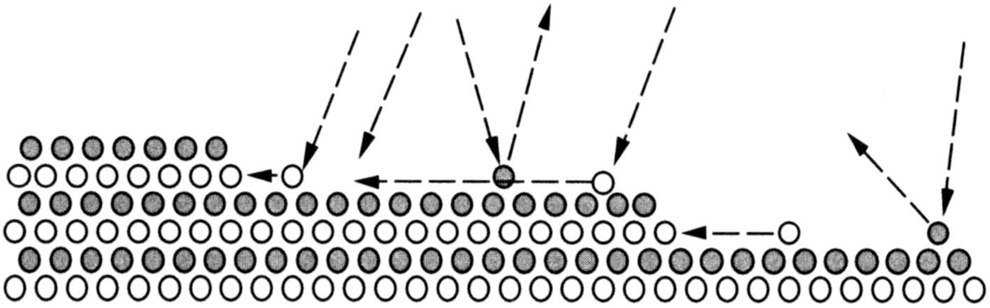

Fig. 9.3. A sketch of GaAs growth. Arsenic (closed circles) re-evaporates when it is not in contact with gallium (open circles).

SOS model (Shitara *et al.* 1992a), where one of the components is simply 'forgotten'–arsenic, in this particular case, because the incorporation of gallium and that of arsenic take place on widely different timescales, and incorporation of gallium is the rate-limiting process during growth.

9.7 Step and kink structure

We have seen (Fig. 1.1) that on a Si(001) surface there are two types of steps, whose equilibrium (the line tension) as well as kinetic (the propagation velocity) properties are different. One might wonder whether these steps will always stay separated, or will join to form double steps–leaving a single kind of terrace on the surface. We will not answer this question here (see for instance Chadi (1994) and references therein). We only remark that, at equilibrium, this answer necessarily requires including elasticity. It is what we will do in chapter 15.

Surface reconstruction may also favour special orientations of the surface itself ('magic vicinals' of Si(111), Takayanagi *et al.* 1987, Bartolini *et al.* 1989). It may also favour a bunching of kinks on each step into 'multikinks' (Wang *et al.* 1990). This is shown in the STM micrographs displayed as the frontispiece of this chapter.

9.8 The (111) face of semiconductors

The (111) face is characterized by a very complicated reconstruction. To understand its nature, consider first a (hypothetical) 1×1 non-reconstructed structure (Fig. 9.4).

Each surface atom has a dangling bond. Compared to (001), it is not much, but, as it happens, one can easily reduce this number by placing one *adatom* on the head of each group of three atoms: a real hat-trick! In fact, this adatom does possess one unsatisfied bond, but the number of dangling bonds *per unit surface* is reduced by a factor three! However, such adatoms

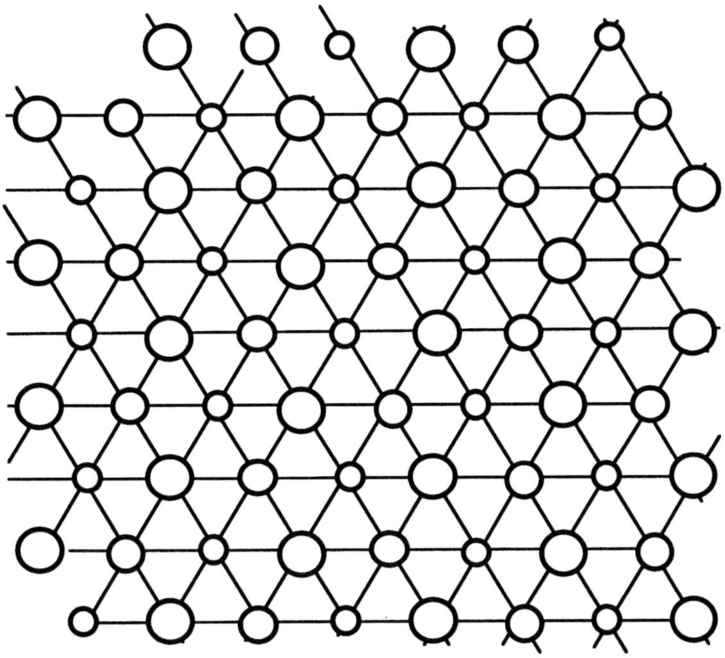

Fig. 9.4. Unreconstructed structure of the (111) face of silicon or germanium. It is unstable at low temperature. The size of the circles is proportional to their proximity to the surface.

deeply perturb the other atoms. On silicon, this perturbation favours the appearence of dimer bonds, stacking faults, and even surface vacancies. The outcome is the celebrated DAS (dimer–adatom–stacking-fault) 7 × 7 model of Takayanagi *et al.* (1985), whose structure is sketched in Fig. 9.5. It was established in the whereabouts of 1985 by combining STM (Binnig *et al.* 1983) and transmission electron diffraction experiments (Takayanagi *et al.* 1985). It includes 12 adatoms, 9 dimer bonds and a stacking fault in each unit cell. Some features of the structure can also be seen in this chapter's frontispiece micrographs.

At about 850 °C, the (111) silicon surface undergoes a 'deconstruction' transition to a 1×1 structure. This does not mean that it takes the structure of Fig. 9.4! On the contrary, it is likely that the adatom density stays near 10%, but that this adatom layer is thoroughly disordered (Latyshev *et al.* 1990, Sakamoto & Kanamori 1993).

9.9 Orders of magnitude

Up to this point, we have discussed several important physical quantities of surface physics: surface free energy, step line tension, diffusion and

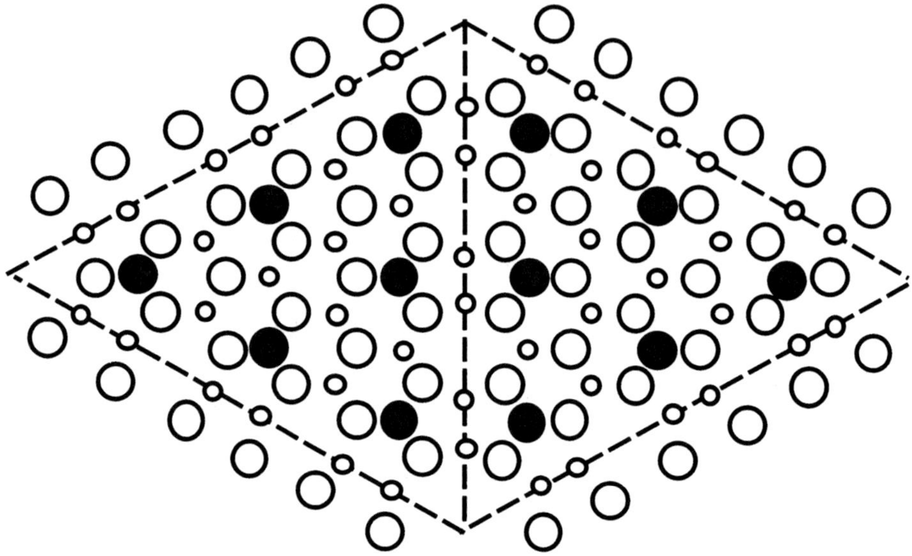

Fig. 9.5. Reconstructed 7 × 7 structure of the (111) face of silicon. Note the adatoms (closed circles), the advacancies (at the cell corners) and the stacking faults (dashed lines).

evaporation barriers, step kinetic coefficients, and so on. We rarely mentioned numerical values for such quantities. This is what we are going to do here for silicon.

Clearly, the order of magnitude for surface energetics is the electron-volt. But energy differences may be much smaller. For instance, take the cleavage (111) plane of silicon. This surface reconstructs after cleavage in a metastable 2×1 structure, which transforms into the 7×7 upon annealing. *Ab initio* calculations have shown that the energy difference between the two structures is only 60 meV per 1×1 cell! (Brommer *et al.* 1992, Štich *et al.* 1992). By an amusing coincidence these two papers neighbour each other in the same *Physical Review Letters* issue, bearing very similar titles, and originating respectively from Cambridge, Massachusetts, and Cambridge, UK! Absolute values of the surface energy are given in these two papers as 1.179 and 1.153 eV per surface atom, respectively.

As we saw in chapter 3, studying the equilibrium shape of a crystal may yield *relative* values of the surface free energies as a function of the surface orientation, via the Wulff construction. Eaglesham *et al.* (1993c) accomplished this task for silicon, measuring the equilibrium shape of bulk voids formed by implantation of He atoms into Si. After annealing at 700 °C, they found equilibrium shapes dominated by $\{111\}$ and $\{001\}$ facets, with relative free energies $\sigma(001) = 1.11\sigma(111)$. They were also able to measure facet sizes, which directly give the step free energy (see problem

2 in chapter 3) for both facets. For steps oriented along [110], the significant step direction, they found $\gamma_{(111)}([110]) = 0.11$ eV per atom for (111), and $\gamma_{(001)}([110]) = 0.023$ eV per atom for (001). The conclusions of this paper have been confirmed by Follstaedt (1993), with the same technique.

Entirely different conclusions have been drawn by Bermond *et al.* (1995), who looked at the equilibrium shape of small silicon columns at high temperature (1050 °C). These authors find a much smaller anisotropy of σ (only a few percent) and they also assign the highest energy to the (111) facet. Moreover, the (001) facet seems not to belong to the equilibrium shape, in accord with theoretical calculations (Mukherjee *et al.* 1994).

These controversies show well how difficult it is to study the equilibrium shape in non-metallic materials.

STM experiments have been designed to measure directly the kink and step formation energy on Si(001) (Swartzentruber *et al.* 1990) at about 600 °C. For low-energy (S_A) steps it was found $\gamma_{(001)}(S_A) = 0.028$ eV per atom, and $\gamma_{(001)}(S_B) = 0.09$ eV per atom for high-energy (S_B) steps. Note, incidentally, that S_A steps are straight, while S_B steps are rough in Fig. 1.1: this may seem strange at first thought, but is consistent with the fact that a kink on an S_A step would expose a 'high-energy' S_B segment–an unfavourable situation–and viceversa. The kink energy of one step would thus equal the formation energy of the other one. For comparison, calculated values by Chadi (1987) are 0.01 and 0.15 eV/atom for S_A and S_B steps, respectively.

Using another microscopy technique, namely reflection electron microscopy (REM), Alfonso *et al.* (1992) measured the stiffness $\tilde{\gamma}$ of steps on Si(111) by directly looking at equilibrium step fluctuations at high temperatures (900 °C). This is the kind of experiment we discussed in section 2.7. Their result (correcting for a factor of 3 in one of their formulae) is roughly $\tilde{\gamma} = 0.5$ eV/atom. Before comparing to Eaglesham *et al.*'s result, note that at this temperature the surface is in the disordered 1×1 phase, and the step structure differs from that of steps on a 7×7 reconstruction.

We saw in section 2.7 that the amplitude of a thermal step fluctuation is proportional to its wavelength L, independently of the physical mechanism responsible for the fluctuations. This is no longer true for the *time* $\tau(w, L)$ needed for the fluctuation to reach the amplitude w: steps fluctuate by exchanging terrace atoms with some atom reservoir, and with a given kinetics. It has been shown by various approaches (Bartelt *et al.* 1992, Saldit & Spohn 1993, Pimpinelli *et al.* 1993b, 1994) that $\tau(w, L)$ indeed varies with L as $L^{\tilde{z}}$, the exponent $\tilde{z}$ in turn depending on the kinetics of the equilibrium processes governing the fluctuations. For instance, atom diffusion along the step edge yields $\tilde{z} = 4$, while adatom surface diffusion from one step to its neighbours yields $\tilde{z} = 2$. The investigation of the temporal behaviour of step fluctuations may thus allow a measurement of kinetic coefficients for surface processes. This has been done on Si(111) (Pimpinelli *et al.* 1994): at 900 °C and under ultra-high vacuum, adatom exchange between steps via surface diffusion was found to be the source of step fluctuations. It was thus possible to estimate the *mass surface diffusion constant* $\rho_0 D_s \approx 10^{-7}\text{cm}^2/\text{s}$. Assuming D_s to be thermally activated (*cf.* chapter 7), $D_s = D_0 \exp(-\beta W_{\text{sd}})$, this translates to an activation energy $W_{\text{sd}} = 1 \pm 0.2$ eV, given an adatom density of the order of 10% (see section 9.8 above) and a prefactor $D_0 = 10^{14}$ s^{-1}. Some assumption on the adatom density is necessary, because this measurement is essentially a macroscopic one, and can only give the Fick diffusion coefficient $\tilde{D} = \rho_0 D_s$ (*cf.* chapter 7).

A more direct method for measuring the tracer diffusion constant D_s has been devised by Mo *et al.* (1992), who exploited it on Si(001). It consists of counting with a tunneling microscope the number of adatom islands nucleated at a given temperature and deposition flux during an MBE experiment. The island density is related to the diffusion constant D_s in a way that we will investigate in detail in chapter 11. In the case of silicon (001) the situation is complicated by the anisotropy of the reconstruction, whose effect is to give two diffusion constants for the two non-equivalent directions. The difficulty can be surmounted by comparing the actual measurements with kinetic Monte Carlo simulations. Mo *et al.* (1992) found thus $W^e_{\text{sd}} \approx 0.7$ eV for the easy direction and $W^h_{\text{sd}} \approx 1$ eV for the hard one. Along the same line, Voigtländer & Zinner (1993) were able to measure W_{sd} on the 7×7 Si(111) (between 400 and 600 °C), finding $W_{\text{sd}} = 0.75 \pm 0.2$ eV.

Finally, we mention that the exploitation of the results of section 6.6, consisting of the measurement of the velocity of step flow during evaporation, has allowed the determination of the evaporation rate for different silicon surfaces. The results are: i) the activation energy for silicon evaporation is 4.2 ± 0.2 eV, independently of the surface orientation ($\{111\}$ or $\{001\}$); ii) the step velocity is linear in the step-step separation for all temperatures (up to 1300 °C) and all separations (up to several microns) tested. According to a simple model of atom detachement from

Table 9.1. *Values of surface and defects energies (eV) for Si(001); see text for symbols.*

γ_A	γ_B	$W_v + W_a$	W^e_{sd}	W^h_{sd}
0.028	0.09	4.2	0.67	1.0

Table 9.2. *Values of surface and defects energies (eV) for Si(111), below and above the* $7 \times 7 - 1 \times 1$ *transition* (T_c)*; see text for symbols.*

T-range	γ	$W_{sd} + W_a$	$W_v + W_a$	W_{sd}
$T < T_c$	0.11			0.75
$T > T_c$	0.5	1.2	4.2	

steps, diffusion and desorption (the Burton, Cabrera and Frank's model of chapter 6), these results imply that the diffusion length of adatoms before desorption is of the order of tens of microns even at very high temperatures (of the order of 90% of the bulk melting temperature).

The numerical orders of magnitude are summarized in Table 9.1 for Si(001) and in Table 9.2 for Si(111).

10
Growth instabilities of a planar front

Deformed beyond deformity, unformed,
Insipid as the dough before it is baked,
They change their bodies at a word.

William Butler Yeats (*The Phases of the Moon*)

In chapter 6, we have described the growth of a stepped surface, assuming that the steps remain straight. Is that correct? The answer will appear as the main conclusion of this chapter.

Similarly, we have described in chapter 4 a model of crystal growth in which crystals keep a convex shape with planar facets; as we know from the observation of snowflakes, it is not always so. When a crystal grows from a liquid, convex shapes are often unstable.

These two phenomena are described by rather similar equations, in which diffusion of matter or energy to or away from the front (or the steps) plays an essential part. The instabilities of a two-dimensional interface are often caused by diffusion in the third dimension, and instabilities of one-dimensional steps are caused by diffusion in the dimension perpendicular to the steps on the surface.

10.1 Diffusion limited aggregation: shape instabilities

In chapter 6, we have studied the evolution of an array of parallel, straight steps. The possibility of certain instabilities (step bunching) has been considered, but the straight nature of steps was not questioned. As will be seen in this chapter, it has to be questioned!

This question already arises in the case of a single step. Away from the step, the adatom density ρ satisfies the following obvious generalization of (6.1):

$$\frac{\partial}{\partial t}\rho(x,y,t) = D\nabla^2\rho(x,y,t) - \frac{1}{\tau_v}\rho(x,y,t) + F\,. \qquad (10.1)$$

The simplest possible boundary condition along the step is obtained by replacing $\mathrm{d}\rho/\mathrm{d}x$ in (6.17) by $\mathbf{n}\cdot\nabla\rho$, where $\mathbf{n}$ is the unit normal to the step. If the step is nearly a straight line parallel to the y axis, an appropriate approximation is provided by equations (6.17), which read

$$D\frac{\partial\rho}{\partial x} = \Delta'(\rho_0 - \rho) \qquad (10.2a)$$

on the upper (left-hand) side of the step, and

$$-D\frac{\partial\rho}{\partial x} = \Delta''(\rho_0 - \rho) \qquad (10.2b)$$

on the lower (right-hand) side of the step. The kinetic coefficients Δ' and Δ'' are defined as in section 6.4, namely

$$a\Delta' = \frac{DD'}{D-D'}, \qquad a\Delta'' = \frac{DD''}{D-D''}\,.$$

Indeed, equations (10.2) are not quite correct, because they ignore the line tension of the step, which will be introduced in section 10.3. The line tension of steps will be seen to be effective only for fluctuations of short wavelength. For the sake of simplicity, we shall first work with equations (10.1) and (10.2). Now, we claim that, if the Schwoebel effect is present, equation (10.1) with the boundary conditions (10.2) may lead to an instability (Bales & Zangwill 1990a). The reason can be very simply explained without equations. In order to discuss the stability of a straight step, we argue as in section 6.4. We assume a weak perturbation to be present, and we ask whether it increases or decreases with time. The perturbation of the step may be a small bump (Fig. 10.1). Because of the Schwoebel effect, the adatoms mainly reach the step from the terrace ahead. Those which are just in front of the bump will of course mainly go to the bump itself, and those which are very far from the bump will not reach it. The important point is that the adatoms which are not just in front of the bump, but not too far, will preferentially diffuse to the nearest point on the step, which lies on the bump. Thus, the bump grows and the straight shape is unstable.

This is very similar to the mechanism which produces fractal aggregates in diffusion limited aggregation (DLA), a model introduced by Witten & Sander in 1981. The true DLA model does not use the deterministic, continuum equation (10.1), but treats diffusion as a random walk of atoms on a lattice. In addition, evaporation is forbidden ($\tau_v = \infty$) and

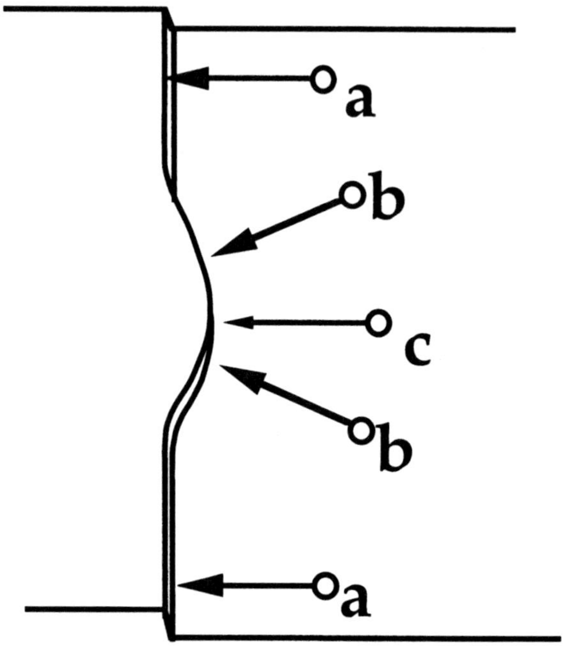

Fig. 10.1. A bump on a growing step. Atoms moving from the forward direction go preferentially to the bump.

the formation of new clusters is avoided by considering only a finite, sufficiently small area. Finally, atoms which stick to the cluster edge are not allowed to detach, so that equations (10.2) are not useful. DLA and its variations have been mainly treated numerically (Jullien *et al.* 1984, Pietronero & Tosatti 1986, Pelcé 1988, Vicsek 1989, Family & Vicsek 1991, Barabási & Stanley 1995) but a partly analytical approach has been proposed in Pietronero *et al.* (1988) and in Pietronero & Schneider (1990).

In the remainder of this chapter, we shall mainly concentrate on the deterministic version of DLA, defined by equation (10.1) with the boundary conditions (10.2) along the step.

10.2 Linear stability analysis: the Bales-Zangwill instability

Equations (10.1) and (10.2) look linear. However, they only are linear if the step is straight. If it is not, they define a non-linear problem, as will be seen, because the (time-dependent) function which gives the step profile determines the boundary conditions (10.2) and must be in turn determined by the solution of (10.1). For a nearly straight step, this problem may be linearized, and the resulting linear equations allow for the investigation of the stability of a straight-step solution. This is called a *linear stability analysis.*

The system of equations (10.1) and (10.2) has indeed a solution (studied in chapter 6) where the step Γ is a straight line. The y axis will be chosen parallel to its direction. Let x be the perpendicular direction in the plane of the surface. The corresponding solution of (10.1) is $\rho(x,t) = \varphi(x - vt)$, where

$$D\varphi'' + v\varphi' - \frac{1}{\tau_v}\varphi + F = 0\,. \tag{10.3}$$

Note that, being familiar with step flow since chapter 6, we do not make the quasi-static approximation yet. Do not worry, we shall do it later! Taking the step at $\xi \equiv x - vt = 0$, the solution can be deduced from problem 6.1:

$$\varphi(\xi) = \varphi_1 \exp(\alpha_1 \xi) + F\tau_v \quad (\xi < 0) \tag{10.4a}$$

$$\varphi(\xi) = \varphi_2 \exp(-\alpha_2 \xi) + F\tau_v \quad (\xi > 0) \tag{10.4b}$$

where

$$\alpha_1 = \frac{1}{2D}\left(\sqrt{v^2 + 4D/\tau_v} - v\right) \tag{10.5a}$$

and

$$\alpha_2 = \frac{1}{2D}\left(\sqrt{v^2 + 4D/\tau_v} + v\right) \tag{10.5b}$$

while the constants φ_1 and φ_2 are determined by (10.2):

$$D\varphi_1\alpha_1 = \Delta'(\rho_0 - \varphi_1 - F\tau_v)$$

$$D\varphi_2\alpha_2 = \Delta''(\rho_0 - \varphi_2 - F\tau_v)\,.$$

Hence

$$\varphi_1 = \frac{\rho_0 - F\tau_v}{\alpha_1 D/\Delta' + 1} \tag{10.6a}$$

$$\varphi_2 = \frac{\rho_0 - F\tau_v}{\alpha_2 D/\Delta'' + 1}\,. \tag{10.6b}$$

We now want to investigate the stability of this solution. As we said above, we now assume that, at $t = 0$, the step has a slightly distorted shape. Since the problem is now linear, we can make a Fourier expansion of the perturbation, and consider only one component, so that the position of the step at $t = 0$ is (Fig. 10.2)

$$\delta x = \eta \cos(qy) \tag{10.7}$$

where η is small. The solution of (10.1) and (10.2) is expected to be equal to (10.4) plus a small perturbation. This perturbation is expected to be

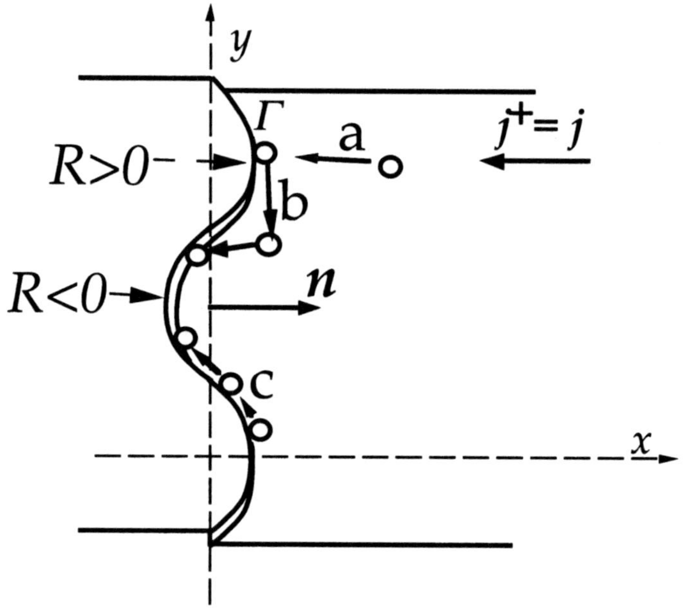

Fig. 10.2. A sinusoidal perturbation on a step: will the amplitude decrease or increase with time? The answer to this question is in the linear stability analysis. Adatoms go preferentially to the tips (a), and this effect is destabilizing. Then, in order to minimize the interface energy, they can diffuse to the holes after being reemitted (b) or slide along the step (c). Effects (b) and (c) tend to stabilize the straight step. They are neglected in the standard DLA model.

a sinusoidal function of y, more precisely $\delta\rho = \psi(\xi)\cos(qy)$. Inserting this into (10.1), and neglecting the y-dependence of the step velocity v, yields $\psi(\xi)$ as an exponential function of ξ. The approximation to neglect the y-dependence of v is comparable to the quasi-static approximation in chapter 6: indeed, we compute here the solution for a y-independent velocity, and we are going to compute soon the y-dependent contribution to v. Finally, the appropriate solution of (10.1) is

$$\rho(\xi) = \varphi_1 \exp(\alpha_1 \xi) + F\tau_v + \eta_1 \exp(\Lambda_1 \xi)\cos(qy) \quad (\xi < \eta\cos(qy)) \tag{10.8a}$$

$$\rho(\xi) = \varphi_2 \exp(-\alpha_2 \xi) + F\tau_v + \eta_2 \exp(-\Lambda_2 \xi)\cos(qy) \quad (\xi > \eta\cos(qy)) \tag{10.8b}$$

where

$$\Lambda_1 = \sqrt{\frac{v^2}{4D^2} + q^2 + \kappa^2} - \frac{v}{2D} \tag{10.9a}$$

and

$$\Lambda_2 = \sqrt{\frac{v^2}{4D^2} + q^2 + \kappa^2} + \frac{v}{2D} \,. \tag{10.9b}$$

We recall that $\kappa^2 = 1/(D\tau_v)$ as in equation (6.6).

The constants η_1 and η_2 are found by inserting (10.8) into (10.2). The expression (10.8) is only acceptable if the terms of second and higher order in η, η_1 and η_2 are neglected. Equating the coefficients of $\cos(qy)$, one obtains within this approximation

$$\eta_1 = -\frac{(\Delta' + D\alpha_1)\varphi_1\alpha_1\eta}{D\Lambda_1 + \Delta'} \tag{10.10a}$$

and

$$\eta_2 = \frac{(\Delta'' + D\alpha_2)\varphi_2\alpha_2\eta}{D\Lambda_2 + \Delta''} \,. \tag{10.10b}$$

Now, the local velocity $v + \delta v(y)$ of the step can be computed:

$$v + \delta v(y) = -\Delta' \left(\frac{\partial \rho}{\partial \xi}\right)_{\text{up}} + \Delta'' \left(\frac{\partial \rho}{\partial \xi}\right)_{\text{down}}$$

or, according to (10.2)

$$v + \delta v(y) = -\Delta'(\rho_0 - \rho)_{\text{up}} - \Delta''(\rho_0 - \rho)_{\text{down}}$$

where the subscripts 'up' and 'down' refer to the upper (left-hand) and lower (right-hand) side of the step, respectively. The excess velocity $\delta v(y)$ is the y-dependent term of the left-hand side. In view of (10.8) and (10.10), one obtains

$$\begin{aligned}
\delta v(y) &= (\Delta'\varphi_1\alpha_1\eta - \Delta''\varphi_2\alpha_2\eta + \Delta'\eta_1 + \Delta''\eta_2)\cos(qy) \\
&= \Big(\Delta'\varphi_1\alpha_1 - \Delta''\varphi_2\alpha_2 - \frac{(\Delta' + D\alpha_1)\Delta'\varphi_1\alpha_1}{D\Lambda_1 + \Delta'} \\
&\quad + \frac{(\Delta'' + D\alpha_2)\Delta''\varphi_2\alpha_2}{D\Lambda_2 + \Delta''}\Big)\eta\cos(qy) \\
&= \left[\frac{(\Lambda_1 - \alpha_1)\Delta'\varphi_1\alpha_1}{D\Lambda_1 + \Delta'} - \frac{(\Lambda_2 - \alpha_2)\Delta''\varphi_2\alpha_2}{D\Lambda_2 + \Delta''}\right] D\eta\cos(qy) \,.
\end{aligned}$$

and, according to (10.6)

$$\begin{aligned}
\delta v(y) = {} & (\rho_0 - F\tau_v) D\delta\xi(y) \\
& \times \left(\frac{\Lambda_1 - \alpha_1}{D\Lambda_1 + \Delta'}\frac{\Delta'\alpha_1}{\alpha_1 D/\Delta' + 1} - \frac{\Lambda_2 - \alpha_2}{D\Lambda_2 + \Delta''}\frac{\Delta''\alpha_2}{\alpha_2 D/\Delta'' + 1}\right) \,.
\end{aligned} \tag{10.11a}$$

In order to avoid uninteresting complications, we now neglect v in (10.5) and (10.9), so that $\alpha_1 = \alpha_2 = \kappa$, and $\Lambda_1 = \Lambda_2 = \Lambda_q$, where

$\Lambda_q = \sqrt{q^2+\kappa^2} > \kappa$. Also, we introduce the two lengths $d' = D/\Delta'$ and $d'' = D/\Delta''$. Thus,

$$\begin{aligned}\delta v(y) &= (\rho_0 - F\tau_v)(\Lambda_q - \kappa)D\kappa\delta x(y) \\ &\quad\times\left[\frac{1}{(1+\kappa d')(1+\Lambda_q d')} - \frac{1}{(1+\kappa d'')(1+\Lambda_q d'')}\right] \\ &= (\rho_0 - F\tau_v)(\Lambda_q - \kappa)D\kappa\delta x(y) \\ &\quad\times\frac{\Lambda_q d'' + \kappa d'' + \kappa\Lambda_q d''^2 - \Lambda_q d' - \kappa d' - \kappa\Lambda_q d'^2}{(1+\kappa d')(1+\Lambda_q d')(1+\kappa d'')(1+\Lambda_q d'')} \\ &= (\rho_0 - F\tau_v)(\Lambda_q - \kappa)D\kappa\delta x(y) \\ &\quad\times(d''-d')\frac{\Lambda_q + \kappa + \kappa\Lambda_q(d''+d')}{(1+\kappa d')(1+\Lambda_q d')(1+\kappa d'')(1+\Lambda_q d'')}\,.\end{aligned} \tag{10.11b}$$

The case of interest is when adatoms do not like to go down the step (Schwoebel effect), i.e. $\Delta' < \Delta''$, or $d'' < d'$. Since growth occurs if $\rho_0 < F\tau_v$, it follows from the formulae (10.11) that

$$\frac{\delta v(y)}{\delta x(y)} > 0\,.$$

Therefore, a straight, isolated step is unstable against meandering, if its line tension is neglected (Bales & Zangwill 1990a). The physics of growth is full of similar instabilities, which may be physically rather different, but which result from equations similar to those of this section, and where the basic mechanism which produces the instability is always the diffusion of matter or energy which eventually reaches a growing interface. As explained in section 10.1, the instability results from the fact that the diffusing objects prefer to go to bumps, thus producing an increase of these irregularities. We shall briefly describe a well-known, three-dimensional, instability of this kind in sections 10.7 and 10.8.

10.3 Stabilizing effects: line or surface tension

As stressed in the preceding section, equations (10.2) do not take the step line tension into account. Neither do the equations (10.11), since they follow from (10.2).

Qualitatively, it is clear that the line tension tends to stabilize the straight step, even though it is not clear that this stabilizing effect will dominate the destabilizing effect of diffusion. Indeed, the free energy tends to decrease, and in particular the step free energy, which is clearly minimal for a straight step, at least if the line tension is approximately isotropic.

Similarly, in the growth of a three-dimensional solid, the surface tension favours the shape of minimal energy, which is often the shape of minimal area, namely the planar shape.

Which one will be the winner? The stabilizing line or surface tension? Or the destabilizing effect? To obtain the answer, one has first to write the correct equations. The modified diffusion equation (10.1) is all right, but equations (10.2) should be modified. We want to express the current of particles near the step (or any moving front) in two manners. First, it is proportional to the gradient of the concentration ρ near the step:

$$\mathbf{j} = -D\nabla\rho \,.$$

Second, it is the difference between the current of atoms sticking to and detaching from the step edge. The former contribution can be written, as in chapter 6, $D'\rho(-a)/a$, where $\rho(-a)$ is the adatom density one atomic distance away from the upper side of the step. It should be compensated by the current from the step, when the step is in equilibrium with a uniform surrounding adatom density ρ_1. Therefore, the current from the step is $-D'\rho_1/a$, and the net current is (for a step parallel to the y axis):

$$j = -\frac{D'}{a}[\rho_1 - \rho(-a)] = -\frac{D'}{a}(\rho_1 - \rho + a\partial\rho/\partial x) \tag{10.12}$$

where ρ is the value that the density $\rho(x)$ would take on the step if the graph of $\rho(x)$ were continued up to the step. The only new feature of the equation (10.12) with respect to section 10.2, is that the equilibrium adatom density ρ_0 has been replaced by ρ_1. The excess density $\delta\rho = \rho_1 - \rho_0$ can be deduced from the excess chemical potential $\delta\mu$ given, as seen in chapter 2, by the Gibbs-Thomson relation $\delta\mu = \gamma/R$. Here R is the radius of curvature of the step, with a positive sign for a protruding step, as shown in Fig. 10.2. Now, formula (5.4) yields $\delta\rho = \beta\gamma/R$. Inserting this into (10.12), one obtains

$$j = -\frac{D'}{a}\left(\frac{\beta\gamma}{R} + \rho_0 - \rho + a\frac{\partial\rho}{\partial x}\right) \,.$$

An invariant form is easily obtained as

$$\mathbf{n}\cdot\mathbf{j} = -\frac{D'}{a}\left(\frac{\beta\gamma}{R} + \rho_0 - \rho + a\mathbf{n}\cdot\nabla\rho\right)$$

where $\mathbf{n}$ is the unit vector normal to the step directed toward the lower side. Writing $\mathbf{j}$ as $-D\nabla\rho$, and using $\Delta' = DD'/[a(D-D')]$, the boundary conditions read

$$\Delta'\left(\frac{\beta\gamma}{R} + \rho_0 - \rho\right) = D(\mathbf{n}\cdot\nabla\rho) \tag{10.13a}$$

at the upper side of the step, and similarly:

$$\Delta''\left(\frac{\beta\gamma}{R} + \rho_0 - \rho\right) = -D(\mathbf{n}\cdot\nabla\rho) \tag{10.13b}$$

at its lower side.

Formulae (10.13) introduce an additional non-linearity because of the curvature $1/R$, which is given by the non-linear expression

$$1/R = -X''\left(1 + X'^2\right)^{-3/2},$$

and because of the product $\mathbf{n} \cdot \nabla = \left[1 + X'^2\right]^{-1/2} [\partial_x - X'\partial_y]$, where $X(y)$ denotes the position of the step and $X' = \mathrm{d}X/\mathrm{d}y$.

10.4 Stability of a regular array of straight steps: the general case

We now want to generalize the linear analysis made in section 10.2 for a single step, neglecting the line tension. We shall give a general formula (which is derived in appendix G) and draw all its consequences.

We consider an array of parallel, straight steps at distance ℓ, perturbed by a weak modulation

$$\delta x_m(y,t) = \epsilon(t)\cos(ky)\exp(im\varphi) \tag{10.14}$$

where m labels the successive steps (Fig. 10.3). The steps are parallel to the y axis, φ is an arbitrary constant and ϵ is a function of t. Arguing as in section 10.2, and taking the line tension into account as explained in section 10.3, one obtains (as shown in appendix G) the following generalization of (10.11)

$$\frac{\partial}{\partial t}\delta x_m = \omega \delta x_m \tag{10.15a}$$

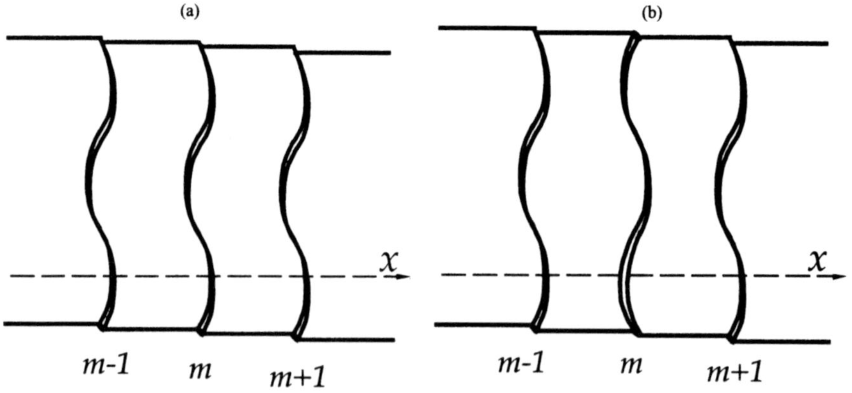

Fig. 10.3. Same problem as in Fig. 10.2, but for an array of steps. The figures correspond to $\varphi = 0$ (a), and to $\varphi = \pi$ (b).

where ω is independent of m and y; this is the advantage of the choice (10.14). It follows that $\epsilon(t)$ varies exponentially with time:

$$\epsilon(t) = \epsilon(0)\exp(\omega t)\,. \tag{10.15b}$$

The regular array of straight steps is therefore stable if $\mathrm{Re}(\omega) \leq 0$ for all ω, and unstable if $\mathrm{Re}(\omega) > 0$ for some ω. The choice (10.14) is clearly not a restriction of the problem since (10.14) is just the general Fourier component of the modulation.

The problem is now to calculate ω. This is done in appendix G. The result includes an imaginary part which will not be reproduced here (and vanishes with φ) and a real part

$$\omega(k,\varphi) = g(k,\varphi) - k^2 f(k,\varphi) \tag{10.16}$$

where the second term describes the effect of the line tension while the first one is independent of the line tension and is proportional to the 'supersaturation' $\Delta F = F - \rho_0/\tau_v$. This definition of the supersaturation is appropriate for MBE, and it has been given at the end of section 5.6 for near-equilibrium conditions. In any case, once the supersaturation is suitably defined, it plays the same role in the theory of step flow (e.g. equation 6.7) and of instabilities.

The first term of (10.16) is (appendix G)

$$g(k,\varphi) = \Omega\Delta F\,(d' - d'')\,\frac{N(k,\varphi)}{D(k)} \tag{10.17a}$$

where $d'' = D/\Delta''$ and $d' = D/\Delta'$ as given by (6.20), Ω is the atomic area, generally set equal to 1 in this book, and the functions $N(k,\phi)$ and $D(k)$ have the following form:

$$\begin{aligned} N(k,\varphi) = {} & \Lambda_k\,(d' + d'') \\ & \times [\cos\varphi + \Lambda_k x_s \sinh(\Lambda_k \ell)\sinh(\kappa\ell) - \cosh(\Lambda_k\ell)\cosh(\kappa\ell)] \\ & + k^2 x_s^2[\cosh(\kappa\ell) - 1]\sinh(\Lambda_k\ell) \end{aligned} \tag{10.17b}$$

and

$$\begin{aligned} D(k) = {} & [(d' + d'')\cosh(\kappa\ell) + (\kappa d'd'' + x_s)\sinh(\kappa\ell)] \\ & \times \left[(d' + d'')\,\Lambda_k\cosh(\Lambda_k\ell) + \left(\Lambda_k^2 d'd'' + 1\right)\sinh(\Lambda_k\ell)\right]\,. \end{aligned} \tag{10.17c}$$

The second term of (10.16) is given by

$$\begin{aligned} f(k,\varphi) = {} & \frac{D\rho_0\gamma}{k_B T}\Lambda_k \\ & \times \frac{\Lambda_k\,(d' + d'')\sinh(\Lambda_k\ell) - 2\cos\varphi + 2\cosh(\Lambda_k\ell)}{(d' + d'')\,\Lambda_k\cosh(\Lambda_k\ell) + (\Lambda_k^2 d'd'' + 1)\sinh(\Lambda_k\ell)}. \end{aligned} \tag{10.18}$$

It is readily seen that f is always positive. Therefore, the term $-k^2 f(k)$ is always negative and therefore stabilizing, either in growth or evaporation conditions. This is in agreement with the intuitive argument of section 10.3: the line tension (if isotropic) is always stabilizing.

The basic question is the sign of (10.17a). Usual Schwoebel effect implies $d' - d'' > 0$. The 'supersaturation' ΔF is positive in growth and negative in evaporation.

The case of a single step

This corresponds to the limit $\ell = \infty$. Of course, the phase shift φ between steps disappears. One finds

$$f(k) = \frac{D\rho_0\gamma}{k_B T}\Lambda_k \frac{\Lambda_k (d' + d'') + 2}{(d' + d'')\Lambda_k + (\Lambda_k^2 d'd'' + 1)}, \tag{10.19a}$$

$$\frac{N(k)}{D(k)} = \frac{\Lambda_k(d' + d'')(\Lambda_k x_s - 1) + k^2 x_s^2}{(d' + d'' + \kappa d'd'' + x_s)\left[(d' + d'')\Lambda_k + \Lambda_k^2 d'd'' + 1\right]}$$

and

$$\begin{aligned} g(k) &= \Omega\Delta F \frac{(d' - d'')\left[\Lambda_k (d' + d'')(\Lambda_k x_s - 1) + k^2 x_s^2\right]}{(d' + d'' + \kappa d'd'' + x_s)\left[(d' + d'')\Lambda_k + \Lambda_k^2 d'd'' + 1\right]} \\ &= \Omega\Delta F \frac{(d' - d'')\left[\Lambda_k (d' + d'')(\Lambda_k - \kappa) + \left(\Lambda_k^2 - \kappa^2\right) x_s\right]}{\left[\kappa(d' + d'') + \kappa^2 d'd'' + 1\right]\left[(d' + d'')\Lambda_k + \Lambda_k^2 d'd'' + 1\right]} \\ &= \Omega\Delta F(d' - d'')\frac{(\Lambda_k - \kappa)\left[\Lambda_k(d' + d'') + \Lambda_k x_s + 1\right]}{(\kappa d' + 1)(\kappa d'' + 1)(\Lambda_k d' + 1)(\Lambda_k d'' + 1)}. \end{aligned} \tag{10.19b}$$

Making use of (6.20) and (6.6), and of the relation $\Delta F = F - \rho_0/\tau_v$, one recovers the result (10.11) of section 10.2.

For small k, $f(k)$ is k-independent and

$$g(k) = \Omega\Delta F \frac{k^2}{2}(d' - d'')\frac{(d' + d'') + 2x_s}{(\kappa d' + 1)^2(\kappa d'' + 1)^2}.$$

Thus, in the case of a single step, both terms of (10.16) are proportional to k^2. The line tension, if large enough, is able to stabilize a single, growing, straight step. There is a critical value of the ratio Supersaturation/Line tension, beyond which the planar step or interface is unstable. The proportionality to k^2 of the surface tension contribution $k^2 f(k)$ is reminiscent of formula (8.9). If evaporation is forbidden, proportionality to a higher power of k is expected in the stabilizing term according to section 8.3, so that the surface tension should be unable to prevent the instability at long wavelengths. This will be checked below.

10.5 The case of MBE growth without evaporation

This corresponds to the limit $\tau_v \to \infty$, $\kappa\ell \to 0$. Then,

$$g(k,\varphi) = \Omega F\,(d' - d'') \times \frac{k\,(d'+d'')\,[\cos\varphi + k\ell\sinh(k\ell) - \cosh(k\ell)] + k^2\ell^2/2\sinh(k\ell)}{(d'+d''+\ell)\left[(d'+d'')k\cosh(k\ell) + (d'd''k^2+1)\sinh(k\ell)\right]}$$

and

$$f(k,\varphi) = \frac{D\rho_0\gamma}{k_B T}k\frac{k\,(d'+d'')\sinh(k\ell) - 2\cos\varphi + 2\cosh(k\ell)}{(d'+d'')\,k\cosh(k\ell) + (k^2d'd''+1)\sinh(k\ell)}\,.$$

We are now going to show that there is an instability for $k\ell \ll 1$. The above formulae become

$$\begin{aligned} g(k,\varphi) &= \Omega F\,(d'-d'')\,\frac{k\,(d'+d'')\left[\cos\varphi - 1 + k^2\ell^2/2\right] + k^3\ell^3/2}{(d'+d''+\ell)\left[(d'+d'')k + (d'd''k^2+1)k\ell\right]} \\ &= \Omega F\,(d'-d'')\,\frac{(d'+d'')\left[\cos\varphi - 1 + k^2\ell^2/2\right] + k^2\ell^3/2}{(d'+d''+\ell)^2} \end{aligned}$$

or

$$\begin{aligned} g(k,\varphi) = \Omega F\,(d'-d'')\,&\frac{(d'+d'')\,[\cos\varphi - 1]}{(d'+d''+\ell)^2} \\ +\Omega F\,(d'-d'')\,&\frac{k^2\ell^2}{2\,(d'+d''+\ell)}\,. \end{aligned} \tag{10.20a}$$

The other quantity which appears in (10.16) is

$$\begin{aligned} f(k,\varphi) &= \frac{D\rho_0\gamma}{k_B T}\frac{(d'+d'')\,k^2\ell + 2(1-\cos\varphi) + (k\ell)^2}{d'+d''+\ell+k^2d'd''\ell} \\ &= \frac{2D\rho_0\gamma}{k_B T}\frac{1-\cos\varphi}{d'+d''+\ell} + k^2\ell\frac{D\rho_0\gamma}{k_B T}\left[1 - \frac{2(1-\cos\varphi)d'd''}{(d'+d''+\ell)^2}\right]. \end{aligned} \tag{10.20b}$$

It is sufficient to consider the case $\varphi = 0$, because for a normal Schwoebel effect ($d' > d''$) $\omega(k,\varphi)$ is maximum at $\varphi = 0$, which is thus the mode which increases most rapidly. The first, destabilizing, term of (10.16) is proportional to k^2, and the second, stabilizing, term is only proportional to k^4 (Fig. 10.4). Thus, in ordinary MBE growth conditions (no adatom desorption) with normal Schwoebel effect, a train of steps growing in step flow mode is *always* unstable to long-wavelength ($k \to 0$), in-phase ($\varphi = 0$) shape fluctuations of all steps (*meandering instability*).

Other cases can be addressed, for instance a single step at equilibrium in the absence of evaporation. Formulae (10.16-18) yield a stabilizing

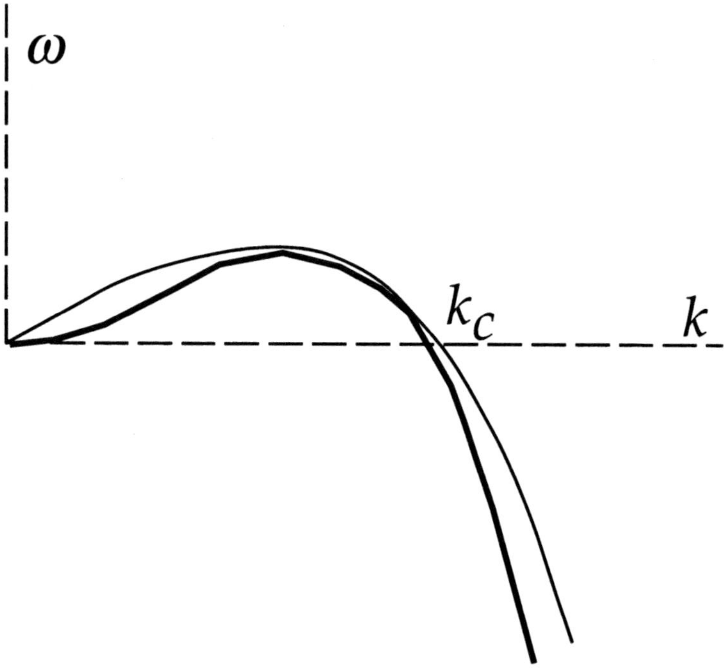

Fig. 10.4. The relaxation rate $\omega(k)$ as given by formulae 10.16 and 10.19 (thick line) and 10.24 (thin line). The straight shape is unstable against any perturbation of wave number smaller than k_c.

term proportional to k^3 (see problem 10.2) in that case. However, at low temperature, this is not necessarily true because ρ_0 is very small and (10.18) vanishes. The healing effect is then the atom diffusion along the steps edge (Fig. 10.2, process c), and the corresponding healing term is proportional to k^4, in agreement with (8.13).

Note that in this section the supersaturation ΔF has been written simply F because of τ_v being very large. In deposition without atom desorption, the adatom density on a terrace of size ℓ is approximately $\rho_0 + F\ell^2/D$. During step flow growth, which concerns us here, the deposited density $F\ell^2/D$ should not exceed very much the equilibrium density ρ_0. If this happens, islands start to nucleate near the terrace centre, as we will discuss in chapter 11, and step flow stops.

10.6 Beyond the linear stability analysis: cellular instabilities

In the previous section, we have shown that a meandering instability of the step edges should occur in MBE growth. However, experimentally, this instability is rarely reported. This may be due to the fact that it does not appear altogether because the sample is too small. It may be alternatively due to the fact that the meanders do not disturb the growing crystal too much, and nobody cares!

'Not too much' means that there could be a stable steady state, and that the steady state is not too far from a straight step (or planar interface or growth front).

An example encountered at the end of section 10.4, is when the instability appears only above some critical value ζ_c of a control parameter ζ (here, the ratio Supersaturation/Line tension or Supersaturation/Surface tension). Then, one can hope that the planar front is not too much disturbed when $\delta\zeta = \zeta - \zeta_c$ is small. To check this statement, one can expand the non-linear equations (10.13) in powers of $\delta\zeta$. The resulting equations are still non-linear (Bena *et al.* 1993, Misbah & Rappel 1993) but then one generally tries to approximate them by some standard equation, familiar to specialists.

Cellular shapes

The simplest such non-linear equation is obtained when, assuming the sinusoidal shape (10.14) with a small amplitude ϵ, one finds (see for references Caroli *et al.* 1991) the following generalization of (10.15), where ω_0 and ω_1 are positive constants:

$$\frac{d\epsilon}{dt} = \omega_0\epsilon - \omega_1\epsilon^3 . \tag{10.21}$$

This is still a non-linear equation, but so simple that many scientists are ready to make the most horrible approximations to arrive at it. Equation (10.21) has a stationary solution $\epsilon = \sqrt{\omega_0/\omega_1}$. If we are lucky, this solution is stable. The 'Landau-Hopf' equation (10.21) can be considered as derived from the Landau free energy $\mathscr{F} = -\omega_0\epsilon^2/2 + \omega_1\epsilon^4/4$ in phase transition problems, where now ω_0 and ω_1 are functions of some control parameters. A continuous or 'second order' transition occurs, if ω_1 is positive, when ω_0 changes sign (Landau & Lifshitz 1959a). In the case of equation (10.21), a similar phenomenon occurs, but it is called 'bifurcation' instead of 'transition', and 'supercritical' instead of continuous or second order. Similarly, if ω_1 is negative, a bifurcation occurs for a particular value of ω_0, which is called 'subcritical' and corresponds to a 'first order phase transition'.

The basic point is that, if ϵ is small, (10.14) may be expected to be a good approximation of the surface profile, and ϵ has a time independent value. Such a structure, where steps, or interfaces, have a steady shape which is periodic in space, is called a cellular structure.

Fig. 10.5. A presumably chaotic array of steps on GaAs(001) observed by large-scale STM after vacumm annealing at 520 °C. The size of the visualized area is 400 nm × 400 nm. The physical situation (vacuum annealing) is different from that addressed in this chapter (MBE growth), but this picture gives an idea of what the Bales-Zangwill instability can lead to: large step fluctuations, somewhat comparable with thermal fluctuations, but no step bunching (from Snyder *et al.* 1994, courtesy of B. Orr).

The Bales-Zangwill instability does not generally lead to cellular patterns, but rather to chaotic structures (Bena *et al.* 1993). Fig. 10.5 show such an apparently chaotic array of steps observed by STM on an evaporating surface (see problem 10.7). In crystal growth, however, solid-liquid interfaces often take cellular structures. The study of solid-liquid interfaces is a very vivid field of contemporary physics, which has been the subject of many elaborate theories and beautiful experiments. Its description will be just briefly sketched below.

10.7 The Mullins-Sekerka instability

While most of this book is devoted to MBE growth, this section addresses crystal growth from the liquid phase. When growing a crystal from the liquid, one generally tries to maintain a planar solid-liquid interface. For instance on can pull the crystal (Fig. 10.6) at a constant velocity V along the x direction between two ovens at positions x_1 and x_2 where well defined temperatures T_1 and T_2 are maintained. These temperatures are such that $T_1 < T_M < T_2$, where T_M is the melting temperature of the material. This type of growth is called *directional solidification.*

Experimentalists have known for a long time that the solid-liquid interface can only remain planar if its velocity V is smaller than some threshold V_{MS}. This threshold depends on the temperature gradient $(T_2 - T_1)/(x_2 - x_1)$ and vanishes with it. When $V > V_{\mathrm{MS}}$, the interface starts meandering. This instability was explained by Mullins & Sekerka (1963, 1964). It is reviewed in Langer (1980), Pelcé (1988), Caroli *et al.* (1991), and Pomeau & Ben Amar (1991).

Like the effect addressed in section 10.1, the Mullins-Sekerka instability is related to diffusion away from the interface (which now plays the same part as the steps in the previous sections). The diffusing objects are not the atoms of the material itself, since those are already present in the liquid. In practice, the diffusing objects are generally impurities, which have a strong influence on melting (the benefic effect of salt to prevent rain from freezing on roads is well known). However, we shall consider the case of a pure material. Then, the diffusing quantity is the energy. Energy conservation is only important in an extremely pure system, because heat (i.e. energy) diffusion is always much faster than matter diffusion. The energy density (or the local temperature which is related to it by the formula $\delta\epsilon = C_P \delta T$) satisfies, away from the interface, the diffusion equation

$$\frac{\partial T}{\partial t} = D_T \nabla^2 T \, . \tag{10.22}$$

This equation is analogous to (10.1). Approximate boundary conditions on the solid-liquid interface, which should replace (10.2) or (10.13), are obtained by assuming that the temperature T is constant on the interface

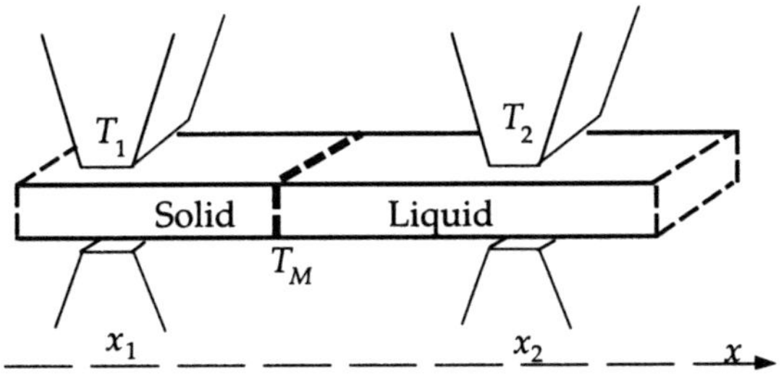

Fig. 10.6. Pulling a crystal. In an actual crystal-growth setting, the x axis is vertical and the solid phase is above.

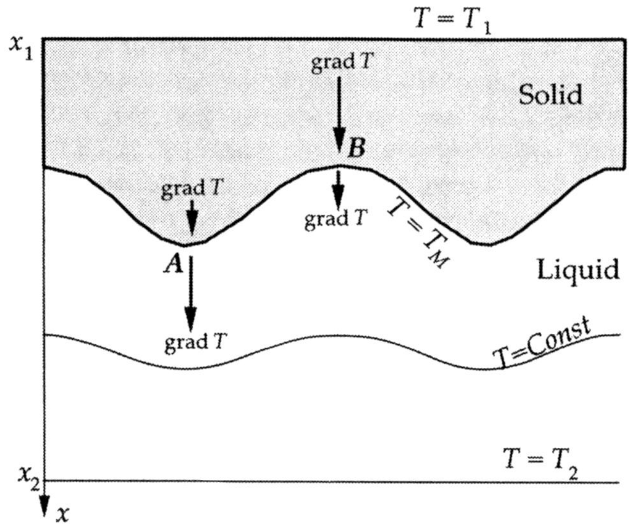

Fig. 10.7. Stabilizing effect of the temperature gradient in directional growth. If the solid-liquid interface has a wavy shape, the temperature gradient is stronger on the tips A, which therefore melt. Consequently, the interface becomes planar again. Thin lines are isothermal surfaces. The Mullins-Sekerka instability appears if undercooling is taken into account.

and equal to the melting temperature T_M:

$$T_{\text{surface}} = T_M \; . \tag{10.23}$$

This relation is only approximate for the same reason as (10.2), since the effect of the interface curvature is neglected. The correct equation should make use of the Gibbs-Thomson relation.

As will now be seen, the Mullins-Sekerka instability is somewhat more complicated than the DLA mechanism outlined in section 10.1. Indeed, if we try to reproduce the argument of sections 10.1 and 10.2, we may assume that the solidification front protrudes in a point A. It is seen in Fig. 10.7 that the temperature gradient is then stronger ahead of A and weaker behind it. The result is that A will receive more heat from ahead and will evacuate less heat from behind. Consequently, freezing will take place more slowly in A and the interface will advance more slowly. An analogous argument shows that if the interface lags behind in B, freezing will be more rapid. Thus, the temperature gradient is stabilizing! The same argument on the solid side of the interface is easily seen to yield stability as well. But then, where does the experimentally observed instability come from?

Mullins and Sekerka realized that the instability is due to undercooling. The liquid right in front of the interface is undercooled, i.e. its temperature

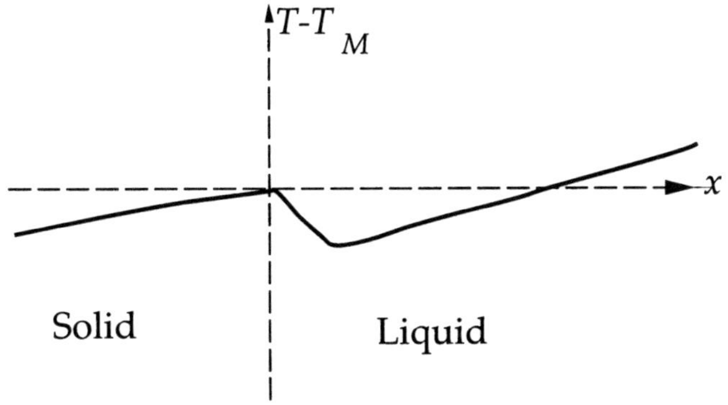

Fig. 10.8. Temperature profile in directional solidification.

is lower than T_M (Fig. 10.8). The temperature gradient just ahead of the advancing interface is seen to have the wrong sign! An argument similar to the above one now suggests that, if the sign of the temperature gradient is inverted, the result is also inverted: that is, undercooling, if strong enough, can destabilize the planar interface. The equations corresponding to this hand-waving argument will be found in the references quoted above. The linear stability analysis is analogous to that of section 10.2 and appendix G, but the final result is rather different. Indeed, the relaxation rate is found to have the form

$$\omega(k) = Av|k| - B|k|^3 - C\Delta T \tag{10.24}$$

where v is the pulling velocity, ΔT the temperature (or impurity concentration) gradient, and the positive constants A, B, and C depend on the thermodynamical properties of the material. The form (10.24), valid for $|\, k \,| \gg v/D_T$, shows the existence of a stabilizing term due to surface tension proportional to k^3. The power 3 has the same origin as the power 3 of formula (8.15), which also comes from diffusion away from the surface (see also problem 10.2). The destabilizing term is proportional to k, rather than k^2 as found in sections 10.3 and 10.4. The reason for that is addressed by problem 10.3. Finally, there is a negative term due to the thermal gradient, which is independent of k and can stabilize the planar interface.

As in the case of steps, a crucial question is to know what happens beyond linear stability analysis. In the simplest cases, the interface eventually takes a steady shape with a cellular structure analogous to those described in section 10.6. These shapes are more easily observed in the growth of transparent crystals growing between two glass plates (Figs. 10.9, 10.10)

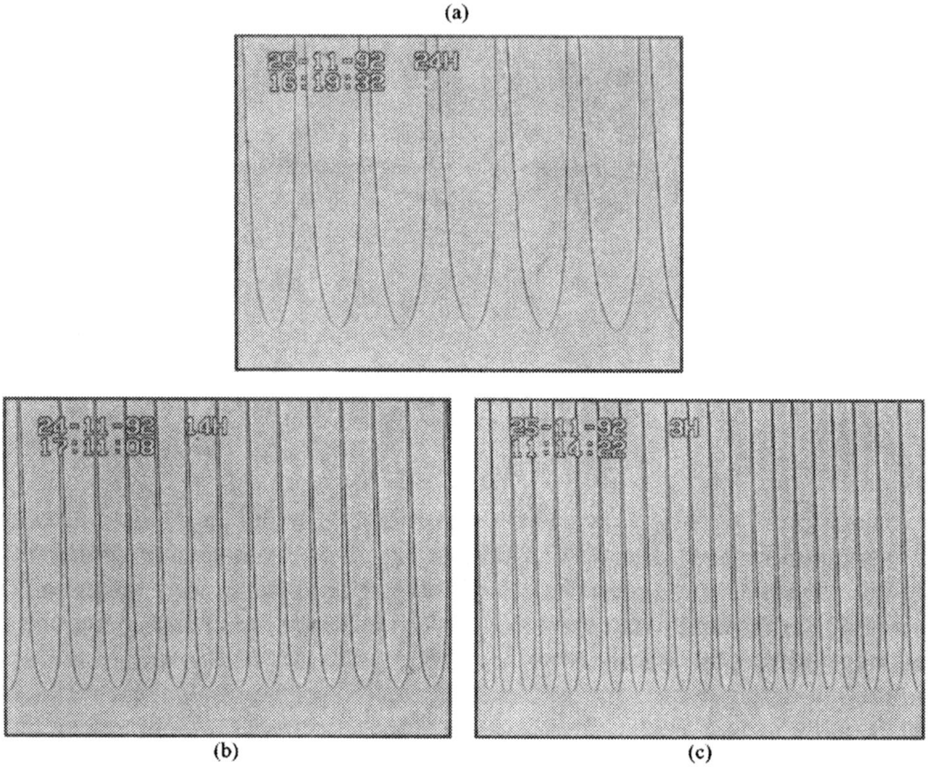

Fig. 10.9. Solid-liquid interface observed during crystallization of pivalic acid in a two-dimensional geometry. The solid phase is above. The sample thickness is 30 microns and the respective pulling velocities are approximately 2, 15 and 50 microns/s (courtesy of P. Oswald, Ecole normale supérieure de Lyon).

rather than in ordinary, three-dimensional crystals (Jamgotchian *et al.* 1993).

10.8 Growth instabilities in metallurgy

Despite the advantages of liquid crystals in fundamental research on growth instabilities, the technological interest of these instabilities in metallurgy should not be left aside. The observation of a cellular structure is only possible after a cut of the sample after cooling (Fig. 10.11a). The section (perpendicular to the growth direction) shows a polycrystalline structure or at least a heterogeneous distribution of impurities (Fig. 10.11b). This structure testifies that the cellular structure of the interface during the growth was very different from the sinusoidal shape (10.7), but

(a) (b)

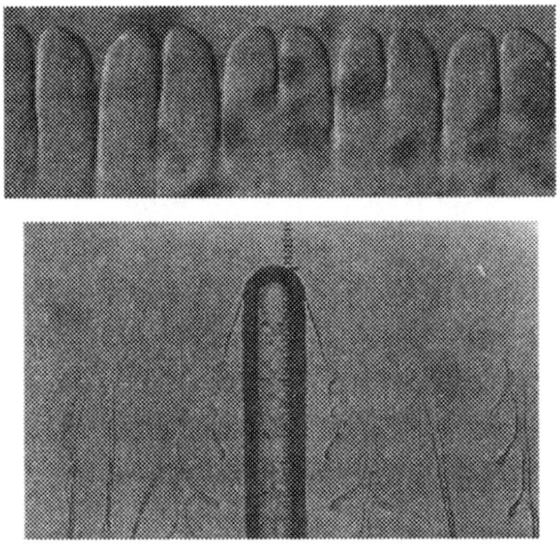

Fig. 10.10. When the growth velocity increases, the simple cellular structure is replaced by more complicated structures. a) Doublet structures in crystallizing succinonitrile. b) Dendritic shapes observed also in succinonitrile (from Jamgotchian *et al.* 1993a,b)

instead had sharp, deep pockets behind the solidification front, as in Fig. 10.9. These holes are due to the following reason: the diffusing objects (subject to equation (6.1) which corresponds to the genesis of the instability) are impurities. These impurities diffuse rather slowly in the liquid, but practically not at all in the solid. Consequently, the impurities which remain trapped in the pockets cannot escape and the pockets become infinitely deep. A similar effect occurs in an advancing step (section 10.1) if the Schwoebel barrier is large.

(a) (b)

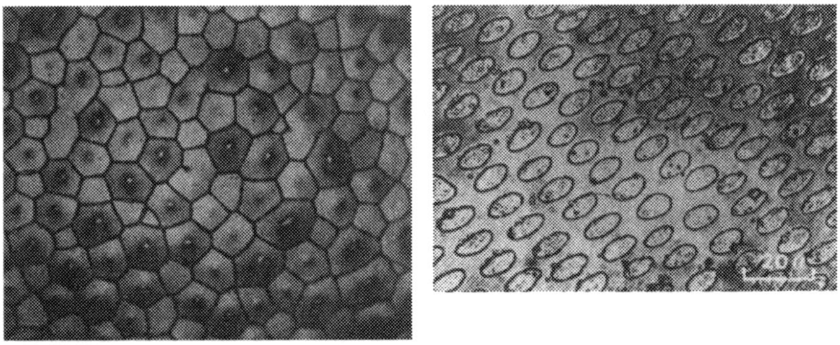

Fig. 10.11. a) Cellular structure in an alloy, due to the Mullins-Sekerka instability during growth, as observed after cooling and cutting the crystal perpendicular to the x axis. The cell boundaries are impurity-rich zones. b) Structure observed in a eutectic alloy (courtesy of Y. Malméjac, Centre d'Etudes Nucléaires de Grenoble).

Growth instabilities are important in the metallurgy of eutectics alloys. A eutectic is an alloy where two phases of different composition can simultaneously crystallize at the same temperature. At equilibrium, the two phases would separate. However, complete separation would require diffusion of matter on macroscopic distances, and therefore a practically infinite time. In practice, one of the phases forms wires inside the other phase (Fig. 10.11b). This local phase separation occurs during growth and in the vicinity of the interface. It follows that the final structure of the solid is determined by equations which are very similar to those which have been studied in this chapter, namely a diffusion equation analogous to (10.22) and boundary conditions analogous to (10.13).

10.9 Dendrites

When one goes beyond the linear stability analysis, the cellular structure (Figs. 10.9, 10.10) is the simplest evolution. However, in many cases, there are no cellular solutions or they are unstable. Snowflakes offer a familiar, and beautiful example of much more complicated structures, which do not evolve to a steady state. Such ramified structures are called dendrites, from the Greek $\delta\epsilon\nu\delta\rho o\nu$, tree.

Ignoring this ethymology, theorists often use the word dendrite to designate non-ramified objects which look like stalagmites rather than trees. Examples (Fig. 10.12) are the finger-like patterns obtained when pushing water into oil between two transparent plates–a problem studied experimentally by Saffmann & Taylor (1958).

A great deal of elaborated theories address the relatively simple case of a non-ramified 'dendrite'. The first one was by Ivantsov (1947) who was able to show that, if surface tension, elasticity, and crystal structure are neglected, the equations in an infinite space possess solutions which have

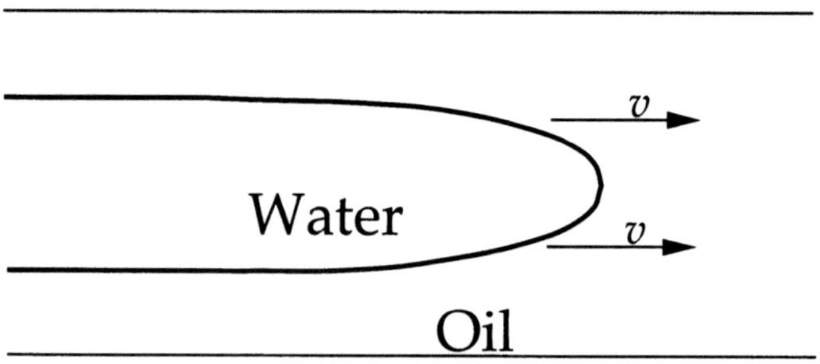

Fig. 10.12. Finger-like pattern formed by water penetrating into oil in a channel (Saffman & Taylor 1958). This system has many analogies with the other instabilities addressed in this chapter and has been studied by many scientists (Pelcé 1988).

the form of parabolae. A simple result, obtained by very clever ways. One problem is that there are an infinity of possible parabolae and it is not clear which one should be chosen. This is the mode selection problem (Combescot & Ben Amar 1991). Other problems are: the stability of the parabolic shape, and the effect of lattice anisotropy and surface tension. These highly non-linear problems are beyond the scope of this book. The reader will find information and bibliographies in Langer (1980), Ben Amar (1991a,b), Pomeau & Ben Amar (1991), Pelcé (1988).

10.10 Conclusion

When introducing this chapter, we asked whether the regular step flow postulated in chapter 6 is stable. The answer is: generally not! For instance, in MBE growth without evaporation and in the presence of the Schwoebel effect, the amplitude of a sinusoidal perturbation of the step shape of wave number k was found in section 10.5 to change with time as $\exp[\omega(k)t]$, where

$$\omega(k) = A\Delta F k^2 - B\gamma k^4 \tag{10.25a}$$

where A and B are positive constants. For small k, ω is always positive and there is an instability.

Only in the case of a single step, it was found in section 10.4 that the line tension γ can stabilize the straight shape, since (10.25a) is replaced by

$$\omega(k) = (A\Delta F - B'\gamma)k^2 \,. \tag{10.25b}$$

It is remarkable that the Schwoebel effect does not produce a meandering instability during evaporation because ΔF is negative. But, during evaporation, it makes step flow unstable with respect to step bunching, which is much worse. Thus, step flow is in principle always unstable.

The reader, whom we tried to motivate at the beginning of this chapter by promising allusions to snowflakes, may be disappointed to have learnt so little about them. As a matter of fact, no theory is able to reproduce quantitatively the complicated shape of a snowflake. The crystal symmetry obviously plays an important part, which is often neglected in simple approaches. Moreover, while much attention has been paid by many authors to the growth of a single 'finger', not as much has been devoted to the fingers which grow on fingers and make it look like a tree. The tree-like shape of snowflakes, in agreement with section 10.7, results from growth from a highly supersaturated medium, which is of course the vapour since the temperature in clouds is far below the triple point (Davis 1974, Hobbs 1974). A discussion taking into account diffusion of water molecules in the vapour and on the crystal surface gives a correct prediction of the growth shape of ice at low supersaturation, when it is not dendritic, and allows to estimate the supersaturation, at which dendrites appear (Mason

1992). But the problem of snowflakes, already tackled by Kepler (1611) is still only qualitatively understood.

Problems

10.1. From equations (10.16-18), rederive the step-bunching instability in the case of evaporation with Schwoebel effect.

Solution – Pimpinelli *et al.* (1994).

10.2. Show that the relaxation time of a sinusoidal perturbation of an isolated straight step at equilibrium in the absence of evaporation is proportional to k^3.

Solution – In the first formula which gives f in section 10.4, one should take the limit $k\ell = \infty$. Then

$$f(k) = \frac{D\rho_0\gamma}{k_B T} k \frac{k\left(d' + d''\right) + 2}{(d' + d'')k + \left(k^2 d' d'' + 1\right)}$$

and

$$k^2 f(k) \simeq \frac{D\rho_0\gamma}{k_B T} k^3$$

for small k. This vanishes with ρ_0, so that, at low temperature, one should take diffusion along the step edge into account.

10.3. Explain why formula (10.24), which describes the growth of a crystal from the liquid phase, looks so different from formulae (10.25) which apply to steps.

Solution – The effect of the surface tension–which determines the third-order term in (10.24)–has been discussed in the text, and is related to the discussion of section 8.3. What will now be discussed is: why are the terms related to diffusion proportional to k in (10.24), instead of k^2 in (10.25)?

First of all, this is only true for $k \gg v/D$, which is the usual physical case. Now, a discussion along the lines of section 10.2 would lead to an equation analogous to (10.11a). However, since the diffusing quantity is conserved, D/τ_v should be replaced by 0 in (10.5) so that $\kappa' = 0$ and $\kappa'' = v/D$. In the limit $k \gg v/D$, the constants Λ' and Λ'' defined by (10.9a) are equal to $|\, k \,|$. Substituting these results into (10.11a) yields $\omega(k) \sim |\, k \,|$.

10.4. Consider a step advancing on a surface growing by MBE. The Schwoebel effect is total and there is no evaporation. Assume a cellular structure as in Fig. 10.9 and 10.10, where deep pockets develop at the rear of the steps. What is the shape of these pockets? As a first approximation, neglect the effect of the other steps. Then, discuss the effect of the other steps and treat a nearly total Schwoebel effect.

Solution – One step and total Schwoebel effect. The point M in Fig. 10.13 will collect all matter falling on the line MK. We are looking for a stationary shape. Let $\partial y/\partial x$ be called y'. Geometry yields

$$y' = \frac{y}{AH} = \frac{MH}{AH} = \frac{MK}{AM} = \frac{HK}{y}$$

where $y = MH$. It follows $AH = y/y'$, $HK = yy'$, hence $AM = \sqrt{AH^2 + MH^2} = \sqrt{(y/y')^2 + y^2} = y/y'\sqrt{1+y'^2}$ and $MK = y'AM = y\left(1+y'^2\right)^{1/2}$.

Therefore $\delta M/\delta x = AM/AH = \sqrt{1+y'^2}$.

The quantity of matter received in time $\mathrm{d}t$ by the segment δM is

$$\left(1 + MK/R\right) F\mathrm{d}tMK\delta M = \left(1 + MK/R\right) F\mathrm{d}ty\left(1+y'^2\right)^{3/2}\delta x\,,$$

where $R = (y'')^{-1}\left(1+y'^2\right)^{3/2}$ is the radius of curvature of the step or interface. On the other hand, this should be equal to $\partial y/\partial t \mathrm{d}t\delta x = v\partial y/\partial x \mathrm{d}t\delta x$.

It follows that

$$\frac{y'}{\left(1+y'^2\right)^{3/2}} = \frac{Fy}{v}\left(1 + \frac{MK}{R}\right) = \frac{Fy}{v}\left(1 + \frac{y''y}{1+y'^2}\right)\,.$$

If y' is small, the solution is $y = \exp\left[F(t - x/v)\right]$.

10.5. Discuss the relaxation rate of a sinusoidal perturbation of an array of steps in the absence of Schwoebel effect as a function of the distance ℓ between steps, the evaporation rate and the wavelength.

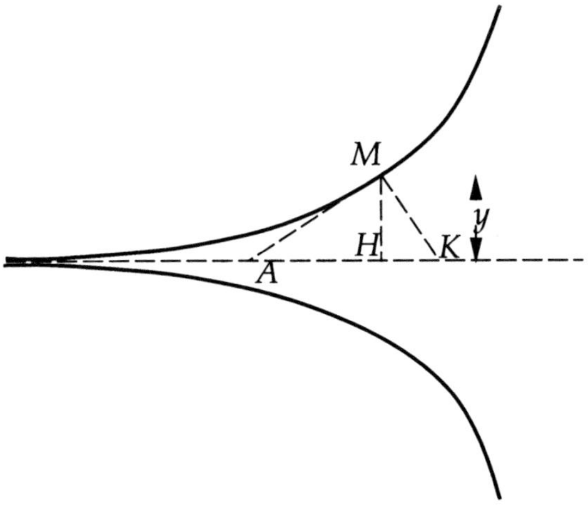

Fig. 10.13. Rear of a cellular structure in the case of a strong Schwoebel effect (or in a growing, solid-liquid interface, with diffusion in the liquid phase only).

Solution – Bartelt *et al.* (1992), Pimpinelli *et al.* (1993b, 1994).

10.6. Consider the motion of an isolated step in MBE growth in the absence of evaporation. Show that equation (10.25a) holds. Prove that the coefficient of the stabilizing term is $k_B T B = \tilde{\gamma}/\tau_k$, where $1/\tau_k$ is the rate of emission of 'gradatoms' (i.e. isolated atoms diffusing along steps) by kinks.

Solution – Pimpinelli *et al.* (1994) and in the French version of this book.

10.7. Discuss why the Bales-Zangwill instability can be held responsible for the finger-like structures of Fig. 10.5, although the morphology of an *evaporating* surface, rather than that of a growing one is shown there.

Solution –In crystal evaporation surface vacancies (advacancies) are produced at a T-dependent rate. At fixed temperature, the production rate may be thought of as an effective advacancy deposition rate. Adatoms are also produced, but they desorb quickly at high enough T. If this is the case, advacancies are the important diffusing entities. For a diffusing advacancy it is likely that attachment to a step is easier from the upper terrace than from the lower one. This Schwoebel effect will lead to the same instability as for adatoms, as the reader should check.

11
Nucleation and the adatom diffusion length

The theoretical villain, however, was what
Dr. Breed called 'a seed'. The seed,
which had come from God-only-knows-where,
taught the atoms the novel way in which
to stack and lock, to crystallize, to freeze.

K. Vonnegut (*Cat's cradle*)

During molecular beam epitaxy, atoms arrive on the surface, and diffuse over a length ℓ_s before being incorporated into the crystal. Can we estimate ℓ_s? The solution to this problem is contained in review articles by Stoyanov & Kashchiev (1981) and by Venables *et al.* (1984).

The simplest case is realized when two diffusing adatoms form an immobile and undissociable pair when they meet, and the resulting cluster grows with a compact shape. Stoyanov & Kashchiev have shown that in this case $\ell_s \sim (D/F)^{1/6}$, where D is the diffusion constant of adatoms and F the deposition flux. More generally, they have shown that $\ell_s \sim (D/F)^\gamma$, where γ is a function of the *critical nucleus size*. The critical nucleus is defined as the largest dissociable atom cluster. The experimental verification is not that easy: the critical nucleus size, and thus γ, depend in an unspecified way on the temperature and the deposition flux. A large host of computer simulation data (Monte Carlo) have anyway confirmed the relation between ℓ_s, D and F.

Orders of magnitude – The lengths should be less than the distances between defects on the surface, i.e. a few microns at the very best. The temperature must be low–which means below 0 °C for metals–otherwise ℓ_s becomes too long.

Experimental techniques – Microscopy (STM, LEEM, REM), atom-beam scattering, high-energy electron diffraction.

11.1 The definition of the diffusion length

We have seen that surface diffusion is, together with the beam intensity, the crucial ingredient of MBE crystal growth. Unlike the beam intensity, which is fixed by the experimentalist, it is not easy to know how big the diffusion constant is. On the other hand, it is very easy, by simply changing the substrate temperature, to vary the diffusion constant over several orders of magnitude; the variation range of the deposition flux in a typical experiment spans on the contrary only two or three orders of magnitude.

In this chapter, we will try to answer the question: 'How far will a freshly deposited adatom diffuse before disappearing?'. It is fundamental to this end to specify the mechanism by which adatoms disappear, that is, what determines the adatom lifetime on the surface. In chapter 6 we investigated step instabilities during crystal sublimation, and we encountered the diffusion length of adatoms before desorption, $x_s = 1/\kappa$. The adatom lifetime was in that case of the order of the reciprocal desorption rate τ_v. At the temperatures of interest for, say, silicon MBE–between 300 and 700 °C–adatom desorption is negligible during the time needed to deposit a monolayer (ML): indeed, the sublimation rate of silicon is of the order of 10^{-14} ML s^{-1} at 500 °C (see chapter 9), and typical deposition rates in MBE range between 10^{-3} and 1 ML s^{-1}. Adatoms will thus diffuse until they find some obstacle to trap them. Such obstacles are in general steps, terrace edges–or other adatoms. The presence of other adatoms was seen in chapter 6 to define a second diffusion lenght, which we called ℓ_s. It measures the average distance travelled by an adatom before meeting another adatom, or, more generally, an adatom cluster.

After an initial transient, we expect the growing surface to reach a stationary state where new steps form and disappear as adatom clusters nucleate and coalesce. The average distance between these steps is clearly of the order of magnitude of the diffusion length ℓ_s. The latter gives also the order of magnitude of the typical cluster (also called terrace or island) size–not necessarily the average size, which may be infinite on a rough surface.

The characteristic length ℓ_s is accessible to measurement, either directly by electron or tunneling microscopy, or by diffraction. For instance, Ernst *et al.* (1992b) measured ℓ_s on Cu(001) between 160 and 200 K by He scattering, and Mo *et al.* (1991, 1992) did the same for Si(001) by STM near room temperature. Since then, many others made similar measurements on a variety of systems.

It may also be that growing terraces do not fit in with each other, and ℓ_s is then simply the distance between misfit defects. On the other hand, it is likely that in many situations the size of monocrystalline domains cannot exceed ℓ_s, especially when surface reconstruction is concerned, as it is for silicon, as well as in the case of growth on a (111) face of an fcc crystal.

We remark finally that the intensity oscillations in RHEED (reflection high-energy electron diffraction) or TEAS (thermal-energy atom scattering), are also related to the characteristic length ℓ_s.

In the following we will evaluate ℓ_s, *neglecting evaporation*. The latter can be accounted for, as done by Stoyanov & Kaschiev (1981). Bulk diffusion will also be neglected.

11.2 The nucleation process

We will consider the growth by molecular beam epitaxy of an infinite, high-symmetry surface of a crystalline element. Let F be the beam intensity, the sticking coefficient being unity. Otherwise, F is the beam intensity times the sticking coefficient. Since evaporation is absent, the time needed for filling a layer is $1/F$ (as usual, we choose to measure lengths in terms of the interatomic distance, so that $1/F$ and $1/D$ both have dimension of times). Freshly landed adatoms diffuse until they find an adatom cluster, or terrace. We will assume that clusters exceeding a critical size i^*, that is, containing more than i^* atoms, are stable and immobile. It means that they do not dissociate, and that their center of mass does not diffuse. At low temperature, i^* is expected to be 1 (Fig. 11.1). In fact, i^* does not only depend on T, but also on F.

The interesting case from a technological point of view is that of a small F compared with the adatom diffusion constant D. The time $1/F$ is in fact of the order of the average time interval between two atoms landing at the same site: if $1/F > 1/D$, an adatom can easily diffuse away before the arrival of the next one, so that the terraces are large and the growing layer is smooth. In practice, for instance in deposition experiments on Cu(001), the flux can be varied between 10^{-4} and 10^{-2} ML s^{-1}, while the diffusion constant is estimated–we will see how–of the order of 10^4 s^{-1} at 250 K (Ernst *et al.* 1992b, Zuo *et al.* 1994, Dürr *et al.* 1995).

Given our 'hit and stick' hypothesis, terrace nucleation looks very much like diffusion limited aggregation (DLA), that we described in chapter 10. One might thus expect clusters to be fractal–as indeed they often are (Hwang *et al.* 1991, Bott *et al.* 1992, Michely *et al.* 1993, Pimpinelli *et al.* 1993a, Brune *et al.* 1994, Bartelt & Evans 1994a,b, Jensen *et al.* 1994). To begin with, however, we assume terraces to be compact rather than fractal (Fig. 11.1). The birth of a new terrace takes place by forming a nucleus on a flat preexisting terrace (Fig. 11.1a). This occurs at a rate $1/\tau_{\mathrm{nuc}}$ per unit time and per site. Then, nucleated terraces grow by adatom capture

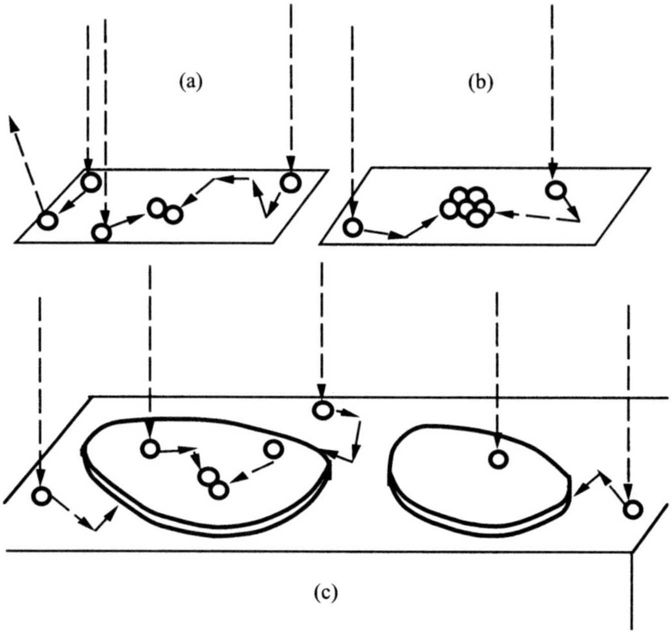

Fig. 11.1. The adatoms landing on a perfect high-symmetry terrace diffuse until they meet and form a critical nucleus. This nucleus is, at low temperature, a single adatom, so that a pair is a stable cluster (a). Then, other atoms join the stable cluster (b), which grows into a terrace by capturing adatoms in competition with neighbouring terraces (c), which eventually coalesce after reaching a size of order ℓ_s.

(Fig. 11.1b). After reaching a size of order ℓ_s, the terraces coalesce (Fig. 11.1c). Such a description implies that terraces reach the size ℓ_s in a time of order $1/F$. The average number $\ell_s^2/(F\tau_{\mathrm{nuc}})$ of nucleation events in an area $\ell_s \times \ell_s$ during a time $1/F$, must be of order 1. In other words,

$$\frac{1}{\tau_{\mathrm{nuc}}} \approx \frac{F}{\ell_s^2}\,. \tag{11.1a}$$

It is easy to extend this argument to the case of fractal terraces of fractal dimension d_f. In this case, the terrace radius reaches the magnitude ℓ_s after a time $t_f \approx (1/F) \times \ell_s^{d_f-2}$. The average number of nucleation events during time t_f inside an area $\ell_s \times \ell_s$ has to be of order 1, which means:

$$\frac{1}{\tau_{\mathrm{nuc}}} \approx \frac{F}{\ell_s^{d_f}}\,. \tag{11.1b}$$

In practice, the appropriate value for d_f is the one which corresponds, as we said, to DLA. Numerical simulations yield $d_f \simeq 1.7$ in two dimensions. Thus, the estimates given by (11.1a) and (11.1b) do not differ very much.

Formulae (11.1) do not depend on i^*, nor on the diffusion constant D. Our next task will therefore be to relate D to ℓ_s and to the adatom density.

11.3 Adatom lifetime and adatom density

Let us call the adatom lifetime τ_1. In the absence of evaporation, the average distance travelled by an adatom is of the order of the distance between island edges, ℓ_s; from elementary diffusion theory we have

$$\tau_1 \approx \frac{\ell_s^2}{D} . \tag{11.2}$$

On the other side, the lifetime τ_1 also fixes the adatom density ρ_1 (number of adatoms per site): it is equal to the beam intensity times τ_1, that is,

$$\rho_1 = F\tau_1 . \tag{11.3}$$

This estimate neglects all the adatoms which are on the surface at equilibrium–which corresponds to setting $F = 0$, since evaporation is negligible. The equilibrium adatom density is, in general, quite small at the temperatures of interest for MBE, compared with (11.3). Anyway, one could easily take it into account, as we did in chapter 6.

From (11.2) and (11.3) it follows:

$$\rho_1 \approx \frac{F\ell_s^2}{D} . \tag{11.4}$$

The reader should compare this result with (6.12). Relation (11.4) only holds on average, since ρ_1 is not uniform. In chapter 6 we investigated its temporal and spatial variation; note, in particular, that (11.4) is only an approximation of (6.12). The merit of the present derivation is to be simple enough to be generalized to the evaluation of the density of dimers, trimers, and so on.

Still, all relations we obtained are independent of i^*. We have to include the details of the nucleation process to make the role of i^* explicit.

11.4 Adatom-adatom and adatom-island collisions

In order to evaluate the nucleation time τ_{nuc} which appears in (11.1), we need to call in the density of critical 'germs' or 'nuclei' (clusters containing i^* atoms). Such germs form when an adatom hits a subcritical cluster of $i^* - 1$ atoms. It is thus necessary to compute the number of such collisions per site and unit time. During a time $t = \tau_1$, a diffusing adatom visits (see chapter 6) a number of distinct sites of the order of

$$v(t) \approx \frac{Dt}{\ln(Dt)} . \tag{11.5}$$

Replacing t by τ_1, we find the probability of a given adatom hitting an n-atom cluster (or n-mer):

$$p_n \approx \frac{D\tau_1}{\ln(D\tau_1)}\rho_n\,, \tag{11.6}$$

where ρ_n is the n-mer density.

11.5 The case $i^* = 1$

The simplest case is $i^* = 1$. This condition ought to be verified at low enough temperature. We are now faced with the following dilemma: we would like to use formula (11.1a), which assumes that islands are compact. However, we know from chapter 10 that at low temperature terraces are not compact. The problem we are going to solve might therefore look like a mere academic exercise, were it not for the three following reasons.

a) The difference between (11.1a) and (11.1b) is very small;
b) The simple case that we are going to treat first is a good training field for more difficult situations. We will actually see that it is very easy to get wrong results;
c) Islands may indeed be compact even at very low temperature where dimers are stable: it is sufficient that bound adatoms have enough mobility along the island edges to be able to look for the most favorable binding site. This is actually the case for the (001) face of fcc metals.

The condition $i^* = 1$ implies that dimers (adatom bound pairs) are stable and immobile: expression (11.1) is the dimer nucleation rate. It is equal to the rate of collisions between adatoms. Thus, it equals the product of p_1 (equation 11.6) times the number F of atoms impinging on the unit surface per unit time. According to (11.2), (11.4) and (11.6), we find:

$$\frac{1}{\tau_{\text{nuc}}} \approx Fp_1 \approx \frac{FD\tau_1}{\ln(D\tau_1)}\rho_1 \approx \frac{F^2\ell_s^4}{D\ln(\ell_s^2)}\,. \tag{11.7}$$

Inserting this result into (11.1a) yields:

$$\frac{\ell_s^6}{\ln(\ell_s^2)} \approx \frac{D}{F}\,. \tag{11.8}$$

This formula is generally written neglecting the logarithm (Stoyanov & Kaschiev 1981). The logarithmic correction is in fact always small, and what one usually finds in literature,

$$\ell_s \approx \left(\frac{D}{F}\right)^{\frac{1}{6}} \tag{11.9}$$

is acceptable, in practice. For instance, if, as in the case of copper mentioned before, $F = 10^{-2}$ ML s^{-1} and $D = 10^4$ s^{-1}, at low temperature, formula (11.9) gives $\ell_s \approx 10$ interatomic distances, while (11.8) gives $\ell_s \approx 12$ interatomic distances. At higher temperature, for instance if $D = 10^{10}$ s^{-1}, (11.9) and (11.8) respectively yield for the same F $\ell_s \approx 100$ and 140 interatomic distances, i.e. the same order of magnitude. Note that neither (11.8) nor (11.9) apply at high temperatures, and the actual distances between steps can be much larger, say 1000 atomic distances.

Several quantities (hopping distances, sticking coefficients, ...) have been omitted from (11.8). Indeed, our purpose is just to show what the basic physics is, without bothering the reader with general but cumbersome formulae. More details can be found in Stoyanov & Kaschiev (1981). It should be noted that surface reconstructions can deeply alter this simple picture. For instance, Mo *et al.* (1991) found that on Si(001) the diffusion and sticking-to-step coefficient of adatoms are quite anisotropic, at least at low enough temperature (see chapter 9). Extreme anisotropy in diffusion is expected to change the exponent 1/6 in (11.9) to 1/4 (Pimpinelli *et al.* 1992, Bartelt & Evans 1993). An experimental test of this prediction has been achieved by Bucher *et al.* (1994).

11.6 Numerical simulations and controversies

The first verification of (11.9) has been attempted through numerical simulations. They are based on the following scheme: one starts from a flat surface, then sends atoms on it at a rate F, and lets them diffuse with a diffusion constant D until they are stopped by another adatom, or by a cluster. The adatom and cluster densities $\rho_1(t)$ and $N(t)$ are continuously monitored, as functions of the deposition time. In the stationary regime one should find $N \approx \ell_s^{-2}$. For small t one can approximately describe the process in the form of rate equations (Venables *et al.* 1984):

$$\dot{\rho}_1(t) = F - 2D\rho_1^2(t) - D\rho_1(t)N(t) \tag{11.10a}$$

$$\dot{N}(t) = D\rho_1^2(t) \tag{11.10b}$$

with the initial conditions $\rho_1(0) = N(0) = 0$.

The ρ_1^2 term describes dimer formation, while $\rho_1 N$ is proportional to the rate of adatom capture by islands. Equations (11.10) completely neglect island coalescence, and are therefore only valid for $Ft \ll 1$, say $Ft < \frac{1}{4}$ for definiteness.

The term proportional to ρ_1^2 is negligible at stationarity with respect to $\rho_1 N$. This is not true, however, for very small t: this is seen by solving (11.10) (Tang 1993), which yields $\rho_1(t) \simeq Ft$ and $N(t) \simeq \frac{1}{3}F^2Dt^3$. The two terms become of the same order at $t \approx t_1 = (DF)^{-1/2}$, in which case one finds $\rho_1(t_1) \approx N(t_1) \approx (F/D)^{1/2}$, whence

$$\ell_s \approx (D/F)^{\frac{1}{4}} . \tag{11.11}$$

If we neglect further terrace formation after the time t_1, we find formula (11.11) rather than (11.9) as we expected. Actually, if the ratio F/D is *not* quite small–$F/D > 10^{-3}$, say–new terraces do not form after $t = t_1$. This is probably the reason why De Miguel *et al.* (1987) and Irisawa *et al.* (1990) gave (11.11) as the correct result in the stationary regime. In fact, Monte Carlo simulations by L.-H. Tang (see Villain *et al.* 1992b, Tang 1993), shown in Fig. 11.2, have established formula (11.9). Note that these simulations do not include any mechanism to allow forming compact islands. Formula (11.1b) should rather be employed, so that instead of (11.9) one gets the following:

$$\ell_s \approx \left(\frac{D}{F}\right)^{1/(4+d_f)} . \tag{11.12}$$

From the same Fig. (11.2) one can tell that the difference between (11.12) and (11.9) is indeed small.

Simulations which disregard the island morphology–islands are treated as pointlike–also confirm (11.9) (Bartelt & Evans 1992).

The situation, however, is not completly settled in the case of fractal islands (Bales & Chrzan 1994).

11.7 Experiments

The main interest of formulae like (11.9) and (11.12) lies in the possibility of measuring D, since F is known. We recall that the diffusion of radioactive tracers does not give D, but the quantity (7.12), which is much smaller.

Mo *et al.* (1992) applied the first this idea to the measurement of D on Si(001) (see chapter 9): in this case, however, the situation is complicated by the reconstruction, and the consequent anisotropy of adatom diffusion and sticking to steps. Equation (11.9) is not applicable in this case, and comparison with numerical simulation is needed. A similar analysis has been performed by Stroscio *et al.* (1993) (see also Stroscio & Pierce 1994) for Fe on Fe(001), where diffusion is expected to be isotropic, and by Bucher *et al.* (1994) for the system Cu on Pd(110), where the transition from anisotropic to isotropic diffusion was investigated. Robinson *et al.* (1996) have measured the island density by X-ray diffraction in the anisotropic system Cu on Cu(110).

The main problem of experiments is that, before using (11.9) for measuring D, one is not sure that it actually applies! In other words, one has

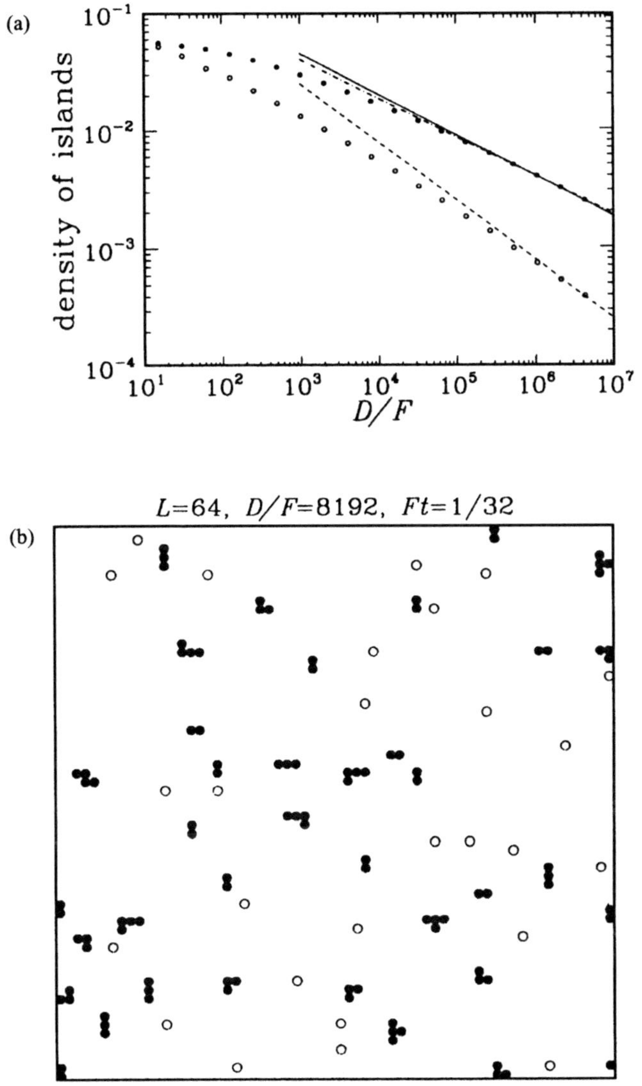

Fig. 11.2. a) The island density $N = 1/\ell_s^2$ as a function of D/F is shown as closed points, according to computer simulations by L.-H. Tang. The continuous, dash-dotted and dashed straight lines represent equations (11.9), (11.12) with $d_f = 1.7$, and (11.11), respectively. Open points represent the adatom density when it is maximum. It is well described by (11.11) (for large values of D/F), while ℓ_s follows (11.9) or (11.12). b) If one starts deposition on an ideal surface with no adatoms, after a time t_1 the number of adatoms and of adatom clusters become equal. At that time, the distance between islands is given by (11.11). This distance must be large enough for (11.9) or (11.12) to be verified at later times.

to check that the power γ in the relation $\ell_s \sim (D/F)^\gamma$, is really 1/6. One possible way to do this is to keep T (and thus D) fixed, and vary the beam intensity. This has been tried first on Cu(001) by Ernst *et al.* (1992b) who indeed measured ℓ_s at constant temperature ($T \sim 220$ K) and varying F. Unfortunately, they found $\ell_s \sim F^{-1/4}$ rather than $\ell_s \sim F^{-1/6}$! The measurement has been later repeated by Zuo *et al.* (1994), with a different technique (high-resolution low-energy electron diffraction instead of He-atom scattering) at various temperatures. At low temperature ($T \sim 223$ K), they have observed a crossover from a low-flux exponent 1/6 to a high-flux value 1/4. At higher temperatures (263 – 305 K) an exponent 3/5 has been seen. As we will see shortly, the latter values agree with a larger critical nucleus than the dimer, i.e. $i^* > 1$. According to Zuo *et al.*, the flux values of Ernst *et al.* lie already in the high-flux region; furthermore, even slightly above 220 K the assumption $i^* = 1$ starts to be questionable. We will now see what happens when dimers are not stable and i^* is no longer 1.

11.8 The case $i^* = 2$

We will now include two effects that were neglected up to this point.

i) When temperature increases, dimers become unstable, which means that their dissociation rate $1/\tau_2$ becomes too large to be neglected.
ii) Dimers start diffusing. Their diffusion constant D_2 must be included.

Trimers are now supposed to be stable. A kinetic equation for the dimer density ρ_2 is obtained equating the rate of change $\dot{\rho}_2$ to the sum of the following five terms, where logarithmic corrections are neglected.

i) The dimer creation rate is equal to the collision rate of adatoms (per site and unit time) $F\rho_1$. According to (11.2) and (11.6) one has

$$(\dot{\rho}_2)_1 = D\rho_1^2 \,. \tag{11.13}$$

ii) The reverse process (dimer dissociation) yields a term:

$$(\dot{\rho}_2)_2 = -\frac{\rho_2}{\tau_2} \,. \tag{11.14}$$

iii) The stable triplets nucleation rate due to adatom-dimer collisions (per site and unit time) is Fp_2. It also contributes to the kinetic equation for the dimer density with a minus sign. According to (11.2) and (11.6), one finds thus:

$$(\dot{\rho}_2)_3 = -D\rho_1\rho_2 \,. \tag{11.15}$$

iv) Dimer disappearance due to capture by an island gives a term

$$(\dot{\rho}_2)_4 = -\frac{\rho_2}{\tau_2^{\text{diff}}}$$

where $\tau_2^{\rm diff}$ is defined as the time needed for a dimer to diffuse to a terrace. To find this time, one replaces D by D_2 into (11.2). Therefore,

$$(\dot{\rho}_2)_4 = -\frac{D_2\rho_2}{\ell_s^2} \,. \tag{11.16}$$

v) Finally, we write the analogue of (11.13), with a minus sign, to account for collisions between diffusing dimers:

$$(\dot{\rho}_2)_5 = -D_2\rho_2^2 \,. \tag{11.17}$$

Adding all the terms above yields:

$$\dot{\rho}_2 = D\rho_1^2 - \frac{\rho_2}{\tau_2} - D\rho_1\rho_2 - \frac{D_2\rho_2}{\ell_s^2} - D_2\rho_2^2 \,. \tag{11.18a}$$

As in the preceding section, we look for a stationary regime, where this equation reads:

$$D\rho_1^2 - \frac{\rho_2}{\tau_2} - D\rho_1\rho_2 - \frac{D_2\rho_2}{\ell_s^2} - D_2\rho_2^2 = 0 \,. \tag{11.18b}$$

The island nucleation rate is provided by the sum of $-(\dot{\rho}_2)_3$ and $-(\dot{\rho}_2)_5$, that is,

$$\frac{1}{\tau_{\rm nuc}} \approx D\rho_1\rho_2 + D_2\rho_2^2$$

and according to (11.1):

$$D\rho_1\rho_2 + D_2\rho_2^2 \approx \frac{F}{\ell_s^2} \,. \tag{11.19}$$

We have now three unknown quantities, ρ_1, ρ_2 and ℓ_s, related by three equations. However, there is some ambiguity in the definition of ℓ_s: is it the average distance between islands *including dimers*, or not? The difference is, in fact, a trifle, because there are many more large terraces than dimers. In other words, the following always holds:

$$\rho_2\ell_s^2 \ll 1 \,. \tag{11.20a}$$

This relation will be proved later. Similarly, one might wonder whether D is the diffusion constant of adatoms on an ideal surface, or whether one should account for the possibility of an adatom being retarded during its random walk by taking temporarily part in a dimer. Again, this problem is not important if (11.20a) holds, because most adatoms reach an island without meeting other adatoms to form dimers.

Our next task is therefore to prove (11.20a). To do this, we show that the last term in (11.18b), as well as the second term in (11.19), can be neglected in an approximate evaluation, provided $D_2 < D$. In fact, it is enough to show that

$$D_2\rho_2 < D\rho_1 \,. \tag{11.20b}$$

Now, formula (11.18b) implies:

$$\rho_2 \approx \frac{D\rho_1^2}{1/\tau_2 + D\rho_1 + D_2/\ell_s^2 + D_2\rho_2} < \frac{D\rho_1^2}{D\rho_1} = \rho_1 \,.$$

Equation (11.20b) follows, provided $D_2 < D$. It is in fact enough that D_2 is not much larger than D.

At this point, (11.4), (11.19) and (11.20b) imply:

$$\rho_2\ell_s^2 \approx \frac{F}{D\rho_1} \approx \frac{1}{\ell_s^2} \,,$$

which proves (11.20a).

Using now (11.20b), and eliminating ρ_2 between (11.18b) and (11.19), yields:

$$\frac{D^2\rho_1^3}{1/\tau_2 + D\rho_1 + D_2/\ell_s^2} \approx \frac{F}{\ell_s^2}$$

and inserting equation (11.4) in this formula we finally find the equation specifying ℓ_s,

$$\boxed{F^2\ell_s^8/D \approx 1/\tau_2 + F\ell_s^2 + D_2/\ell_s^2 \,.} \tag{11.21}$$

The main merit of this equation is to be an algebraic one! One can even find simple power laws in certain limit cases (Villain *et al.* 1992a,b).

(a) $F\ell_s^2$ is larger than both $1/\tau_2$ and D_2/ℓ_s^2.

In this case one finds again the relation (11.9). If we now insert the latter into the two inequalities implied by (a), we find the following two conditions on the validity of (11.9):

$$DF^2\tau_2^3 > 1 \tag{11.22a}$$

$$D^2F > D_2^3 \,. \tag{11.22b}$$

This regime is always attained if F is large enough. Thus, if F is large enough the assumption $i^* = 1$ should always hold. This seemingly disagrees with the experimental finding of Zuo *et al.* (1994) mentioned in section 11.7.

(b) $1/\tau_2$ is larger than both $F\ell_s^2$ and D_2/ℓ_s^2.

In this case the relation (11.21) gives

$$\ell_s \approx \left(\frac{D}{F^2\tau_2}\right)^{\frac{1}{8}} . \quad (11.23)$$

Inserting (11.23) into the inequalities implied by (b) yields the following two conditions on the validity of (11.23):

$$DF^2\tau_2^3 < 1 \quad (11.24a)$$

$$F^2D_2^4\tau_2^5 < D . \quad (11.24b)$$

In principle, such conditions are always met if the beam intensity F is small. Again, this seems to contrast with Zuo *et al.* (1994).

(c) D_2/ℓ_s^2 is larger than both $F\ell_s^2$ and $1/\tau_2$.

In this case, (11.21) becomes:

$$\ell_s \approx \left(\frac{DD_2}{F^2}\right)^{\frac{1}{10}} . \quad (11.25)$$

The relation $\ell \sim F^{-1/5}$ agrees with Monte Carlo simulations by Jensen *et al.* (1994).

Inserting (11.25) into the inequalities implied by (c) yields the following two conditions on the validity of (11.25):

$$D^2F < D_2^3 \quad (11.26a)$$

$$F^2D_2^4\tau_2^5 > D . \quad (11.26b)$$

These are only possible if

$$D_2\tau_2 > \frac{D}{D_2} . \quad (11.27)$$

The relation (11.23) is given in Stoyanov & Kaschiev (1981) in a different form. First, they introduce a number of parameters which we have thoroughly put to 1. Second, and most important, they do not use explicitly D_2 and τ_2, but rather the activation energy for surface diffusion, W_{sd}, and the bond energy E_2 of a dimer. The energy W_{sd} is defined through

$$D = D_0 \exp\left(-\frac{W_{sd}}{k_BT}\right) \quad (11.28)$$

where D_0 is a constant which in our units has the dimension of a frequency. Our notations are recovered if we note that in (11.18b), the first two terms must satisfy detailed balance at equilibrium,

$$D\tau_2 = (\rho_2/\rho_1^2)_{eq} = \exp\left(\frac{E_2}{k_BT}\right) . \quad (11.29)$$

Note that the dimer bond energy E_2 is positive. We stress that (11.29) relates E_2 to the product $D\tau_2$, and not to τ_2 only, as one could believe at first sight.

11.9 Generalization

The preceding argument has been extended to the general case where i^* takes any integer value (Villain *et al.* 1992a,b). The details are given in appendix H. The only assumption concerns neglecting coalescence of clusters of less than i^* atoms. The following results are found:

(a) the order of magnitude of ℓ_s is given by an algebraic equation;
(b) as a function of temperature, $\ell_s(T)$ increases faster than an Arrhenius exponential;
(c) inside appropriate regions of the (T, F) plane, ℓ_s can be approximated by a power law $\ell_s \sim F^{-\gamma}$, where γ lies between 1/6 and 1/2. Within such regions, ℓ_s depends on T as an Arrhenius exponential, $\ell_s \sim \exp[-E/(k_B T)]$, *via* the diffusion constant of adatoms, D, those of n-mers, D_n, and their lifetimes τ_n.

Table 11.1. contains a summary of (twice) the exponents γ and the correspondent activation energies E which may be computed with the assumption of the existence of a critical nucleus of size i^*, for different physical situations. They are functions of the size i^* of the critical nucleus, of its formation energy E_{i^*}, and of the barriers for adatom diffusion, W_{sd}, and desorption W_v, and for cluster diffusion, $W_{sd}^{clusters}$. The notation '3D' indicates that three-dimensional islands are considered. References are Stoyanov & Kaschiev (1981), Venables *et al.* (1984), Villain *et al.* (1992a,b), Bartelt & Evans (1992,1993) and Jensen *et al.* (1997). Note that strictly speaking, on a square lattice the critical nucleus size is a well

Table 11.1. *Possible values of the exponent 2γ and activation energies E for different physical regimes.*

Regime	2γ	E
iso. adatom diffusion	$i^*/(2+i^*)$	$(E_{i^*} + i^* W_{sd})/(2+i^*)$
fractal islands	$2/(4+d_f)$	$2W_{sd}/(4+d_f)$
aniso. adatom diff.	$i^*/[2(1+i^*)]$	$(E_{i^*} + i^* W_{sd})/[2(1+i^*)]$
adatom desorption	$2i^*/3$	$[2E_{i^*} - W_{sd} + (2i^*+1)W_v]/3$
vicinal surface	$i^*/2$	$(E_{i^*} + i^* W_{sd})/2$
diffusing clusters	$2/5$	$(W_{sd} + W_{sd}^{clusters})/5$
iso. adatom diff. (3D)	$2i^*/(5+2i^*)$	$2(E_{i^*} + i^* W_{sd})/(5+2i^*)$
adatom desorption (3D)	$2i^*/3$	$2[E_{i^*} - W_{sd} + (i^*+1)W_v]/3$

defined concept only for $i^* = 1$ (Ratsch *et al.* 1994, Bartelt *et al.* 1995) as we discuss in section 11.11 below. In a real experiment, which seldom spans more than two or three orders of magnitude in F/D, one is likely to observe the apparent power laws mentioned in point (c) above.

11.10 Diffraction oscillations in MBE

As we anticipated in chapter 5 (see Fig. 5.4a), ideal layer-by-layer growth is a cyclical process: terraces nucleate, grow, coalesce, nucleation takes place again, and so on. Since the early 1970s, well before widespread use of MBE, crystal growers were well aware of the oscillating behaviour of growth rates, and of its relation to layer-by-layer growth (Bostanov *et al.* 1975).

The first report of oscillatory behaviour in MBE was, not by chance, by a group working on growth properties of GaAs (Harris *et al.* 1981a), whose superior optical properties make it a good candidate for applications in optoelectronics. The experiment, however, was complicated by the presence of a dopant, and the relation of the oscillations with the actual morphology of the growing surface was overlooked, even though the authors established the equality between the frequency of the oscillations and the growth rate. But, oscillations of what? Experimentalists employed, as they still do, the reflection high-energy electron diffraction (universally known as RHEED) pattern, as a probe of the surface reconstruction, and therefore of the surface quality. Textbooks tell us that the diffraction from an arrangement of three-dimensional objects is made of spots, whereas that of two-dimensional objects is made of rods; the length of these rods was seen to oscillate during MBE depositions. What makes RHEED a very useful tool for MBE study is the experimental geometry: high-energy electrons are required for observing at grazing incidence, which allows observations *in situ*, i.e. during deposition by the molecular beam.

However, unlike X-ray and He atom scattering, high-energy electron diffraction, as it turns out, is far from being described by the kinematic approximation, multiple diffractions being non-negligible. The kinematic approximation would amount to saying that electrons diffracted by terraces at different levels above the surface interfere, so that a flat surface reflects specularly the incident beam, while from a multilevel (rough) surface the diffracted beams cancel and the signal falls.

An important contribution to understanding the experiment was given by Van Hove *et al.* (1983), who showed that RHEED oscillations are most pronounced in the out-of-phase condition, which is the most sensitive to surface steps. This suggests that electrons are randomly diffracted by step edges, and that the important oscillating quantity is in fact the total step length per unit surface (or 'step density'). Incidentally, Van Hove *et al.* also made the interesting observation of an increase of the oscillation period at high temperature, which they correctly ascribed to adatom desorption. Later Van Hove & Cohen (1985) and Kojima *et al.* (1985) reported RHEED oscillations for evaporating GaAs, showing that evaporation may also take place in a layer-by-layer mode.

Whatever the precise origin of RHEED oscillations–destructive interference from different surface levels, or scattering from surface steps–it was clear that they are the signature of layer-by-layer growth. In a typical experiment some ten oscillations can be seen before they are damped away with passing time (i.e. increasing thickness of the deposited layer). Special growth conditions allow to see hundreds or even thousands of them. However, increasing the temperature usually leads to a complete disappearence of the oscillations, which are not visible since the very beginning of growth. This phenomenon is taken to mean that the growth mode is changing from island nucleation (oscillating RHEED pattern) to step flow (no oscillations) (Neave *et al.* 1985). This behaviour was reproduced with computer simulations (Monte Carlo) by Clarke & Vvedensky (1987). In the simulations, the oscillating quantity is the step density, whose variations match very accurately those of the RHEED pattern.

Further evidence supporting the close relation between step density and RHEED comes from the recent work by Johnson *et al.* (1993), which couples reciprocal and direct space techniques. A GaAs(001) crystal is grown by MBE and its RHEED oscillations monitored: in correspondence of maxima and minima the deposition is stopped, and after quenching to preserve the surface morphology, the surface is looked at with an STM. The conclusion is that the step density oscillates in the same fashion as the RHEED intensity, and that the eventual damping and disappearence of the oscillations at long times is determined by the setting in of a stationary (i.e. time-independent) step density. This is interpreted as a dynamical transition to a kind of step flow mode of growth, where the surface width (a function of the number of exposed layers) does not increase, or does it only very slowly.

11.11 Critical nucleus size and numerical simulations

In section 11.6 we described computer simulations aimed to very low temperature, where dimers are stable. On a square lattice, and with nearest-neighbour interactions, this means that the thermal energy is not high enough to break a single atomic bond. As a consequence, all adatom clusters are stable, and i^* has a well-defined meaning. On the contrary, when the dimer becomes unstable, the trimer and all other clusters containing singly bonded atoms become unstable as well. The tetramer and all clusters containing doubly bonded atoms are instead stable, and the concept of a critical size above which all clusters are undissociable loses its meaning.

Therefore, on a square lattice one expects of observing only three regimes: a low-temperature regime where an atomic bond cannot be broken and $i^* = 1$; a crossover region where singly bonded atoms dissociate, but doubly bonded ones do not; a high-temperature regime where doubly bonded atoms dissociate, and no stable island exists (Ratsch *et al.* 1994). According to Bartelt *et al.* (1995), the crossover regime can lock to a kind of '$i^* = 3$' behaviour when the tetramer is the smallest stable island; larger islands, e.g. a pentamer, which contain singly coordinated atoms, are unstable, however the adatom created by the dissociation of such an island has a greater probability of making again a bond with its mother island than with any other cluster. This allows the existence of a temperature region where exponents corresponding to $i^* = 3$ should be observed, if the nearest-neighbour bond energy is larger than the surface diffusion barrier (Bartelt *et al.* 1995). This energy is at best an effective one in the case of fcc metals, which nevertheless seem to provide a confirmation of these results.

11.12 Surfactants

In MBE growth, island nucleation in the first layer is not the end of the story. Smooth, layer-by-layer growth requires nucleation of the second and further layers to take place at a sufficiently late stage, compared to coalescence in the previous layers. This is usually the case at high temperature (Figs. 11.3 and 11.4a). At lower temperatures, non-smooth growth, with nucleation of three-dimensional, pyramid-like island (Figs. 11.3 and 11.4b) is often observed (Kunkel *et al.* 1990, van der Vegt *et al.* 1992, Ernst *et al.* 1994).

The birth of pyramids is a kinetic phenomenon. It is probably due to the Schwoebel effect introduced in chapter 6: once the 'first floor' of an island is formed, atoms falling on top of it are trapped there by the step-edge barrier and the probability of nucleating the second floor before coalescence of the first one, is very high. The higher the temperature the less effective the barrier, which explains the occurence of layer-by-layer growth at high T. More surprising is the character of growth at very low T. The appearance of (strongly damped) diffraction oscillations (Fig. 11.3) points to a quasi-2D growth mode ('reentrant' layer-by-layer growth). This is confirmed by STM imaging and computer simulations (Fig. 11.4c), which show the surface covered by many, small, two-dimensional islands. The smallness of the islands, which implies a high perimeter-to-area ratio, should allow falling atoms to overcome the

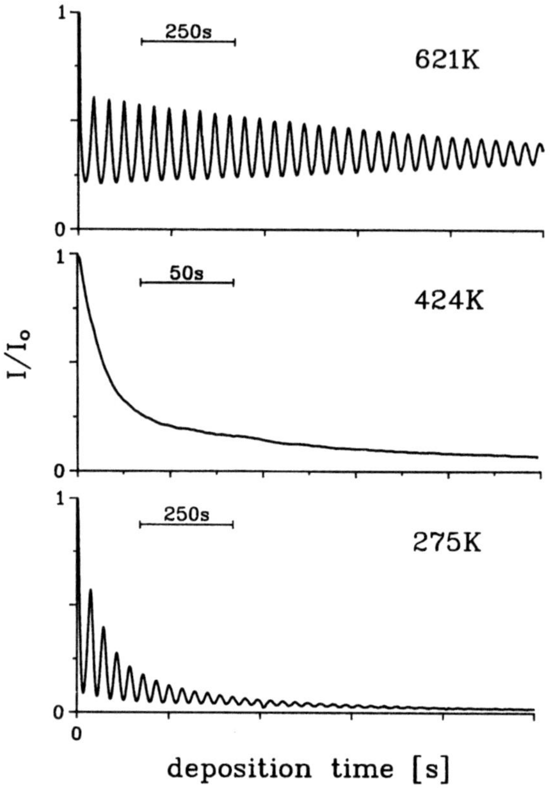

Fig. 11.3. Diffracted intensity in thermal atom scattering during deposition of Pt on a Pt(111) surface at three temperatures. At high and at low T oscillations can be seen, while they disappear at intermediate temperature (from Kunkel *et al.* 1990, with permission of the authors).

Schwoebel barrier, if one assumes that atoms hitting near an island edge are easily incorporated into it, e.g. by exchange with a step atom (Šmilauer *et al.* 1993a).

Much more effective in inducing layer-by-layer growth at low T are the so-called *surfactants*. Steigerwald *et al.* (1988), who investigated growth of pseudomorphic fcc-Fe films on Cu(100) (iron is normally bcc), noted that 'under the influence of adsorbed gases' a 3D Fe film could 'be induced to lie flat.' In fact, such 'surfactants' are adsorbed impurities that exert a smoothing action on a growing surface, and keep floating at the surface all along the growth process.

A well-known case study is Ag on Ag(111) (van der Vegt *et al.* 1992). Three-dimensional growth is observed from 225 up to 575 K at a deposition rate of 0.02 ML s^{-1}. Deposition of approximately 0.15 ML of Sb previous to silver deposition makes the film to grow layer-by-layer even

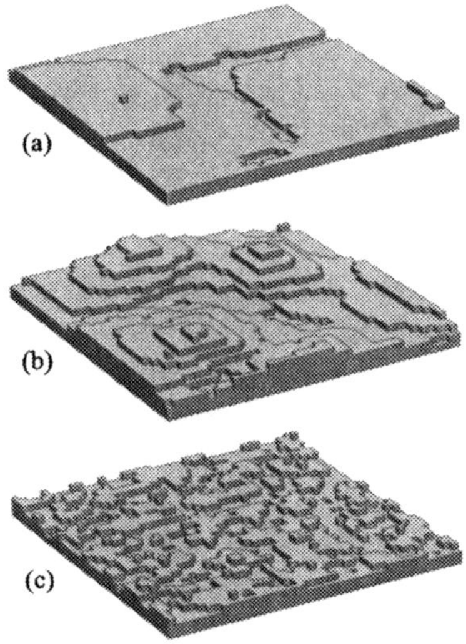

Fig. 11.4. Crystal surface during atom deposition, according to computer simulations by Šmilauer *et al.* (1993a). a) High temperature. b) Intermediate temperature. c) Low temperature. Note that the amplitude of spatial height fluctuations is maximum in panel (b) (with permission by the authors).

at 225 K (van der Vegt *et al.* 1992). It is likely that antimony favours interlayer transport in two ways: i) by acting as a germ for heterogeneous nucleation of Ag islands, it may keep their size small and allow overcoming of step-edge barriers as discussed above in the case of reentrant 2D growth. ii) By binding at island edges, Sb may at the same time reduce the size of the critical nucleus and favour incorporation of incoming Ag atoms by exchange, effectively lowering the step-edge barrier. The latter scenario has been investigated both theoretically and experimentally by Tersoff *et al.* (1994). The smoothing action of an artificially increased density of nuclei has been directly shown by Rosenfeld *et al.* (1993).

The term surfactant refers thus here to something quite different from standard use; we know surfactants to be chemical agents that by sitting at an interface (e.g. detergents at the oil-water interface) affect interfacial free energies and properties, like wetting. In the case of MBE, 'surfactants' rather affect kinetics than thermodynamics. It is true that this term was first used in connection with Ge/Si heteroepitaxy, where such impurities as As were shown to allow wetting of the silicon substrate by a germanium film (Copel *et al.* 1989, 1990). But even in that case, the 'surfactants' do not stay at the interface between the two crystals, and must float at the growth surface to exert their action. In any case, the term 'surfactant' is now common usage, and we will conform.

Some light on surfactant behaviour has been shed by an extensive study by Voigtländer *et al.* (1995). By monitoring the island density during silicon deposition in the presence of various surfactants, they were able to correlate an increased island density in Si homoepitaxy, with the suppression of 3D islanding in Ge/Si. The situation can be summarized as follows.

1) In Ge/Si deposition without surfactant, one reaches the equilibrium state (the substrate temperature is typically 700 K), namely: three layers of Ge, then 3D islands.
2) With 'good' surfactants (As or Sb), one can grow 30 layers of Ge before 3D islands appear. In Si homoepitaxy, As and Sb increase the 2D-island density.
3) With 'bad' surfactants (In or Ga), the Ge/Si system behaves as in 1. In Si homoepitaxy, In and Ga decrease the 2D-island density.
4) The effect of a good surfactant is presumably to reduce the attachment probability of adatoms on islands and therefore (because of the detailed balance condition) the detachment rate of adatoms from islands.

In fact, it is likely that reducing detachment from the island edge also explains the reduced size of the critical nucleus in homoepitaxy, and thus the effectiveness of such impurities on smoothing the growth front.

In the simplified picture presented in this chapter, some possibly important mechanisms with a smoothing action have been neglected. In particular (Evans 1991): i) The 'knock out' effect, in which an atom from the beam hits an atom at a terrace edge, pushes it aside and ends up inside the terrace (Fig. 11.5). ii) The 'funnel' effect, in which an atom from the beam lands on an inclined facet and goes down because the site with highest coordination is at the bottom of the facet. Both effects allow the Schwoebel barrier to be bypassed and facilitate filling crevasses and healing roughness.

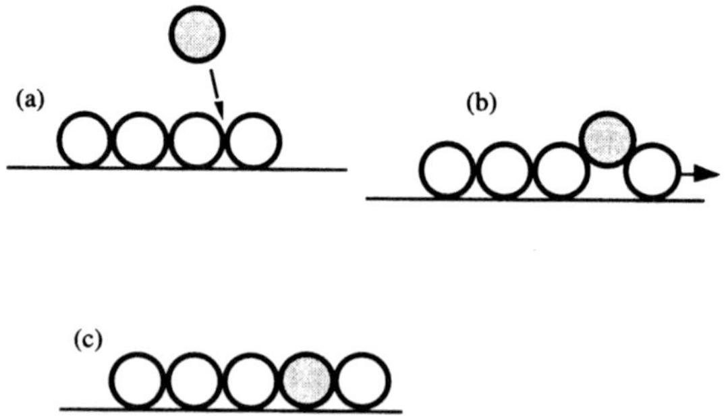

Fig. 11.5. Schematic view of the 'knock out' effect leading to non-activated incorporation of a deposited atom.

12
Growth roughness at long lengthscales in the linear approximation

Grau, teurer Freund, ist alle Theorie,
Und grün des Lebens goldner Baum.

Grey, dear friend, is any theory and green is the golden tree of life.

J. W. Goethe (*Faust*)

A surface can be rough, as seen in chapter 1, because of thermal fluctuations. It can also be rough as an effect of the fluctuations of the beam intensity. In order to study this kind of roughness, one should make averages over beam fluctuations instead of thermal fluctuations. However, while in statistical physics, the probability p of any configuration is given at equilibrium by the Gibbs formula $p = (1/Z)\exp(-\beta E)$, there is no such magic formula, nor any variational principle, in the new statistical mechanics which we shall now discover. In the present chapter, a linear model will be exactly solved, which is a good approximation in certain cases. Certain theorists would probably say that a linear problem is uninteresting, and therefore tedious. However, there are linear problems which are very difficult, e.g. the localization of non-interacting electrons. Then, if by chance we have the opportunity to solve a problem exactly, we should not be ashamed to do it!

12.1 What is a rough surface?

As in the preceding chapter, we consider here a crystal limited by a high-symmetry plane ($\{001\}$ or $\{111\}$) below its roughening transition temperature. If such a crystal is made to grow (for instance by MBE), steps form on its surface as explained in the preceding chapter, where the order of magnitude ℓ_s of their average distance has been calculated. In this chapter and in the following two, we want to study what happens at distances much larger than ℓ_s.

At the beginning of this book, growth was described as a deterministic process. The main purpose of the present chapter is to go beyond this deterministic view. We have already done so in chapter 11, but the consequences of fluctuations on structural disorder on the surface were not derived. Growth by molecular beam epitaxy will be specifically addressed, and the fluctuations which will be taken into account are those of the beam intensity. These beam fluctuations are often called 'shot noise' in the specialists' jargon, because true noise (the one you can actually hear, especially in electronic devices), is associated with random processes which are similar to the beam fluctuations of interest here.

The main effect of these beam fluctuations is the following: a growing surface becomes rough, even if it is below its roughening transition temperature and if it is initially smooth. The word 'rough' is used here in a sense analogous to chapter 1. That is, the height-height correlation function

$$G(\mathbf{r},t) = \left\langle \left(z(\mathbf{r}',t) - z(\mathbf{r}'+\mathbf{r},t)\right)^2 \right\rangle \tag{12.1}$$

goes to infinity when r and t go to infinity. In this formula, $\mathbf{r}$ and $\mathbf{r}'$ denote vectors of a fixed plane (say xy) and z denotes the height of the surface in the direction perpendicular to this plane. The mean value represented by $\langle ... \rangle$ in (12.1) is not a thermal average as in chapter 1, but an average on all random processes: beam intensity, sticking, diffusion, nucleation of new terraces. The initial state of the surface is assumed to be planar:

$$z(\mathbf{r},0) = 0, \quad G(\mathbf{r},0) = 0 \,. \tag{12.2}$$

If $G(\mathbf{r},t)$ diverges when $\mathbf{r}$ and t go to infinity, the surface will be said to be kinetically rough. Note that this expression has been used by the Dutch school, and in section 5.7, with a different (and less precise) meaning, appropriate to growth in a weak supersaturation.

Note that (12.1) is an equal-time correlation function. One might also introduce the correlation function $\Gamma(\mathbf{r},\tau;t)$ between heights at $\mathbf{r}$ and $\mathbf{r}'+\mathbf{r}$ at the respective times t and $t+\tau$. This complication would not be very useful, however.

12.2 Random fluctuations and healing mechanisms

In the present chapter, we are mainly interested in the effect of random fluctuations of the beam intensity $f(\mathbf{r},t)$ in MBE growth. The beam intensity will be assumed to be the sum of a constant part F and a random part $\delta f(\mathbf{r},t)$, which has no correlations in space and in time. Thus, if $\mathbf{r} = (x,y)$ denotes coordinates perpendicular to the beam direction (the length unit being as always the atomic distance), one has

$$\langle \delta f(\mathbf{r},t)\delta f(\mathbf{r}',t')\rangle = F\delta_{\mathbf{r},\mathbf{r}'}\delta(t-t') \,. \qquad (12.3)$$

The factor F appears because the total mean square fluctuation in time t at any site is equal to the average number of atoms falling on this site, as it should be:

$$\int_0^t \mathrm{d}t' \int_0^{t'} \mathrm{d}t'' \,\langle \delta f(\mathbf{r},t')\delta f(\mathbf{r},t'')\rangle = Ft \,. \qquad (12.4)$$

It is of interest to consider briefly the case of an SOS model without diffusion or evaporation: atoms fall randomly onto sites $\mathbf{r}$ of a two-dimensional lattice, form piles (Fig. 1.8) and do not move any more. Such a model has some resemblance with what happens at very low temperatures, when there is no evaporation and very little diffusion: it would lead to an extremely rough surface. Indeed, the correlation function (12.1), for $\mathbf{r} \neq 0$, is just equal to (12.4) multiplied by a factor 2 (because the fluctuation in $\mathbf{r}+\mathbf{r}'$ is to be added to the fluctuation in $\mathbf{r}'$):

$$G(\mathbf{r},t) = \left\langle \left(z(\mathbf{r}',t) - z(\mathbf{r}'+\mathbf{r},t)\right)^2 \right\rangle = 2Ft\left[1-\delta_{0,\mathbf{r}}\right] \,. \qquad (12.5a)$$

Fortunately, if the temperature is appropriately chosen, the formation of a smooth surface is warranted by matter transport which carries the atoms where they should be. These healing mechanisms are mainly those discussed in chapter 8: a) surface diffusion, b) evaporation, c) bulk diffusion. In addition, if vacancies form, their incorporation in the bulk may help the smoothing of the surface, even though the resulting crystal is no longer ideal. This mechanism will be neglected although it is an essential feature of a theoretical model by Meakin *et al.* (1986), which has the great interest of being exactly soluble. Diffusion in the bulk will also be neglected.

In the growth of elements (Si, Ge, metals), surface diffusion is the main healing mechanism. Evaporation may be important for the growth of binary or ternary systems like GaAs, GaAlAs or CdTe.

Before forgetting formula (12.5a), one should note that it may capture some of the physics of columnar growth, which often takes place in the manufacturing of tapes for magnetic recording (Abelmann 1994). We may imagine that the first layer forms islands as in chapter 11, but they are not in registry with the substrate (which is not crystalline anyway). Thus, the islands do not fit together at coalescence, and the subsequently added material forms columns. Speculatively, one may imagine that diffusion is not easy from column to column, so that a kind of non-correlated roughness similar to formula (12.5a) develops. Of course, in contrast with (12.5a), there are correlations inside each column. When the beam is not perpendicular to the surface, shadowing effects appear, whose genesis would be difficult to understand in the absence of hindrance to diffusion. A theory of the late stages of shadowing has been formulated by Bales & Zangwill (1990b).

12.3 A subject of fundamental, rather than technological, interest

We now give an evaluation of the order of magnitude of the effect of beam intensity fluctuations in a realistic case. This evaluation is only correct in certain limits, but it has the advantage of being simple. We consider a deposition time such that an average number of h atomic layers have been grown on the initially flat surface. We assume that diffusion is able to heal the surface on a distance ℓ_s and has no effect on longer distances (this is an approximation). The distance ℓ_s depends on temperature and has been evaluated in chapter 11. The average number of atoms deposited on a square of size $\ell_s \times \ell_s$ is $\ell_s^2 h$ and, since it is a sum of independent, random numbers equal to 0 or 1, its mean square deviation is also $\ell_s^2 h$. The fluctuation of h is therefore

$$\delta h \simeq \sqrt{\ell_s^2 h}/\ell_s^2 = \sqrt{h}/\ell_s\,. \tag{12.5b}$$

Assume for instance that, at some temperature, diffusion is able to heal the surface on a length $\ell_s \approx 100$ atomic distances. Then, the height fluctuation is not larger than 1 atomic distance, except for samples thicker than 10 000 atomic layers–a huge thickness in MBE. Now, a healing length of 100 atomic distances is quite small. In usual MBE conditions, ℓ_s can be much larger and δh will be even smaller. We conclude that, in usual MBE, beam fluctuations are presumably not important. The roughness observed in certain experiments, such as in Si growing on Si at low temperature (Eaglesham & Gilmer 1992) is most likely due to the instability caused by the Schwoebel effect as discussed in section 6.7.

However, kinetic roughness is a fascinating subject. It corresponds indeed to a new branch of statistical mechanics, which developed in the last three decades of the twentieth century. This is a statistical mechanics where thermal fluctuations are replaced by fluctuations of a different nature (beam intensity, diffusion...). In the new statistical mechanics, there is no free energy to minimize, and the steady state (if there is one) is not given by any magic formula similar to the Gibbs formula $p = (1/Z)\exp(-\beta E)$ of equilibrium statistical mechanics. It will be seen in the next chapters, however, that certain methods of usual statistical mechanics, e.g. the renormalization group, can be applied–and even succesfully sometimes...!

Thus, we persist in studying kinetic roughness because of its fundamental interest, although its technological importance is presumably minor. This is the program of the remainder of this chapter and of the following two.

12.4 The linear approximation (Edwards & Wilkinson 1982)

In the calculation presented in the preceding section, the effect of diffusion at long distances was not properly accounted for.

If one wants to do that, the simplest idea is to use the same equations as in chapter 8. In the case of a non-singular surface (i.e. above T_R) the relevant equation was (8.14). In the case of growth, we assume that the surface is always rough, even if it is below T_R. This assumption will be checked to be consistent with the final result. Thus, we just add the beam intensity $f(\mathbf{r},t)$ to the right hand side of (8.14), obtaining

$$\boxed{\dot{z}(\mathbf{r},t) = f(\mathbf{r},t) + \nu\nabla^2 z - K\nabla^2(\nabla^2 z)\,.} \tag{12.6}$$

We shall see that this equation is in general not correct, but does apply in special cases. The model defined by this equation is called the *Edwards-Wilkinson model.*

Equation (12.6) is linear and invariant under translation, and can therefore be solved by introducing the Fourier transform $z_q(t)$ of $z(\mathbf{r},t)$. One obtains

$$\dot{z}_q(t) = f_q(t) - (\nu q^2 + Kq^4)z_q(t)\,. \tag{12.7}$$

The solution of this equation for $t > 0$ is

$$z_q(t) = z_q(0)e^{-(\nu q^2+Kq^4)t} + \int_0^t dt' f_q(t')e^{-(\nu q^2+Kq^4)(t-t')}\,.$$

Since $z_q(0) = 0$, the correlation function is

$$\langle z_q(t)z_{-q}(t)\rangle = \int_0^t \mathrm{d}t' \int_0^t \mathrm{d}t''\,\langle f_q(t')f_{-q}(t'')\rangle\,\mathrm{e}^{-(\nu q^2+Kq^4)(2t-t'-t'')}$$

or, using (12.3),

$$\begin{aligned}\langle z_q(t)z_{-q}(t)\rangle &= F\int_0^t \mathrm{d}t' e^{-(\nu q^2+Kq^4)(2t-2t')}\\ &= F\frac{1-\exp\left[-2(\nu q^2+Kq^4)t\right]}{\nu q^2+Kq^4}\,.\end{aligned} \tag{12.8}$$

The ordinary space correlation function (12.1) is obtained by inverting the Fourier transformation. It will be useful to assume that the space has d dimensions, where d is not necessarily equal to 3. The surface, as well as the space where $\mathbf{r}$ and $\mathbf{r}'$ live, is then $(d-1)$ dimensional, and the inverse Fourier transformation introduces integrals in a $(d-1)$-dimensional reciprocal-space. With the notation $d' = d - 1$, one has

$$G(\mathbf{r}, t) \simeq 2F \int \mathrm{d}^{d'} q \sin^2\left(\frac{\mathbf{q} \cdot \mathbf{r}}{2}\right) \frac{1 - \exp[-2(vq^2 + Kq^4)t]}{vq^2 + Kq^4} \tag{12.9}$$

where $d' = d - 1 = 2$ in the experimentally interesting case. This equation solves the problem. Some consequences will be derived in the next sections, and a more complete analysis is done in appendix I.

12.5 Lower and upper critical dimensions

The denominator in (12.9) vanishes for $q = 0$. This implies a divergence of the integral in the limit $r, t \to \infty$, for certain values of d. Indeed, for $r = \infty$, the sine squared can be replaced by $1/2$ and the exponential by 0, so that

$$\left\langle [z(0, \infty) - z(\infty, \infty)]^2 \right\rangle \simeq 2F \int \mathrm{d}^{d'} q \frac{1}{vq^2 + Kq^4} \, .$$

The integral diverges for $d \leq d_c^u$, where

$$d_c^u = 2 + 1 \quad (\text{if } v \neq 0) \tag{12.10a}$$

$$d_c^u = 4 + 1 \quad (\text{if } v = 0) \, . \tag{12.10b}$$

We write 2+1 instead of 3 because many authors, when talking of the space dimension, refer to the surface dimension d' (which they call d) instead of the true ambient space dimension, that we call d. In order to avoid confusion, we make explicit the value of the surface dimension d' and add the information '+1'.

The dimension d_c^u is called the *upper critical dimension.* It is the dimension above which the surface is smooth in the sense given in section 12.1. Above d_c^u, the model (12.6) would not be self-consistent, but in our 3-dimensional space, this does not matter since (12.10) is larger than or equal to 3.

The consistency of the model (12.6) does not only require that the surface be rough, but also that the fluctuation (12.9) be finite for any finite r as $t \to \infty$. Replacing again the exponential in (12.9) by 1, but now the sine by its argument, one obtains in order of magnitude

$$\left\langle [z(\mathbf{r}', t) - z(\mathbf{r}' + \mathbf{r}, t)]^2 \right\rangle \approx Fr^2 \int \mathrm{d}^{d'} q \frac{1}{v + Kq^2} \, .$$

The integral over the Brillouin zone is always finite if $\nu \neq 0$. However, if $\nu = 0$, the integral diverges when $d' \leq 2$. The convergence of the integral for finite r and infinite t defines the *lower critical dimension* d_c^ℓ,

$$d_c^\ell = 2 + 1 \qquad (\text{if } \nu = 0)\,. \tag{12.11a}$$

Since the physical dimension is $d = 3$, the model (12.6) is seen to be unphysical, at least at long times, when ν vanishes.

For $\nu \neq 0$, the integral which should be finite is $\int d^{d'}q \sim \int dq q^{d'-1}$, which diverges for $d' \leq 0$. Although this value has no physical meaning, it may be of interest to note that

$$d_c^\ell = 0 + 1 \qquad (\text{if } \nu \neq 0)\,. \tag{12.11b}$$

The interesting case is when $d_c^\ell < d < d_c^u$. In the rest of this chapter, this condition is assumed to be fulfilled.

12.6 Correlation length

If $d < d_c^u$, the correlation function (12.9) goes to infinity in the limit $r, t \to \infty$. However, it is finite for any finite time t and has the form sketched in Fig. 12.1. One defines the *correlation length* $\xi(t)$ as the value of r beyond which the quantity (12.9) is close to its asymptotic value

$$\begin{aligned}\left\langle [z(0,t) - z(\infty,t)]^2 \right\rangle &\simeq F \int d^{d'}q \frac{1 - \exp\left[-2(\nu q^2 + K q^4)t\right]}{\nu q^2 + K q^4} \\ &\approx F \int_{\nu q^2 + K q^4 > 1/t} \frac{d^{d'}q}{\nu q^2 + K q^4}\,.\end{aligned} \tag{12.12}$$

A straightforward calculation shows that ξ is given by

$$\nu \xi^{-2} + K \xi^{-4} \approx 1/t\,. \tag{12.13}$$

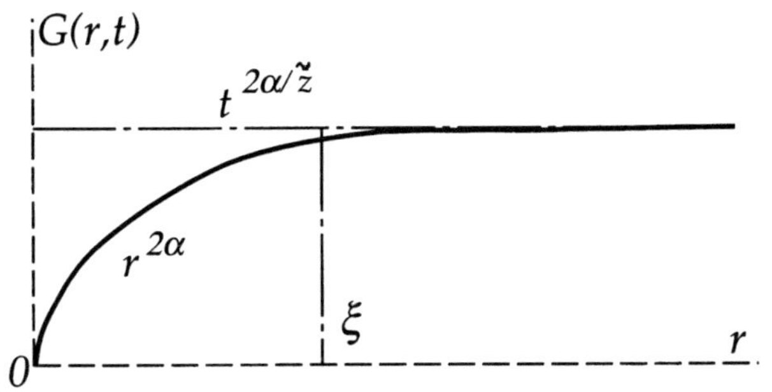

Fig. 12.1. Behaviour of the height-height correlation function $G(\mathbf{r}, t)$ as a function of the distance r parallel to the surface at a given time t.

In the long time limit, ξ behaves algebraically as a function of time, namely

$$\boxed{\xi(t) \sim t^{1/\tilde{z}}} \tag{12.14}$$

where the exponent $\tilde{z}$ is independent of d:

$$\tilde{z} = 2 \quad (\text{if } v \neq 0) \tag{12.15a}$$

$$\tilde{z} = 4 \quad (\text{if } v = 0)\,. \tag{12.15b}$$

Such power laws will be encountered, with different exponents, in the more complicated models addressed in the next two chapters. Moreover, the correlation function (12.1) also varies as a power of r at infinite time, and as a power of time at infinite distance, as we will now show in the case of the model (12.6).

12.7 Scaling behaviour and exponents

Consider the correlation function (12.9) at infinite time. It can be approximated as follows:

$$\begin{aligned}
\left\langle \left[z(\mathbf{r}',\infty) - z(\mathbf{r}'+\mathbf{r},\infty)\right]^2 \right\rangle &\simeq 2F \int \mathrm{d}^{d'}q \sin^2(\mathbf{q}\cdot\mathbf{r}/2) \frac{1}{vq^2+Kq^4} \\
&\approx 2F \int_{qr<1} (\mathbf{q}\cdot\mathbf{r}/2)^2 \frac{\mathrm{d}^{d'}q}{vq^2+Kq^4} + F \int_{qr>1} \frac{\mathrm{d}^{d'}q}{vq^2+Kq^4} \\
&\approx 2Fr^2 \int_0^{1/r} \frac{q^{d'-1}\mathrm{d}q}{v+Kq^2} + F \int_{1/r}^{1} \frac{q^{d'-1}\mathrm{d}q}{vq^2+Kq^4}\,.
\end{aligned} \tag{12.16}$$

In the limit of long distances r, this formula yields for $v \neq 0$:

$$\left\langle \left[z(\mathbf{r}',\infty) - z(\mathbf{r}'+\mathbf{r},\infty)\right]^2 \right\rangle \approx \frac{2F}{v} r^{2-d'} = \frac{2F}{v} r^{3-d}\,. \tag{12.17}$$

This formula is easily seen to be also correct at finite times if $r < \xi$. Quite generally, one can find for most growth models

$$\boxed{\left\langle \left[z(\mathbf{r}',t) - z(\mathbf{r}'+\mathbf{r},t)\right]^2 \right\rangle \approx r^{2\alpha}} \tag{12.18}$$

if $r < \xi(t)$.

This behaviour will also be obtained for the more complicated models introduced in the next two chapters. For the linear model (12.6), comparison of (12.18) and (12.17) yields for $\nu \neq 0$:

$$\alpha = (3 - d)/2 \,. \tag{12.19}$$

The physical case $d = 2 + 1$ is the upper critical dimension, as we already remarked. It is said to be 'marginal'. Formula (12.6) yields in this case (Edwards & Wilkinson 1982)

$$\left\langle \left[z(\mathbf{r}', t) - z(\mathbf{r}' + \mathbf{r}, t)\right]^2 \right\rangle \approx \frac{\pi F}{2\nu} \ln r \qquad (r < \xi) \,. \tag{12.20}$$

According to the definition of ξ, the correlation function (12.1) should be given for $r \geq \xi$ by

$$\left\langle \left[z(\mathbf{r}', t) - z(\mathbf{r}' + \mathbf{r}, t)\right]^2 \right\rangle \approx t^{2\beta} \qquad \text{if} \quad r \geq \xi(t) \tag{12.21}$$

where

$$\beta = \alpha/\tilde{z} \,. \tag{12.22}$$

Relation (12.22) is easily obtained by equating (12.21) to (12.18) for $r = \xi = t^{1/\tilde{z}}$. Relation (12.21) is easily proved (see appendix I) in the case of the Edwards-Wilkinson model (12.6) for $\nu \neq 0$ and $d < 2 + 1$. For $d = 2 + 1$, one finds

$$\left\langle \left[z(\mathbf{r}', t) - z(\mathbf{r}' + \mathbf{r}, t)\right]^2 \right\rangle \simeq \frac{F}{2\nu} \ln(\nu t) \qquad (r > \xi) \,. \tag{12.23}$$

Within the linear approximation (12.6), power laws are obtained only for non-physical values of the space dimension d. It is not so in the non-linear problems considered in the next two chapters, as shown by Monte Carlo simulations by Family & Vicsek (1985, 1991).

Formulae (12.18) and (12.21) are similar to the scaling laws which apply at equilibrium at a critical point. They are typical of a 'self-similar' system, i.e. a system where the correlation functions have forms independent of the length unit. More precisely, if the length unit in the xy plane is changed, and if (12.18) and (12.21) hold, it is possible to find a length unit in the z direction, such that (12.1) is the same. At equilibrium, one

Table 12.1. *Critical dimensions and exponents for the linear model (12.6).*

	d_c^u	d_c^ℓ	$\tilde{z}$	$\alpha = \chi = \zeta$	β
$\nu \neq 0$	$2+1$	$0+1$	2	$(3-d)/2$	$(3-d)/4$
$\nu = 0$	$4+1$	$2+1$	4	$(5-d)/2$	$(5-d)/8$

has to force the system to the critical state, through an appropriate choice of the temperatures and external fields. In the present case, the system evolves spontaneously towards a self-similar state. This is 'self-organized criticality' (Bak & Chen 1991).

The results of this chapter, and those of appendix I, are summarized in table 12.1.

13
The Kardar-Parisi-Zhang equation

"Two added to one–if that could be done"
It said, "with one's fingers and thumbs!"
Recollecting with tears how, in earlier years,
It had taken no pains with its sums.
"The thing can be done," said the Butcher, "I think,
The thing can be done, I am sure.
The thing shall be done! Bring me paper and ink,
The best there is time to procure".

Lewis Carrol (*The hunting of the Snark*)

In 1986, Kardar, Parisi & Zhang (KPZ) realized that the equation which rules the growth of anything (crystalline or not) at large lengthscales is in principle non-linear. The generic equation written by KPZ is still unsolved in 2+1 and larger dimensions, although the roughness exponents α, β, $\tilde{z}$ defined in the previous chapter, can be determined by simulations.

Another surprising point is that, although the KPZ equation is in principle generic, there is no clear experimental evidence of its validity in the case of molecular beam epitaxy. Experiments are often performed in regions where the non-linear term is quite small; the relevant lengthscales are often larger than the sample size or than the size for which gravity becomes important.

13.1 The most general growth equation

In the previous chapter, we obtained an equation for growth, formula (12.6), by simply adding a term (the beam intensity) to the equation (8.14) derived in the absence of growth. This procedure is not justified, since growth may well introduce new terms. This is actually what happens, as will be seen.

Equation (12.6) suggests a more general approach. In this equation, the local growth rate $\dot{z} = \partial z/\partial t$ is the sum of two terms: the first term $\dot{z}_1$ is the effect of the beam, and the second one, $\dot{z}_2$, is due to further restructuring of the surface by diffusion and evaporation. In analogy with (12.6), $\dot{z}_2$ will be assumed to be a function of the local shape of the surface, more precisely an *analytic* function of the surface height $z(x, y)$ and of its partial derivatives $\partial^n z/(\partial x^m \partial y^p)$.

This analytic form is only acceptable for the description at long length-scales and for a rough surface, when the discrete nature of the lattice can be neglected. With these hypotheses, the growth rate will even be independent of z if there is no external field. The resulting equation is

$$\dot{z}_2 = \sum_{\alpha} A_\alpha \partial_\alpha z + \sum_{\alpha\gamma} A_{\alpha\gamma} \partial_\alpha z \partial_\gamma z + \sum_{\alpha\gamma\zeta} A_{\alpha\gamma\zeta} \partial_\alpha z \partial_\gamma z \partial_\zeta z + \ldots \,. \tag{13.1a}$$

The indices α, γ, ζ, take the values x and y, or more generally 1, 2,..., d', where $d' = d - 1$ as in section 12.4. The coefficients A are constants. These coefficients are discussed in appendix J.

We have now to write the part $\dot{z}_1$ due to the deposition of atoms from the beam. As in the previous chapter, the beam will be assumed normal to the surface so that $\dot{z}_1$ is just the beam intensity $f(\mathbf{r}, t)$. The general case, when the beam makes an angle α with the z direction, is studied in appendix K. In MBE, the sample is usually rotated at constant velocity around the z direction, and in this case it is shown in the appendix K that $\dot{z}_1 = f(\mathbf{r}, t) \cos\alpha$. The factor $\cos\alpha$ will be ignored from now on for simplicity, but may be easily taken into account.

Adding the beam intensity to the right hand side of (13.1a), one obtains

$$\dot{z}(\mathbf{r}, t) = f(\mathbf{r}, t) + \sum_{\alpha} A_\alpha \partial_\alpha z + \sum_{\alpha\gamma} A_{\alpha\gamma} \partial_\alpha z \partial_\gamma z + \sum_{\alpha\gamma\zeta} A_{\alpha\gamma\zeta} \partial_\alpha z \partial_\gamma z \partial_\zeta z$$
$$+ \sum_{\alpha\gamma\zeta\xi} A_{\alpha\gamma\zeta\xi} \partial_\alpha z \partial_\gamma z \partial_\zeta z \partial_\xi z + \ldots + \sum_{\alpha\gamma} B_{\alpha\gamma} \partial^2_{\alpha\gamma} z + \ldots \,. \tag{13.1b}$$

The reader is reminded that this equation is correct only at long distances, when the discrete nature of the lattice can be neglected. The discrete lattice can be taken into account by introducing into the right hand side of (13.1b) a z-dependent, periodic term, e.g. proportional to $\cos 2\pi z$. This can be done (Tang & Nattermann 1991, Hwa *et al.* 1991, Rost & Spohn 1994) and one can check that these terms are not important (or 'renormalize to zero') at long wavelength if the surface is rough.

The assumption (13.1) is analogous to Landau's treatment of equilibrium critical phenomena from a free energy functional which is assumed to be analytic in the order parameter. This analogy (already noticed in chapter 12) suggests a renormalization group treatment, analogous to the now usual technique in critical phenomena. This technique will be very briefly addressed in section 13.5 and, more in detail, in appendix L.

13.2 Relevant and irrelevant terms in (13.1): the KPZ equation

The expansion (13.1b) is only useful if higher order terms can be neglected, at least in some limit. The limit in which (13.1b) can be useful is, as said above, that of long distances. A term of (13.1b) is said to be irrelevant if it is negligible at large lengthscales. Otherwise, it is said to be relevant.

It will first be shown that the linear terms of (13.1b), without being really irrelevant, can be eliminated by a 'Galilean transformation', i.e. by choosing a moving frame of reference deduced from the fixed one by a uniform translation. The axes can first be rotated in such a way that a single coefficient $A_\alpha = A_x = A$ is not vanishing. One can then define X by

$$x = X - At\,.$$

For any function $\varphi(x,t)$, one has

$$\mathrm{d}\varphi = (\partial\varphi/\partial x)_t\,(\mathrm{d}X - A\mathrm{d}t) + (\partial\varphi/\partial t)_x \mathrm{d}t$$

and therefore $(\partial\varphi/\partial x)_t = (\partial\varphi/\partial X)_t$ and $(\partial\varphi/\partial t)_X = (\partial\varphi/\partial t)_x - A(\partial\varphi/\partial x)_t$.

Replacing φ by z and inserting this result into (13.1b), the linear terms proportional to ∂_α are seen to vanish.

As in chapter 12, we write the beam intensity f as the sum of its average value and its fluctuation:

$$f\,(\mathbf{r},t) = F + \delta f\,(\mathbf{r},t)\ .$$

If z is replaced by $z - Ft$, i.e. if a Galilean transformation is carried out for z also, the term in F disappears, and equation (13.1b) can be written as

$$\begin{aligned}\dot z(\mathbf{r},t) = \delta f(\mathbf{r},t) &+ \sum_{\alpha\gamma} A_{\alpha\gamma}\partial_\alpha z\partial_\gamma z + \sum_{\alpha\gamma\zeta} A_{\alpha\gamma\zeta}\partial_\alpha z\partial_\gamma z\partial_\zeta z \\ &+ \sum_{\alpha\gamma\zeta\xi} A_{\alpha\gamma\zeta\xi}\partial_\alpha z\partial_\gamma z\partial_\zeta z\partial_\xi z + \ldots + \sum_{\alpha\gamma} B_{\alpha\gamma}\partial^2_{\alpha\gamma} z + \ldots\end{aligned} \qquad (13.1c)$$

where δf satisfies (12.3).

The higher order terms will now be discussed, having simplicity in mind rather than mathematical rigour. Assume that we are interested in height fluctuations with a particular spatial extension R. Their order of magnitude δh is such that $\left\langle \delta h^2(R,t)\right\rangle = G(R,t)$, defined by (12.1). This implies that, for a typical configuration, the average value of $\partial_\alpha z$ on an area of order $R \times R$ is of order $h(R,t)/R$ (Fig. 13.1). Hence

$$\partial_\alpha z \partial_\gamma z \approx \delta h^2(R,t)/R^2\,. \qquad (13.2a)$$

Similarly, the order of magnitude of $\partial^2_{\alpha\gamma} z$ can be roughly estimated to be

$$\partial^2_{\alpha\gamma} z \approx \delta h(R,t)/R^2 \,. \tag{13.2b}$$

This quantity is smaller than (13.2a) for large values of R and t if the surface is rough, as it is assumed to be. Thus, we have no right to neglect the term (13.2a), as we did in chapter 12! This remark was made for the first time by Kardar, Parisi & Zhang in 1986.

This discussion can be extended to the relevance of the general term of (13.1), of order p in z and q with respect to the derivatives ∂_α. This term is of order $\delta h^p/R^q$. Now, since only first and higher derivatives appear in (13.1), not z itself, q is larger than or equal to p. Therefore, any term can only be relevant if $p = q$ (except if terms with $p = q$ vanish identically, but this is generally not so). Finally, above the lower critical dimension, the fluctuation $\delta h(r, t = \infty)$ diverges more slowly than r. Therefore, the only relevant terms are generally those with $p = q = 2$. Neglecting irrelevant terms and diagonalizing the quadratic form containing the coefficients B, (13.1c) reads

$$\dot{z} = f(\mathbf{r}, t) + \sum_\alpha B_{\alpha\alpha} (\partial_\alpha z)^2 \,. \tag{13.3}$$

However, the linear term, although it is irrelevant according to the above argument, is usually kept. A good reason for this is that the quadratic term in (13.3) is often very small at realistic distances and times (Balibar & Bouchaud 1992). Another good reason is that we want an equation which includes the smoothing problem of chapter 8 as a special case. In chapter 8, the case of a vanishing growth rate F was addressed, and the linear equation (8.7) was obtained. After the preceding section, the reader might have some doubt about this equation, but it will be checked in the

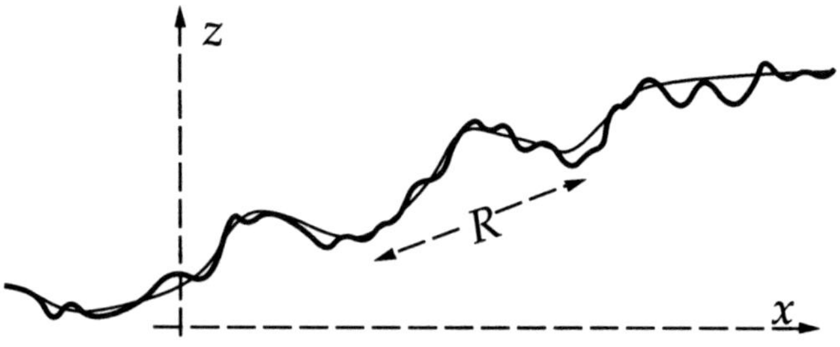

Fig. 13.1. The estimation in (13.2) replaces all terms in (13.1) by those corresponding to a smooth surface resulting from coarse-graining over all fluctuation wavelengths smaller than R.

next section that this doubt is not justified. Therefore, we shall write an equation which contains both (8.7) and (13.3) as special cases.

For the sake of simplification, the growing surface will be assumed to have a high-symmetry orientation, so that $B_{xx} = B_{yy}$. Defining $Z(\mathbf{r}, t) = z(\mathbf{r}, t) - Ft$ as in chapter 12, the following equation of motion is obtained:

$$\dot{Z}(\mathbf{r}, t) = \delta f(\mathbf{r}, t) + \tfrac{\lambda}{2}(\nabla Z)^2 + \nu \nabla^2 Z \ . \tag{13.4}$$

This equation was introduced by Kardar, Parisi & Zhang (KPZ) in 1986 to describe a model of non-crystalline growth proposed by Eden (1961). It is indeed clear that the crystal structure plays no part in the above argument. The result (13.4) may therefore be expected to be rather general. However, it is not quite general: in the case of a vicinal surface, the $B_{\alpha\alpha}$ in (13.3) can be explicitly calculated, and are found (Villain 1991) to be of opposite signs! The resulting roughness is completely different from that of the KPZ model (Wolf 1991) as shown in the appendix L.

Equation (13.4) is invariant under the transformation

$$\lambda \to -\lambda, \quad z \to -z, \quad \delta f \to -\delta f \ .$$

The properties of the KPZ equation (13.4) are therefore independent of the sign of λ.

13.3 Upper critical dimension and exponents

Even though experimental systems are three-dimensional, the upper critical dimension defined in the previous chapter is an important property. For instance, as seen in chapter 1, the weakness of the divergence of the equilibrium correlation function in 2+1 dimensions is related to the fact that $d = 3$ is the upper critical dimension for thermodynamic (equilibrium) roughness.

Now, the upper critical dimension of the linear model (12.6) is 2+1 (provided that $\nu \neq 0$). For $d > 2 + 1$, the effect of the ν term in (13.4) is to make the surface smooth, and there is no obvious reason why the λ term should make it rough. Therefore, the upper critical dimension of (13.4) is expected to be the same as for (12.6),

$$d_c^u = 2 + 1 \ . \tag{13.5}$$

At the critical dimension 2+1, a weak divergence of $\delta h(r, t = \infty)$ is therefore expected, with an exponent $\alpha = 0$. As a matter of fact, numerical simulations show that it is not so (Kim & Kosterlitz 1989, Amar & Family

1990, Forrest & Tang 1990). The roughness exponent α defined by (12.18) is found to be of order

$$\boxed{\alpha \simeq 0.4\,.} \tag{13.6}$$

Comparing with (12.19) and (1.12), we conclude the following.

The surface roughness in the KPZ model is much more pronounced than equilibrium roughness, and than the roughness of the linear model (12.6) *as well.*

The reason why non-trivial exponents are obtained at the upper critical dimension (13.5) is presumably the existence of two fixed points in the renormalization group treatment (Halpin-Healey 1989).

Kim & Kosterlitz (1989) have made simulations for several space dimensions $d < 3$, and found that their results are well reproduced by the following empirical formulae:

$$\alpha = \frac{2}{d+2} \tag{13.7a}$$

$$\beta = \alpha/\tilde{z} = \frac{1}{d+1} \tag{13.7b}$$

$$\tilde{z} = \frac{2(d+1)}{d+2}\,. \tag{13.7c}$$

The corresponding behaviour of the roughness correlation function (12.1),

$$G(\mathbf{r}, t) = \left\langle \left[z(\mathbf{r}', t) - z(\mathbf{r}' + \mathbf{r}, t)^2 \right] \right\rangle$$

is displayed by Fig. 13.2. A heuristic derivation of the exponents in (13.7) has been proposed by Hentschel & Family (1991).

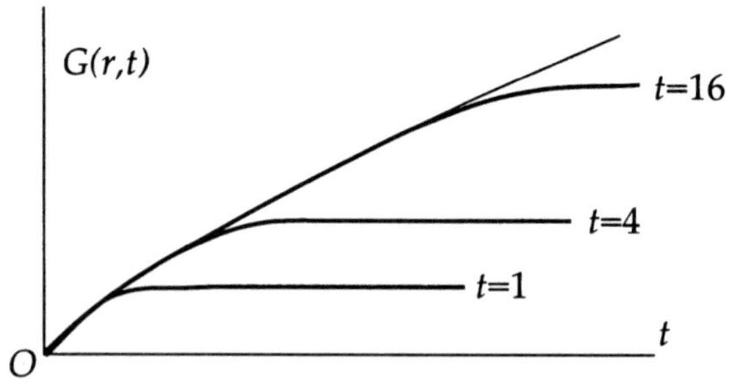

Fig. 13.2. The height-height correlation function (12.1) as a function of the distance for 3 chosen times in the KPZ model. The units are arbitrary.

It turns out that very few experimental checks of (13.6) are available. There are three reasons for this:

i) The coefficient λ of the non-linear term is often small, as we will see below in the next section, and in the next chapter.
ii) In the special case of MBE, which is often addressed in this book, it has been seen in section 12.3 that beam fluctuations are averaged out by surface diffusion if ℓ_s is large, which is generally the case in technological application.
iii) The third problem is that experimentalists usually try to obtain (and succeed in obtaining) as smooth a surface as possible, where α cannot be measured. One of the few experimental checks of the KPZ exponents (actually of β) is by Chevrier *et al.* (1991), but on a fairly restricted range of distances.

13.4 Behaviour of λ near solid-fluid equilibrium

We now consider a solid in contact with a fluid, when the chemical potential is the same in the bulk solid and in the fluid phase. An equivalent statement in the language of this chapter is that evaporation is exactly compensated by deposition, so that $F = 0$. We are going to prove that, in this case, λ vanishes. This is the situation considered in chapter 8, although the beam fluctuations δf were not explicitly taken into account there. However, irreversible thermodynamics, used in chapter 8, is supposed to take thermal fluctuations into account, and this can be done explicitly through the Langevin equation formalism. What we are going to do here is to reconsider the formulae of irreversible thermodynamics used in chapter 8.

In order to discuss the value of λ, it is sufficient to consider the case of a planar interface, so that the term containing ν in (13.4) vanishes. If the interface is planar, the chemical potential on the interface is the same as in the solid and in the fluid, according to the Herring-Mullins formula (2.9). Therefore, the system is in equilibrium, *whatever the orientation of the interface*! It follows that the average value of the left hand side of (13.4) vanishes. The average value of the right hand side of (13.4) should also vanish. Therefore, λ vanishes too. It can be checked (appendix L) that the renormalization group treatment does not introduce a finite λ if its 'bare' value is zero, but it is not even necessary, because the model with $\lambda = 0$ is exactly solvable as shown in chapter 12.

The above argument has been illustrated by a discussion of the solid-superfluid interface of ^{4}He by Balibar & Bouchaud (1992). The authors reexamined experiments on the roughening transition, in which the experimenters tried to stay as close as possible to thermal equilibrium. The conclusion was that indeed λ is extremely small in that case.

Instead of invoking the Herring-Mullins formula, it is possible to consider the change of free enthalpy (Gibbs free energy) due to a translation of the solid-fluid interface at constant pressure. The free enthalpy of the interface is not modified by this translation, and the free enthalpy difference due to melting or solidification vanishes if the chemical potential in both phases is the same. Therefore, the free enthalpy is independent of the position of the interface, and there is equilibrium independently of the orientation of the interface.

13.5 A relation between the exponents of the KPZ model

A relation between the exponents α and $\tilde{z}$ can be obtained by the following argument (which, as discussed later, is heuristic rather than rigorous ... the result is correct, however!). Assume that as an effect of the fluctuations δf a bump or a hole of size ξ and height $\delta h(\xi)$, has formed. The height $\delta h(\xi) \approx \xi^\alpha$. Let τ be the time necessary for the formation of the bump. It is of the same order of magnitude as its lifetime. The decay of the bump may be expected to be given by (13.4) without the beam f. By an argument analogous to that of section 13.2, we make in (13.4) the following replacements:

$$\dot{z} \to \frac{\delta h}{\tau}\,, \quad (\nabla z)^2 \to \frac{\delta h^2}{\xi^2}\,, \quad \nabla^2 z \to \frac{\delta h}{\xi^2}\,.$$

Formula (13.4) now reads

$$\frac{\delta h}{\tau} \approx \frac{\lambda}{2}\frac{\delta h^2}{\xi^2} + \nu\frac{\delta h}{\xi^2}\,.$$

In this relation we insert the following expressions resulting from the definitions (12.14) and (12.18)

$$\xi \sim \tau^{1/\tilde{z}}\,, \quad \delta h \sim \tau^{\alpha/\tilde{z}}\,.$$

For large ξ and $\lambda \neq 0$, one obtains

$$\boxed{\alpha + \tilde{z} = 2\,.} \tag{13.8}$$

The values (13.7) do satisfy this property.

The preceding argument, as well as the one of section 13.2, is not rigorous. Indeed, the relaxation of a defect of size ξ might be modified by smaller defects inside this defect. In this case, one says that λ and ν are renormalized.

It is shown in appendix L that λ is not renormalized and that (13.8) is indeed true (Medina *et al.* 1989).

In the same appendix, a brief description of the renormalization group method is given. In the case of the KPZ model, this method is not very successful except for $d = 1 + 1$. However, for other models to be seen in the next chapter, the exponents can be obtained by the renormalization group.

13.6 Numerical values of the coefficients λ and ν

The easiest case is the Eden model (Eden 1961), in which the average growth rate normal to the surface is independent of the orientation. This model is an appropriate description of the growth of a tumour. In the reference frame (x, y, z), the growth rate normal to the surface is

$$\frac{\partial z}{\partial t}\left[1 + z_x^2 + z_y^2\right]^{-1/2} .$$

The growth equation is

$$\frac{\partial z}{\partial t}\left[1 + z_x^2 + z_y^2\right]^{-1/2} = F + \delta f$$

or

$$\frac{\partial z}{\partial t} = \left[1 + z_x^2 + z_y^2\right]^{1/2}(F + \delta f) .$$

Assuming $|z_x| \ll 1$, $|z_y| \ll 1$ and small fluctuations of the beam, this equation reads

$$\frac{\partial z}{\partial t} \simeq F + \frac{F}{2}\left[z_x^2 + z_y^2\right] + \delta f .$$

The constant term is eliminated as before by a Galilean transformation $Z = z - Ft$, and (13.4) results, with $\nu = 0$ and $\lambda = F$. A finite value of ν would be generated by the renormalization group, or by taking evaporation into account. Balibar & Bouchaud (1992) have applied the method above to the specific case of He, and found λ to be very small.

In the case of MBE growth, the calculation of λ is still an open problem. Attempts have been made by Villain (1991) and by Zangwill *et al.* (1992).

13.7 The KPZ model without fluctuations ($\delta f = 0$)

We consider an initially non-planar, but analytic surface, with an average direction parallel to the z axis, and we assume that its evolution is ruled by (13.4) with $\delta f = 0$. It is of interest to check whether the non-linear term in (13.4) leads to a planar surface. It is actually so, but the process is

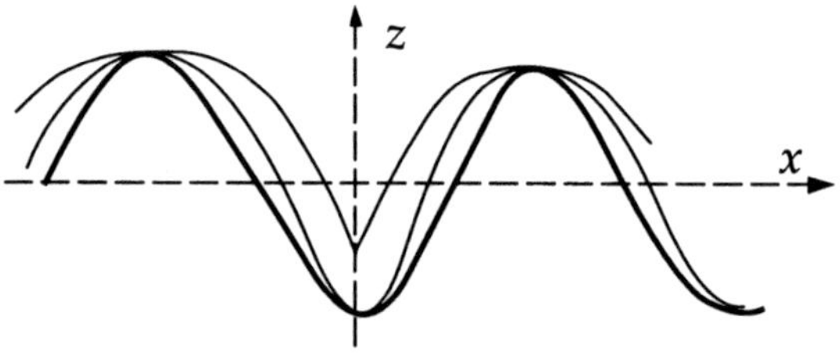

Fig. 13.3. Formation of a kink. The bold line represents the initial state of a surface evolving according to the equation (13.4) with $\delta f = \nu = 0$. The thin lines give the position of the surface at later times.

rather complicated and, if $\nu = 0$, there is a transient, non-analytic shape with kinks (Burgers 1974, Medina *et al.* 1989).

Only the case $d = 1 + 1$ will be considered. Let the one-dimensional surface have an extremum $z = z_0$ at $x = x_0$. If the surface is described by an analytic function (which is true at short times), the quantities $\delta x = x - x_0$ and $\delta z = z - z_0$ verify a quadratic equation near the extremum:

$$\delta z(x,t) = C + \frac{1}{2}A(t)\delta x^2 .$$

Inserting this equation into (13.4) with $\delta f = \nu = 0$, one sees that C is t-independent as long as A satisfies $\mathrm{d}A/\mathrm{d}t = \lambda A^2$, whence:

$$\frac{1}{A(t)} = \frac{1}{A(0)} - \lambda t .$$

As a consequence, $A(t)$ becomes infinite at $t = 1/\left[\lambda A(0)\right]$ if this quantity is positive. Singularities appear, at minima if $\lambda > 0$ (Fig. 13.3) or at maxima if $\lambda < 0$. Note that, when λ is negative in (13.4), it can be made positive by changing z into $-z$, as remarked in section 13.2.

To close this chapter, we mention that in the 1+1-dimensional KPZ model, the exponents can be exactly computed and are correctly given by (13.7) with $d = 2$. The simplest way to obtain α is to use a particular model solved by Ramanlal & Sander (1985). The exponent $\tilde{z}$ can then be derived from formula (13.8).

14
Growth without evaporation

Nous avons l'exposant -6	We have the exponent -6
ou l'exposant -5	or the exponent -5
au lieu de l'exposant -2,	instead of the exponent -2,
mais c'est toujours un exposant.	but it is always an exponent.

Henri Poincaré (*La valeur de la science*)

The KPZ model studied in the preceding chapter is generic and applies in principle to any growing crystal surface of high symmetry, if there is no instability. 'Generic' means, so to speak, that it is applicable in any case unless there is a good reason not to apply it. Indeed, in molecular or atomic beam epitaxy (MBE) there are often exact or approximate conservation laws which make the KPZ model unapplicable. This is the topic of this chapter. The conservation laws apply if there is no evaporation, if the sticking coefficient is unity and if bulk vacancy formation is negligible, or is not affected by the surface. Such conditions are probably realistic in many cases. These conservation laws give rise to models which turn out to be more easily solved than the KPZ model.

14.1 Where λ is shown to vanish in the KPZ equation

We consider the infinite and initially planar surface of a crystal which is growing under deposition by an atom beam. It will be assumed that the beam direction z is normal to the initial direction of the surface. The general case is discussed in the last section of the preceding chapter and in appendix K, and no complication is introduced if the sample is rotated around z.

Since, in the present chapter, evaporation and vacancy formation in the bulk are assumed to be absent, the growth rate $\dot{z}$ in the z direction is equal to the average beam intensity F. It is therefore independent of the local surface orientation, so that $\lambda = 0$ in (13.4) (Kariotis 1989). This result can alternatively be obtained by writing that $\dot{z}(\mathbf{r}, t) - f(\mathbf{r}, t)$ satisfies a conservation equation, namely

$$\dot{z}(\mathbf{r}, t) = f(\mathbf{r}, t) - \operatorname{div}\mathbf{j}(\mathbf{r}, t) \tag{14.1}$$

where the current density $\mathbf{j}$ is perpendicular to z. It is in fact the projection of the intrinsic current density on the xy plane, as we discussed in section 7.4. The term containing λ in (13.4) is not a divergence, so that λ must vanish. The random function f is the beam intensity as in the preceding two chapters. It can be written as the sum of a constant term F and a fluctuation $\delta f(\mathbf{r}, t)$. Beam fluctuations are supposed uncorrelated in space and time.

14.2 Diffusion bias and the Eaglesham-Gilmer instability

If λ vanishes in (13.4), it is essential to calculate the coefficient v of the linear term. For a vanishing beam intensity ($F = 0$), comparison of (13.4) and (8.12) shows that v should vanish too. In the general case $F \neq 0$, it is convenient to remark that (13.4) and (14.1) yield

$$\mathbf{j}(\mathbf{r}, t) = -v\nabla z\ . \tag{14.2}$$

From this equation, one can again check that $v = 0$ for $F = 0$. This can be seen if one considers an infinite, planar surface, where ∇z is constant but does not necessarily vanish. We have seen in the preceding section that, for such a surface, $\partial z/\partial t = 0$. But, in addition, the chemical potential μ on this surface is uniform according to chapter 2, and therefore the current density $\mathbf{j} = -\nabla\mu$ should vanish everywhere. Comparison with (14.2) yields $v = 0$.

If $F \neq 0$, the current on the surface does not vanish if $\nabla z \neq 0$. This is mainly due to the Schwoebel effect discussed in chapter 6: adatoms prefer to be incorporated into the upper ledge than into the lower one. This implies that there is a current parallel to ∇z and with the same sign. Comparison with (14.2) implies, for a high-symmetry surface, $v < 0$. The case of a vicinal surface is different and is discussed in appendix J.

The result $v < 0$ implies that a planar surface is unstable. This is obvious if fluctuations are neglected, since the growth equation $\partial z(\mathbf{r}, t)/\partial t = F + v\nabla^2 z(\mathbf{r}, t)$ yields $\partial z_q/\partial t = -vq^2 z_q$ after Fourier transformation. Taking fluctuations into account in the complete equation

$$\frac{\partial z(\mathbf{r}, t)}{\partial t} = f(\mathbf{r}, t) + v\nabla^2 z \tag{14.3}$$

is quite an easy matter, as seen in chapter 12, and also leads to an instability as long as $\nu < 0$.

Is there eventually a steady state? The answer depends on the details of the model. In the following, we present an argument which should be an adequate description of the growth of a cubic metal initially limited by a (001) surface (Siegert & Plischke 1994, Šmilauer & Vvedensky 1995). A natural generalization of (14.3) is

$$\frac{\partial}{\partial t} z(x,t) = f(r,t) + \operatorname{div}(\nu \nabla z)$$

where ν is now a function of ∇z. This equation means that the current density, projected onto the (001) plane, is given by (14.2). Now, the current density should vanish on the $\{111\}$ planes. Indeed, these planes cannot be considered as stepped surfaces, but as high-symmetry planes on which the diffusion is even faster than on the (001) plane (Stoltze 1994), and isotropic. It then follows from (14.2) that $\nu = 0$ on the $\{111\}$ orientations. Therefore, the initial stage of the evolution drives the surface toward a state made of pyramids with $\{111\}$ facets (Siegert & Plischke 1994). These pyramids would be stable with respect to deformation, but not with respect to coalescence. Therefore, the final state of the surface would be a single pyramid. Formation of pyramids is indeed experimentally observed in growth at sufficiently low temperature (Johnson *et al.* 1994, Ernst *et al.* 1994). The facets are *not* oriented along $\{111\}$, presumably because coalescence takes place before this orientation can be reached (Johnson *et al.* 1994, Hunt *et al.* 1994).

Other one-dimensional models have been studied (Elkinani & Villain 1994, Krug & Schimschak 1995) in which ν does not vanish for any orientation. Then, deep cracks form during growth. These cracks are unphysical, but these models are adequate to study the initial stage of the instability, which is the most interesting for practical applications–at least if one wants to prevent it! Before the instability takes place, there are very few steps and the continuum equation (14.2) is not applicable. On the other hand, the Monte Carlo method is hard to apply in practical cases where the instability sets in at very long times. The 'Zeno model' of Elkinani & Villain (1994) is an attempt to overcome this computational difficulty by treating surface diffusion (which is very time-consuming in Monte Carlo simulations) as a deterministic process along the lines of chapters 6 and 10.

As emphasized by Johnson *et al.* (1994) the Eaglesham-Gilmer instability is expected to disappear for a sufficiently strong miscut, in one-dimensional models. In some of these models, however, miscut surfaces are only metastable, and cracks do form after a sufficiently long time (Krug & Schimschak 1995).

Experimentally, the growth instability observed by Eaglesham *et al.* (1990, 1991) and by Chevrier *et al.* (1994) in the homoepitaxy of semiconductors seems to originate from the Schwoebel mechanism (Eaglesham & Gilmer 1992). The final stage turns out to be in this case an amorphous material.

This does not mean that excellent Si crystals cannot be grown: the crystal-amorphous instability in silicon growth is strongly temperature-dependent, and films up to 100 nm thick or more can be grown at temperatures as low as 300 °C (Eaglesham *et al.* 1990, 1991). Also, one may choose to use as substrate a stepped surface, rather than a high-symmetry one. As seen in section 6.7, the Schwoebel effect stabilizes the step-step separation during step flow growth. Thus, the Schwoebel effect seems destabilizing for a high-symmetry surface, but stabilizing for a vicinal one. Reality is more complicated. The Schwoebel effect damps the fluctuations of the terrace widths on a vicinal surface growing by step flow. However, the same Schwoebel effect is responsible for the Bales-Zangwill instability of section 10.2 (*cf.* also equation (J.5) in appendix J), i.e. it increases the fluctuations of the step shape (meandering), leading to a 'ripple' structure in the step train. Rost *et al.* (1996) have shown that a *secondary instability* of the ripple structure takes place at long deposition times, leading to formation of mounds as on a singular surface. The morphology of a vicinal surface is then indistinguishable from that of an unstable singular surface (Rost *et al.* 1996). Nonetheless, it should be kept in mind that these instabilities are of a kinetic nature, so that their practical importance depends on the time scale over which they show up. Moreover, they are the less effective the higher the growth temperature, and acting on the substrate temperature allows their consequences on the quality of crystals to be made less severe.

14.3 A theorist's problem: the case $\lambda = \nu = 0$

We shall now address the cases where λ and ν are equal to 0 in the KPZ equation (13.4). This can be true only if the Schwoebel effect vanishes, and in general it has no reason to do so. However, there are not many situations where the magnitude of the Schwoebel effect is known, and it may be of interest to consider the case when it is absent.

Moreover, the case $\lambda = \nu = 0$ has a great theoretical interest because the resulting equation is easier than the KPZ problem with $\lambda, \nu \neq 0$, without being trivial as the Edwards-Wilkinson problem $\lambda = 0$, $\nu \neq 0$ solved in chapter 12.

In order to write the growth equation, we look at the general equations (13.1). The terms of order 2 in ∂_α are assumed to vanish as explained above. A high-symmetry orientation will be assumed–which implies that no preferred slope is present–so that by symmetry all terms of order 1 and 3 vanish. Therefore, the dominant terms are presumably those of order 4 with respect to ∂_α. A power-counting analysis similar to that of section 13.2 shows that it is indeed so. The term in $|\nabla z|^4$ should also vanish because it has not the form (14.1). A discussion similar to that of section 13.2 then shows that the dominant term should be of order 3 in z. According to (14.1), this term should be the divergence of a current density $\mathbf{j}$ which should be a homogeneous expression of degree 3 with respect to both z and ∂_α, invariant under any translation $z \longrightarrow z + z_0$. This current should therefore be the sum of products of the form $\Lambda_{\alpha\gamma\zeta}(\partial_\alpha z)(\partial_\gamma z)(\partial_\zeta z)$. It should therefore be an odd fuction of ∇z, which has to vanish if there is no Schwoebel effect. Since v has been assumed to vanish, the Schwoebel effect should vanish and the coefficients $\Lambda_{\alpha\gamma\zeta}$ should also vanish. The dominant term of (13.1a) should therefore be of second order in z. An acceptable expression is $\dot{z}_2(\mathbf{r},t) = \sigma\nabla^2(\nabla z(\mathbf{r},t))^2$. However, comparison with (8.12) shows that $\sigma = 0$ for $F = 0$. It is therefore convenient to keep also the linear term (8.12). The final equation of motion is (Wolf & Villain 1990b, Villain 1991, Golubović & Bruinsma 1991, Das Sarma & Tamborenea 1991)

$$\dot{z}(\mathbf{r},t) = \delta f(\mathbf{r},t) - K\nabla^2\left(\nabla^2 z(\mathbf{r},t)\right) + \sigma\nabla^2(\nabla z(\mathbf{r},t))^2 \ . \tag{14.4}$$

The model (14.4) was introduced for the first time by Sun *et al.* (1989) with special assumptions which define what will be called the 'Montréal model'. Sun *et al.* assumed in fact fluctuations which conserve the number of particles, i.e.

$$\delta f(\mathbf{r},t) = \mathrm{div}\delta\mathbf{j}(\mathbf{r},t) \tag{14.5}$$

where $\delta\mathbf{j}$ is a random current.

The Montréal model can be expected to correspond to a situation when the roughness is due to the fluctuations in the diffusion of adatoms.

14.4 Calculation of the roughness exponents

The exciting feature for the theorists in the model (14.4) is that the exponents α, β and $\tilde{z}$ defined by (12.18), (12.21) and (12.22) can be calculated. Moreover, for non-conserved fluctuations δf, these exponents have, in the physical, three-dimensional space, 'non-trivial' values (opposed to the 'trivial' value $\alpha = 0$ of thermal roughness seen in chapter 1). Whether these values are exact or just approximate will be discussed later.

The first step in the calculation is the following evaluation, similar to that of section 13.5.

Healing time of a defect

As in section 13.5, it will be assumed that, if a bump or a valley of radius ξ and height δh has formed because of the beam fluctuations δf, it will heal after a time τ. The latter can be evaluated from (14.4) if $\dot{z}$ is replaced by $\delta h/\tau$, each derivative ∂_α by $1/\xi$ and each factor z by δh. The fluctuation δf is ignored at this stage as in section 13.5. One obtains

$$\delta h/\tau \approx \left(K\delta h + \sigma\delta h^2\right)/\xi^4$$

or

$$\tau \approx \xi^4/(K + \sigma\delta h) \; . \tag{14.6}$$

Replacing ξ by $\tau^{1/\tilde{z}}$ and δh by $\tau^{\alpha/\tilde{z}}$, one obtains for large ξ, if $\sigma \neq 0$, the equality

$$\alpha + \tilde{z} = 4 \tag{14.7}$$

instead of 2 as for the KPZ model (formula 13.8).

Creation time of a defect

We are now going to study the reverse process: we will evaluate the time needed for the beam fluctuations δf to create a hill or a valley of radius ξ and height δh. The argument is the same as in section 12.3: the number n of atoms landing into an area ξ^{d-1}–where, as usual, $d = 3$ in experimental physics–during a time τ is of order $\bar{n} \approx F\xi^{d-1}\tau$. Since fluctuations are statistically independent, the mean square fluctuation δn of n is of the order of $\sqrt{\bar{n}}$. Now, δn can be identified with the volume $\xi^{d-1}\delta h$ of the defect (with our usual length units where the atomic distance is 1). It follows

$$\tau \approx (1/F)\xi^{d-1}\delta h^2 \; . \tag{14.8}$$

Replacing ξ by $\tau^{1/\tilde{z}}$ and δh by $\tau^{\alpha/\tilde{z}}$, one obtains for large ξ the equality

$$\tilde{z} - 2\alpha = d - 1 \; . \tag{14.9}$$

Combining (14.7) and (14.9), one finds for $\sigma \neq 0$

$$\alpha = (5-d)/3, \;\; \tilde{z} = (7+d)/3 \; . \tag{14.10}$$

In particular, in the physical case $d = 3$, one finds $\alpha = 2/3$. In the case $\sigma = 0$, (14.6) and (14.8) yield $\alpha = (5-d)/2$, in agreement with the result of appendix I.

For the KPZ equation of chapter 13, there is no relation analogous to (14.8). Indeed, the KPZ equation applies to MBE when evaporation is allowed. Now, since a part of the deposited atoms evaporate, the time needed to create a defect cannot be related to the beam fluctuations only.

The above derivation of (14.10) is heuristic rather than strictly correct. Indeed, as in section 13.4, any possible renormalization has been ignored. However, formulae (14.10) have been confirmed by analytic, renormalization group calculations by Lai & Das Sarma (1991) and by Tang & Nattermann (1991), at least in the 'one-loop approximation' similar to that described in appendix L for the KPZ model.

14.5 Numerical simulations

Simulations do not use continuous models like (14.4), but discrete models which have the invariance properties of (14.4), and should hopefully have the same roughness exponents in the limit of long distances and times. The following story shows that such a program may hide traps.

The first models which were studied were 1+1-dimensional (Wolf & Villain 1990b, Das Sarma & Tamborenea 1991, hereafter called WV and DST, respectively). Evaporation was forbidden, so that λ has to vanish in (13.4). No Schwoebel effect was introduced, so that the authors naively expected $\nu = 0$ too. It was later pointed out by Siegert & Plischke (1992) that ν can have a non-vanishing value even without Schwoebel effect, and indeed Kessler *et al.* (1992) confirmed that the WV and DST models are consistent with (14.3) at very long times. Nevertheless, the first simulations of WV and DST, performed on too small samples and too short times, were consistent with (14.4) with $\sigma = 0$. We suggest the following conclusion (which may be provisional) to this story: i) In the absence of Schwoebel effect, ν is really very small, and this is probably even more true in real systems. ii) Non-linearities are often very small in practice.

In 2+1 dimensions, numerical simulations by Kotrla *et al.* (1992) have confirmed the exponents (14.10).

14.6 The Montréal model

The Montréal model corresponds to fluctuations δf which conserve the total number of atoms. The formula (14.7), which is not related to fluctuations, is therefore still valid. How should (14.8) be modified? We have to calculate the fluctuation δN of the number of atoms entering during time τ into the domain $\mathscr{D}$ of area ξ^{d-1} occupied by the defect. Now, if ρ_k is the Fourier transform of the adatom density, $\mathbf{j}_k$ the Fourier transform of the current density and $\mathcal{N}$ the number of sites on the surface, we have

$$\begin{aligned}\frac{\partial}{\partial t}N &= \mathcal{N}^{-1/2}\sum_k\sum_{r\in\mathscr{D}}\dot{\rho}_k\exp(-i\mathbf{k}\cdot\mathbf{r}) \approx \mathcal{N}^{-1/2}\xi^{d'}\sum_k\dot{\rho}_k \\ &= \mathcal{N}^{-1/2}\xi^{d'}\sum_{k<1/\xi} i\mathbf{k}\cdot\mathbf{j}_k\,.\end{aligned}$$

In the Montréal model, the fluctuations of $\mathbf{j}_k$ are statistically independent, and they have magnitude δj and relaxation time τ_0. It follows

$$\left\langle \delta N^2 \right\rangle = \mathcal{N}^{-1} \xi^{2d'} (\tau/\tau_0) \sum_{k<1/\xi} k^2 \left\langle | j_k |^2 \right\rangle = \xi^{d'-2} (\tau/\tau_0) \delta j^2 \,.$$

The square root of this quantity can be identified with the volume $\xi^{d-1}\delta h$ of the defect. Replacing ξ by $\tau^{1/\tilde{z}}$ and δh by $\tau^{\alpha/\tilde{z}}$, one obtains for large ξ

$$\tilde{z} - d - 1 = 2\alpha \,.$$

Combination with (14.7) yields

$$\alpha = (3 - d)/3 \,. \tag{14.11}$$

Thus, the upper critical dimension of the Montréal model is 2+1, which means that the surface would be smooth above the physical dimensions 3. The roughness in 3 dimensions corrresponds to $\alpha = 0$ and is therefore weak, possibly logarithmic.

An interesting extension of the Montréal model has been proposed by Sun *et al.* (1992). These authors take into account the discrete structure of the lattice by adding to the right hand side of (14.4) a term $K\nabla^2$ [$\sin(2\pi z)$] which favours integer values of z. This modulated Montréal model turns out to have a roughening transition when K changes.

14.7 Conclusion

The following remarks can be made.

a) In all the models which have been reviewed (apart from the modulated Montréal model mentioned in section 14.6) the upper critical dimension is at least 3, so that, in the physical, three-dimensional space, the surface is rough in 3 dimensions at all temperatures. The continuous description by formula (13.1b) is therefore consistent.

b) All the kinetic equations which have been written from (12.6) to (14.4) reduce to (8.7) or (8.12) in the absence of growth or evaporation (i.e. for $F = 0$). However, the values of the coefficients (e.g. v in equation 13.4) are very different far from equilibrium ($F \neq 0$) and near equilibrium ($F = 0$)

c) Many questions addressed in chapters 13 and 14 are still open. In the near future, some renormalization group method will hopefully be devised, which will provide an analytic solution of the KPZ model and confirm, or contradict the results (13.7) of Kim & Kosterlitz. The validity of the values (14.10) for the model (14.4) beyond the 'one loop' approximation will be clarified. Finally a systematic investigation of the Eaglesham-Gilmer instability is still to be made.

Problem

14.1. Formula (14.4) yields the growth rate parallel to a given plane. Write the formula analogous to (14.4) for the healing rate normal to the surface, assuming the diffusion constant to be independent of the surface orientation.

Solution – Siegert & Plischke 1992.

15
Elastic interactions between defects on a crystal surface

Je préfère explorer les forêts vierges
que cultiver un jardin de curé.

Louis Néel (*Un siècle de Physique*)

I prefer to explore virgin forests
than cultivate a parson's garden.

So far we have neglected the complications which result from elasticity. Elasticity is essential in the epitaxy of a crystal on another one, because it introduces long range effective interactions between adatoms, steps, and other surface defects. In the present chapter these interactions will be derived without calculations from simple elementary considerations.

A surprising feature of elastic interactions between adatoms is that they are repulsive, while the short range, chemical interaction is of course attractive– and much stronger. This conflict between strong, short range and weak, long range interactions can give rise to long period reconstructed superstructures. These long period features appear because elastic interactions are too weak to create many defects, but can add their effects to create a few ones.

We shall use the linear, continuum approximation of elasticity, which is sufficient for most problems. However, even a linear problem can become complicated in a system which is not infinite in all directions, and therefore, in particular, for a medium limited by a surface. These complications will be avoided in the present chapter and only appear in the next one and in the appendices. The continuum approximation is not completely satisfactory because it does not allow us to describe surface relaxation, which is an atomic scale phenomenon.

Note that in this chapter and in the following one the unit length is not set equal to 1. All relations are thus dimensionally correct.

15.1 Introduction

The structure of a crystal surface (either clean or with an adsorbate) partly depends on elastic mechanisms. For instance, if one considers 2 adatoms, each of them produces a strain of the substrate in its neighbourhood. This strain extends to long distances and influences the position of the other adatom. In other words, the interaction free energy of the 2 adatoms is a function $W(r)$ of their distance r, which depends partly on elastic mechanisms.

The elastic contribution to $W(r)$ is usually weak in comparison with the energy of a chemical bond, which is typically a few electronvolts. Consider for instance a single adatom on the (111) face of an fcc metal (Fig. 15.1). It is not able to change very much the atomic distance of its neighbours, since they have already 9 neighbours which do not like to change their atomic distance. Moreover, in a model based on pair interactions between nearest neighbours, the strain produced by adatoms identical to those of the substrate is strictly zero as shown in appendix M. In the case of a 'foreign' adatom, elastic effects are more important. The order of magnitude suggested by Lau & Kohn (1977) is 0.1 eV for chemisorption.

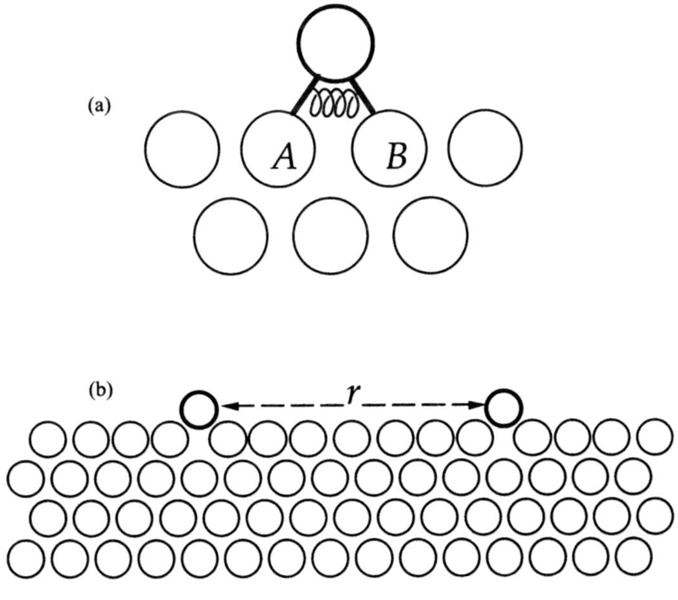

Fig. 15.1. a) An adatom tries to change the distance between its neighbours, and thereby exerts a stress. b) Two adatoms on a planar surface. The energy is proportional to $1/r^3$. An adatom which tends to expand the pair AB below it produces a compression of the neighbouring, more distant, pairs. This discourages other adatoms from approaching the first one. The interaction is therefore repulsive.

It is of some interest to recall first the features of elastic effects in an infinite solid. A point impurity produces at a distance $\mathbf{r}$ a displacement $u(\mathbf{r})$ which satisfies the following equation (rederived in the next chapter as equation 16.11):

$$(\lambda + \mu)\nabla(\mathrm{div}\mathbf{u}) + \mu\nabla^2\mathbf{u} = 0 .$$

Since this equation looks like the Poisson equation of electrostatics, and since the displacement is a vector, it is reasonable to ask whether

$$\mathbf{u}(\mathbf{r}) = \mathrm{Const} \times \mathbf{r}/r^3 \qquad (15.1)$$

(which corresponds to the electric field in electrostatics) is a solution. Actually, it is, since $\mathrm{div}\mathbf{u} = \nabla^2\mathbf{u} = 0$. This decay proportional to r^{-2} is slow in comparison with that of the chemical interactions (which are roughly exponential): *elastic effects are usually long ranged.*

This long range nature of the displacement field generates long range, effective interactions between defects: for instance the energy per unit length of two parallel dislocations of length L and of opposite sign at distance r is proportional to $\ln r$ (Friedel 1956). It does not even decay for large r, but instead it goes to infinity!

It will be argued in the next section that a long ranged, elasticity-mediated interaction is also present between adatoms on a surface. A naive microscopic interpretation of these effects is given in appendix M.

15.2 Elastic interaction between two adatoms at a distance r

Consider first a single adatom (Fig. 15.1a). It tries to modify the positions of the neighbouring atoms of the substrate. In more precise words, it exerts forces on the substrate. The total force is opposite to the sum of the forces exerted by the solid on the adatom, which vanishes as a consequence of mechanical equilibrium. By analogy with electric dipoles, it is natural to introduce the *force dipole moments*, defined as the mechanical moments of a set of forces $\mathbf{F}_R$ acting at points $\mathbf{R}$, namely:

$$m_{\alpha\gamma} = \sum_R R_\alpha F_R^\gamma \qquad (\alpha, \gamma = x, y, z) . \qquad (15.2)$$

A force dipole moment has the dimension of an energy. In this and in the following chapter the length unit is not chosen equal to the atomic distance, and all formulae keep their correct dimensions. The expressions (15.2) are independent of the choice of the origin, since $\sum_R \mathbf{F}_R = 0$ has been assumed.

The strain resulting from the force dipoles of the adatoms is associated with an elastic energy (or, at finite temperature, a free energy) W_0^{el}. The displacement at distance r from the adatom on the surface will be assumed to have still the form (15.1). We show in appendix N that this assumption is indeed true. The components of the strain are the derivatives of $u(r)$ with respect to r, which are proportional to $1/r^3$.

If we now want to deposit another adatom at a distance r from the first one, we have to pay an elastic energy which would be equal to W_0^{el} in the absence of strain, but which is slightly different because of the strain produced by the first adatom. The difference is expected to be proportional to the strain itself, and therefore to $1/r^3$. This expectation is confirmed by a detailed calculation done in section 16.6. Thus, the energy of a pair of adatoms at distance r is equal to twice the energy of a single adatom, plus a correction $W_{\mathrm{int}}(r)$, which will be called *elastic interaction energy*. Its expression will only be given for the symmetry associated to an adatom on a (001) or (111) face of a cubic lattice. In this case,

$$m_{\alpha\gamma} = \delta_{\alpha\gamma}\left[(\delta_{\alpha x} + \delta_{\alpha y})m + \delta_{\alpha z}m_{zz}\right] .$$

The interaction energy of two force dipoles m and m', which is derived in appendices J and L (except for the cross term which is taken from Duport 1996), reads:

$$\begin{aligned} W_{\mathrm{int}} &= \frac{1-\zeta^2}{\pi E r^3}\left[mm' - \frac{\zeta}{1-\zeta}\left(mm'_{zz} + m_{zz}m'\right) + \left(\frac{\zeta}{1-\zeta}\right)^2 m_{zz}m'_{zz}\right] \\ &= \frac{1-\zeta}{2\pi\mu r^3}\left[mm' - \frac{\zeta}{1-\zeta}\left(mm'_{zz} + m_{zz}m'\right) + \left(\frac{\zeta}{1-\zeta}\right)^2 m_{zz}m'_{zz}\right] , \end{aligned} \tag{15.3}$$

where E is the Young modulus and ζ the Poisson coefficient. The moments m and m' are a measure of how uncomfortable the adatoms feel on the surface. In the case of homoepitaxy, one can expect the adatoms to be fairly happy and $m = m'$ to be small.

Since the Poisson ratio ζ is always smaller than $1/2$ (see section 16.3), the interaction (15.3) is seen to be repulsive between atoms of the same kind. This repulsive interaction can be intuitively understood as follows (Fig. 15.1): the main effect of each adatom is to create a local strain. Assume for instance that the first adatom produces an expansion of the atom pair AB just below it. The result is a compression on both sides of the pair AB, which decreases (proportionally to $1/r^3$) with the distance. The second adatom will try to sit in less compressed places, i.e. as far as possible from the first one. Actually, the situation is more complicated because the strain due to an adatom is not a uniform compression or dilatation, but a compression in one direction and a dilatation in the other. The resulting interaction is the sum of a repulsive and an attractive contribution. The net result for two *identical* adatoms is always a repulsion. This is shown in appendices J and L. On the other hand, the presence of competing interactions (in the bulk) is attested by the following astonishing fact: the elastic interaction between two point defects in an infinite solid vanishes at order $1/r^3$ (see problem 16.1 in chapter 16).

A remarkable feature is the conflict between the long range elastic repulsion and the short range chemical attraction. The ground state of a system of two adatoms is generally a pair, but if this pair is broken, what remains is the elastic repulsion. This repulsion is usually much weaker than the attraction because m and m' are usually small, especially in the case of homoadsorption. In fcc metals, for instance, the energy (of chemical nature) necessary to break a pair is of order 0.5 eV (twice the cohesion energy, divided by the number of neighbours of a given atom). On the other hand, the Young modulus and the other elastic constants, listed in appendix P, are of order 10 to 100 GPa. Since 1 GPa = 6.24 eV/nm^3, the order of magnitude of the elastic constants is a few millielectronvolts per atom.

The electric dipole interaction between adatoms is also proportional to $1/r^3$. It is not necessarily small with respect to the elastic interaction, but the latter is more important in heteroepitaxy because it becomes an insurmountable obstacle to layer-by-layer growth if the misfit is large.

15.3 Interaction between two parallel rows of adatoms

The problem is again represented by Fig. 15.1, which has now to be understood as a cross-section of the material.

We first consider the interaction between an adatom row and an isolated adatom at a distance r from this row. The interaction can be obtained from the result of the previous section, by integration of the repulsive contribution in $1/r^3$ over the row. The adatom-row interaction is therefore *repulsive and decays as* $1/r^2$. The interaction between two rows is obtained next by multiplying by the number of atoms per row. Therefore, the interaction energy per unit length between two rows of adatoms is *repulsive and proportional to* $1/r^2$. If L_y is the length of the row, and m, m' are the dipole moments per atom on the respective rows, the interaction energy obtained by integration of (15.3) is

$$W_{\rm int}(r) = \left[mm' - \frac{\zeta}{1-\zeta}\left(mm'_{zz} + m_{zz}m'\right) + \left(\frac{\zeta}{1-\zeta}\right)^2 m_{zz}m'_{zz}\right] \times \frac{L_y}{a}\frac{1-\zeta^2}{\pi E}\int_{-\infty}^{\infty}\frac{dy}{a\left(r^2+y^2\right)^{3/2}} .$$

Introducing the dipole moment per unit length, $dm/dy = m/a$, the energy per unit length, $\mathrm{d}W_{\mathrm{int}}/\mathrm{d}y = W_{\mathrm{int}}/L_y$ is found to be

$$\mathrm{d}W_{\mathrm{int}}(r)/\mathrm{d}y = 2\frac{1-\zeta^2}{\pi E r^2}\left[\frac{\mathrm{d}m}{\mathrm{d}y}\frac{\mathrm{d}m'}{\mathrm{d}y} - \frac{\zeta}{1-\zeta}\left(\frac{\mathrm{d}m}{\mathrm{d}y}\frac{\mathrm{d}m'_{zz}}{\mathrm{d}y} + \frac{\mathrm{d}m_{zz}}{\mathrm{d}y}\frac{\mathrm{d}m'}{\mathrm{d}y}\right) + \left(\frac{\zeta}{1-\zeta}\right)^2 \frac{\mathrm{d}m_{zz}}{\mathrm{d}y}\frac{\mathrm{d}m'_{zz}}{\mathrm{d}y}\right]. \tag{15.4}$$

In this formula, we have omitted an r-independent term which is the sum of the interactions between adatoms of the same row. Note that m and m' are the dipole moments of the atoms in the two rows, assumed to be identical inside each row. Thus, the definition of m and m' is the same in (15.3) and (15.4) ... however their values are not the same! The force dipole exerted by an adatom on the substrate is not the same if this

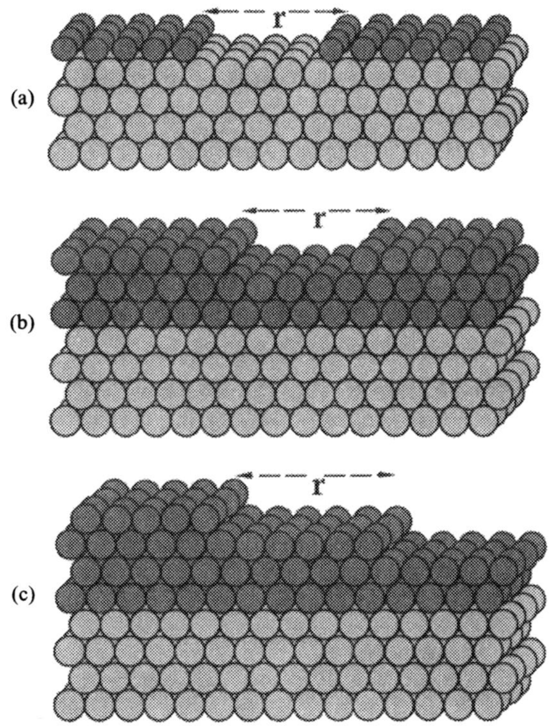

Fig. 15.2. a) Two half-layers adsorbed on a planar crystal surface in commensurate position. b) The same surface, with a finite number of complete adsorbed layers on the substrate. c) Two steps on a commensurate adsorbate. In all three cases, the elastic interaction energy is proportional to $\ln r$. It is repulsive in cases (a) and (b), attractive in case (c).

atom is isolated or embedded inside a cluster. We shall come back to this important caveat in section 15.5.

Formula (15.4) is not restricted to rows of adatoms, but also applies to other cases, which are more realistic because adatoms have no reason to form rows. Other examples of linear defects are discommensurations (Villain & Gordon 1983) or cracks (Berréhar *et al.* 1989, 1992). Another instance is that of steps on a clean surface. This is a more complicated case which will be addressed in section 15.5.

15.4 Interaction between two semi-infinite adsorbed layers

What is addressed in this section is *commensurate* heteroadsorption (Fig. 15.2).

Consider first the interaction between a half-layer and a row at a distance r (Fig. 15.3). It is obtained by integrating the interaction (15.4), which varies as $1/r^2$, over the half-layer. The result is proportional to $1/r$. The interaction between two half-layers is obtained by integrating this result over all the rows of the second half-layer. Ignoring an r-independent term which is the interaction between the atoms of the same monolayer, the interaction energy between two half-layers is thus found to be *proportional to* $\ln r$:

$$W(r) \simeq -\text{Const} \times \ln r \,. \tag{15.5a}$$

The logarithmic behaviour is analogous to the elastic interaction between two parallel dislocations at a distance r inside an infinite solid.

An important case is realized when the strain is solely due to the 'misfit' $\delta a/a$ between the 'natural' lattice constant a of the substrate and that $a + \delta a$ of the adsorbate (where δa can be positive as well as negative). Then, it will be seen in section 16.7 that an adsorbed layer of thickness h is equivalent to a density of force dipoles per unit area

$$\frac{\mathrm{d}m}{\mathrm{d}S} = \frac{E}{1-\zeta}\frac{\delta a}{a}h \,.$$

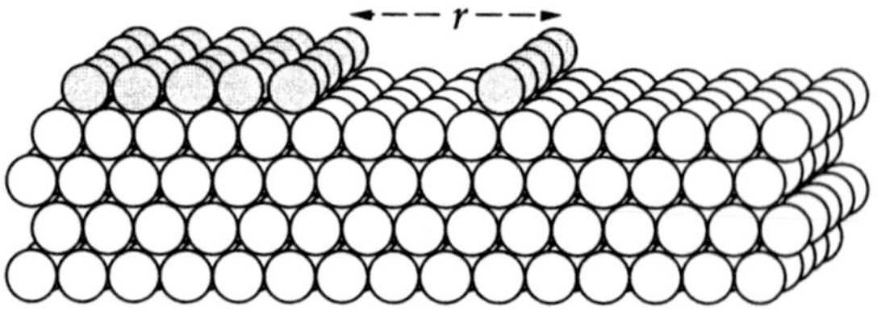

Fig. 15.3. A half-layer and a row adsorbed on a planar surface. The elastic interaction energy is proportional to $1/r$.

The adsorbed layer can be decomposed in stripes of width dx, each having a dipole moment per unit length equal to

$$\frac{\mathrm{d}m}{\mathrm{d}y} = \frac{E}{1-\zeta}\frac{\delta a}{a}h\,\mathrm{d}x\,.$$

Insertion of this expression for dm/dy into (15.4) and integration yield

$$\mathrm{d}W_{\mathrm{int}}(r)/\mathrm{d}y = 2\left(\frac{E}{1-\zeta}\right)^2\left(\frac{\delta a}{a}\right)^2\int_{-\infty}^{0}\mathrm{d}x\int_{r}^{\infty}\mathrm{d}x'\frac{1-\zeta^2}{\pi E(x-x')^2}\,,$$

or

$$\mathrm{d}W_{\mathrm{int}}(r)/\mathrm{d}y = -\frac{2E}{\pi}\frac{1+\zeta}{1-\zeta}(\delta a)^2\ln r\,, \tag{15.5b}$$

where we let $h = a$. A term independent of r has been omitted. For a system with a finite size, this term would be a constant proportional to the logarithm of the size. Being a constant, it can be incorporated into the step energy and ignored in the interaction. In the limit of infinite size, this constant is infinite but can be ignored as well.

This interaction is repulsive since $W(r)$ is minimum and equal to $-\infty$ for $r = \infty$. This is not surprising, since the interaction (15.5) is again the sum of repulsive interactions between adatoms. Thus, this interaction favours the splitting of a monolayer. On the other hand, a monolayer can only split if chemical bonds are broken, and this costs a big amount of energy. It turns out that the constant in (15.5) is small in comparison with chemical bonds. However, the logarithm diverges with r.

As will be seen in the next section, this divergence has an important consequence. Indeed, a monolayer must in principle split into parts. This splitting will produce an energy gain, even if the constant is small in (15.5). The number of pieces just decreases with the constant. This effect is related to the Volmer-Weber growth process mentioned in chapter 5. However, the occurrence of the Volmer-Weber type of growth may be due to a high interface free energy (which may be present even if the growth is incommensurate and there is no strain) rather than to elastic effects.

The logarithmic repulsion (15.5) between two half-monolayers is present if these half-monolayers are directly deposited on the substrate (Fig. 15.2a) but also if they are deposited on any number of complete adsorbed layers in commensurate positions (Fig. 15.2b). This can be seen from the calculation of section 16.7. The particular case where the adsorbate and the substrate have the same elastic constants is of interest because of its simplicity. In that case, the interaction is independent of the thickness of the adsorbate.

The case of two steps of identical 'sign' on an adsorbate (Fig. 15.2c) can be treated in an analogous way. The interaction is still logarithmic, but it is now attractive! The intuitive reason is that the adatoms feel a constraint because of the misfit between the lattice parameters. Near a step, this constraint is partially relieved. Multiple steps correspond to a more complete relief. Hence an attractive interaction between monoatomic steps.

The logarithmic interaction derived in this section is peculiar to commensurate adsorption on a substrate which is chemically different from the adsorbate. The case of homoepitaxy will be discussed in the next section.

15.5 Steps on a clean surface

It is time to be concerned with steps (Fig. 15.4). Arguing as in the previous section and integrating three times the elementary interaction (15.3), one might be led to believe that the interaction energy between two parallel steps is proportional to the logarithm of their distance r on any surface. This is not true in general, as will now be seen. Two cases should be distinguished.

a) Case of a Bravais lattice

A Bravais lattice is a crystal lattice, whose unit cell contains a single atom.

The argument of section 15.4, which leads to a logarithmic interaction, relies on the idea that the adsorbed layer imposes a fixed strain or stress per atom. However, if the adsorbate is identical to the substrate, it is intuitive that there is no such strain. More precisely, this strain is localized near steps. Even more precisely, we can expect that a step behaves as a line defect and produces a strain which decays as $1/r^2$ with the distance r, as seen in section 15.3. The logarithmic interaction appears if the strain is proportional to $1/r$.

The interaction between two steps in the case of a Bravais lattice can be expected to be of the same kind as between two line defects. As seen from section 15.3, it should be proportional to $1/r^2$ (Marchenko & Parshin 1980, Nozières 1991). This point will be discussed in more detail in section 16.11. If the steps have the same sign (Fig. 15.5b) their interaction is

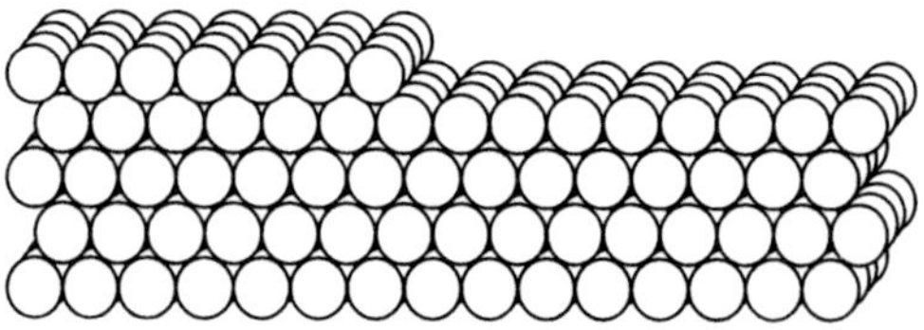

Fig. 15.4. A step on a planar surface.

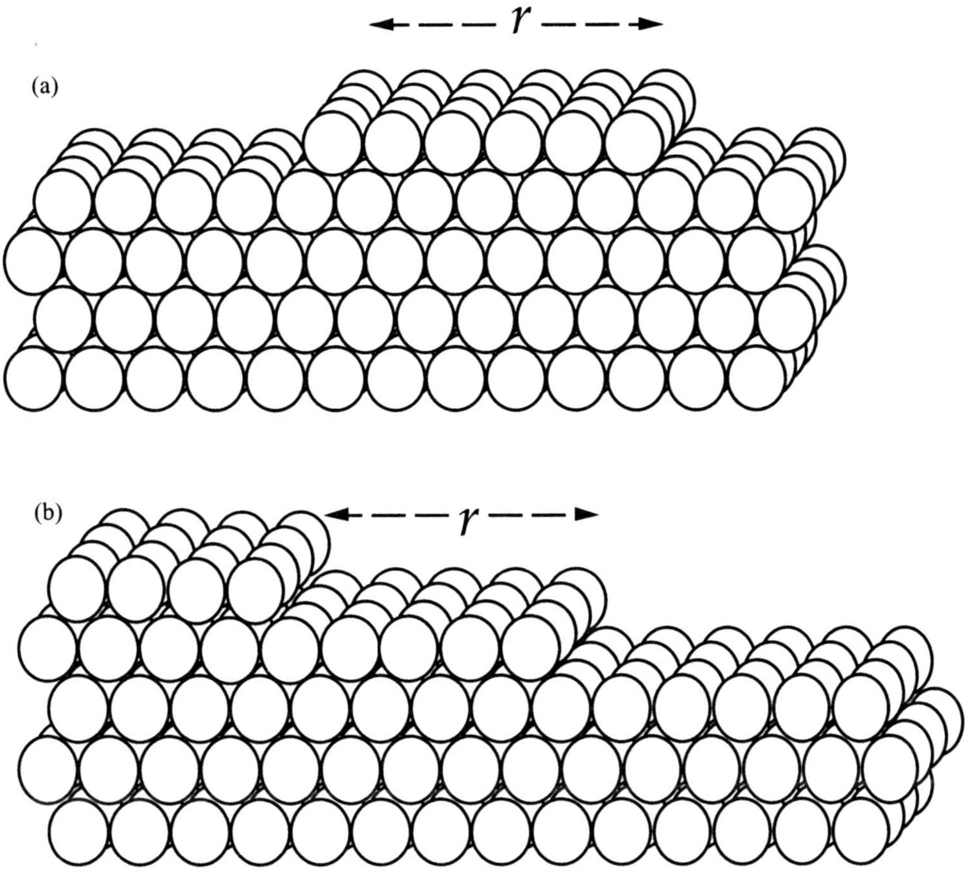

Fig. 15.5. Two steps on a planar surface. a) Steps of opposite sign. b) Steps of identical sign.

repulsive. In the case of two steps of opposite sign (Fig. 15.5a) there is no simple rule to deduce the sign of their interaction.

b) The (100) surface of silicon (Alerhand et al. 1988)

As seen in chapter 9, silicon has two atoms per unit cell, therefore its lattice is not Bravais'. The (100) surface has an orthorhombic symmetry which tends to produce a strain parallel to the surface. For a given state of the surface, this strain corresponds to an expansion in one direction (say x) and a contraction in the other direction, y. If one more layer is added, the surface tends to produce an expansion in the y direction and a contraction in the other direction, x. If two more layers are added, the directions of expansion and contraction are not changed.

Thus, in contrast with the case of a Bravais lattice, an additional layer does exert a stress, though two additional layers produce no stress.

However, in the case of a thick crystal limited by a complete (001) layer, the stress exerted by the surface has no effect: the superficial layer competes with the whole crystal, which does not like to be strained.

The situation is different if the upper layer splits into stripes (Fig. 15.6). Under the stripes, the crystal is expanded in the x direction and compressed in the y direction. Between the stripes, the crystal is expanded in the y direction and compressed in the x direction. Far from the surface, the strain is negligible.

The actual calculation is similar to that of the previous section. Assuming the last layer to be exactly half-filled, and to split into parallel stripes of width ℓ at distance ℓ (Fig. 15.6), the elastic free energy is easily seen to be equal to an ℓ-independent term, plus a logarithmic term analogous to (15.5), namely

$$\delta \mathcal{F}_{\text{el}} \approx -\mathcal{A}\frac{C}{\ell}\ln \ell \tag{15.6}$$

where $\mathcal{A}$ is the total surface area and C is a constant.

The total free energy $\delta\mathcal{F}$ (counted from that of the flat surface) is obtained by adding a term proportional to the number of broken chemical bonds, and therefore to $1/\ell$:

$$\delta\mathcal{F}/\mathcal{A} \approx \frac{w_0}{\ell} - \frac{C}{\ell}\ln \ell\,,$$

where w_0 is the energy of a chemical bond. The minimization of $\delta\mathcal{F}$ yields $\ell \simeq \exp(w_0/C)$ if $w_0 \gg C$. At equilibrium, the surface splits into stripes, even if the elastic energy is very small. If it is very small, the time necessary to reach equilibrium may be unphysically long. However, stripe formation on Si surface, as theoretically predicted by Alerhand *et al.*, has been observed experimentally (Men *et al.* 1988). However, more complicated superstructures have been observed (Tromp & Reuter 1992,

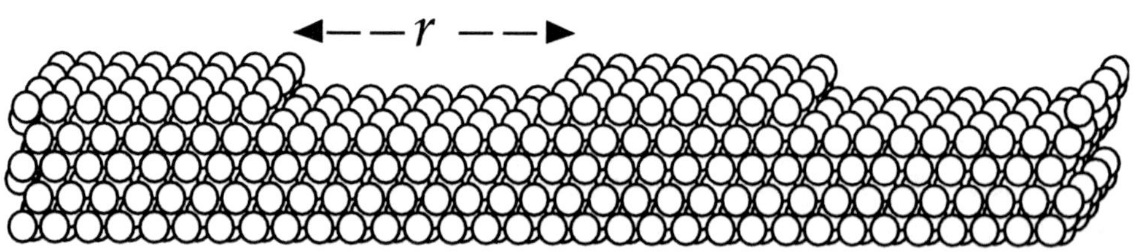

Fig. 15.6. Superstructure of the (001) surface of silicon according to Alerhand *et al.* (1988). The upper layer has an orthorhombic distortion, and the expanded directions are orthogonal in the two types of stripes.

1993) and theoretically investigated (Mukherjee *et al.* 1993). Kinetic effects may play an important role.

The elasticity-driven reconstruction of Si(001) has some analogy with the long period magnetic structures observed in ferromagnetic thin films. In these systems, there is a competition between weak, but long range magnetic dipole interactions and a strong, short ranged exchange interactions (Gehring & Keskin 1993). This is analogous to the competition between weak, but long ranged elastic interactions and strong, short ranged chemical interactions on the Si surface. Note that magnetic thin films, as well as silicon, have a technological importance! In both cases the energy has the form (15.5).

Other long period structures appear when one tries to cut a crystal along an unstable orientation, forbidden by the Wulff construction or the double tangent construction (see chapter 3). The surface takes then a superstructure (formed in the simplest case by alternating facets) whose period is determined by elasticity (Marchenko 1981, Cahn 1982, Phaneuf *et al.* 1988).

15.6 More general formulae for elastic interactions

Up to this point, in this chapter, we have calculated the interactions between simple surface defects. We would like to be able to calculate the energy of any structure. The general equations will be derived in the next chapter, but one can already treat a few simple cases.

i) First case (Fig. 15.7): an adsorbate is in epitaxy on a substrate, i.e. it is commensurate with it (without misfit dislocations). The substrate-adsorbate interface is planar and parallel to the xy plane. The substrate is infinitely thick. The adsorbate surface is assumed to be almost planar and parallel to the xy plane, except for small height variations $\delta z(x, y)$, which are smooth functions of x and y. The field $\delta z(x, y)$ can be assumed to vanish at infinity.

Then, one can just add the contributions (15.3). The stress exerted on the substrate by a column of height z is just z times the stress p_0 exerted by a single adatom. Therefore, if z is small enough everywhere, the elastic free energy (counted from the dislocation-free state with a flat interface) is

$$F_{\mathrm{el}} = K \int \mathrm{d}^2 r z^2(r) + \frac{1-\zeta^2}{\pi E} p_0^2 \int \int \mathrm{d}^2 r \mathrm{d}^2 r' \frac{z(r)z(r')}{|\mathbf{r}-\mathbf{r}'|^3} . \qquad (15.7)$$

The first term is the sum of the energies of isolated adatoms, plus the interaction of adatoms stacked at the same location. It is often

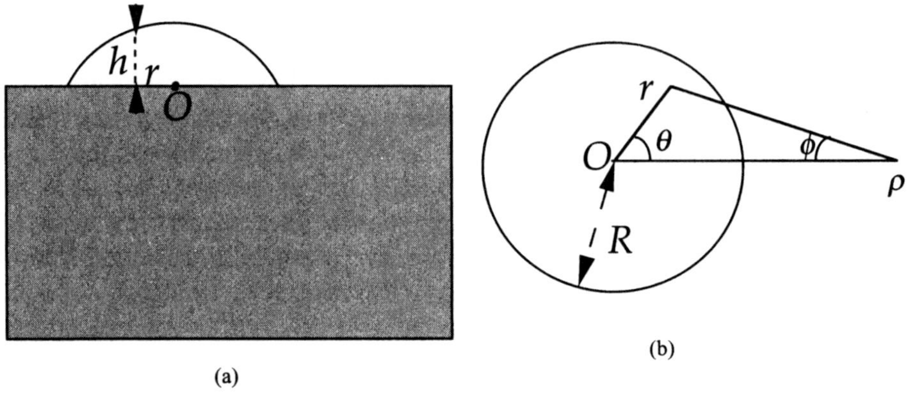

Fig. 15.7. An adsorbed bubble on a surface. a) Cross section. b) Top view.

neglected in the remainder of this chapter, but it is not always possible to do it. Since the elastic free energy for constant $z(r)$ is by definition zero,

$$K = -\frac{1-\zeta^2}{\pi E} p_0^2 \int d^2r \frac{1}{r^3} .$$

It follows from this relation that (15.7) is not modified by changing the origin of z, which is therefore arbitrary and not necessarily situated on the subtrate.

If z is not small, an algebraic complication appears, due to the different elastic constants in the two media. For the sake of simplification, it will be assumed that the adsorbate and the substrate are isotropic solids with *identical values of the Poisson ratio* ζ *and of the Young modulus* E. Then, the above formula is also applicable to a thick adsorbate.

The stress p_0 is related to the strain δa (the misfit), which is known since it is the difference between the lattice parameter of the substrate a and that of the adsorbate, a'. For small δa, Hooke's law (see next chapter) states that the stress-strain relation is linear, namely

$$p_0 = \frac{E}{1-\zeta} \frac{\delta a}{a} . \tag{15.8}$$

Relations (15.7) and (15.8) imply that the interaction between steps on an adsorbate depends only on the misfit and on the elastic constants, in contrast with the interaction (15.3) between adatoms, which also depends on the dipole moments m and m'. This difference shows that the remarkably simple results (15.7) and (15.8) hold only for modulations of long wavelengths, for instance for large terraces of monoatomic thickness. Strictly speaking, the energy contains, in addition to (15.7), a correction due to the edges of these terraces.

ii) What happens now in the case of a clean surface (without an adsorbate)? If one tries to apply (15.7) and (15.8), one has $\delta a = 0$ and therefore no interaction between steps. However, the correction term remains. Stewart & Goldenfeld (1992) have proposed the following formula for the elastic free energy of a undulating solid surface, if its local orientation is not too far from a high-symmetry direction $z = \text{Const}$:

$$\mathscr{F} = \frac{C}{2}\int \mathrm{d}^2r \left(\nabla^2 z\right)^2 + C'\int\int \frac{\mathrm{d}^2r\mathrm{d}^2r'}{\mid \mathbf{r}-\mathbf{r}' \mid}\left(\nabla^2 z(\mathbf{r})\right)\left(\nabla^2 z(\mathbf{r}')\right) . \tag{15.9}$$

C and C' are constants, which cannot be deduced from elasticity theory. One can for instance apply this formula to two linear defects at a distance r. One easily finds that their interaction energy is inversely proportional to $1/r^2$, in agreement with the argument given in subsection 15.5a.

Generally, one can consider a homogeneous elastic solid bounded by a plane and infinite in all other directions. Such a solid is said to be semi-infinite. If external forces $f_\alpha(x, y)$ are applied to the boundary plane (the surface of the solid body), the elastic free energy has the form

$$\mathscr{F} = \int \mathrm{d}^2r \int \mathrm{d}^2r'\Gamma_{\alpha\beta}(\mathbf{r}-\mathbf{r}')f_\alpha(\mathbf{r})f_\beta(\mathbf{r}') \tag{15.10}$$

where the tensor $\mathbf{\Gamma}(\mathbf{r}-\mathbf{r}')$ is called a Green function. The determination of Green functions in elastic media is the subject of many pages in several textbooks: a useful reference is chapter 13 of Morse & Feshbach (1953). A simplified version is given in next chapter and in the appendices. Despite the algebraic complications, it should be emphasized that the quadratic form (15.10) is very simple. This simplicity stems from the fact that, in linear elasticity, there is a linear relation between the forces (or the stresses) and the atomic displacements (or the strains).

However, formula (15.10) does not completely solve the problem since the forces f are not known, except for certain limits, e.g. when (15.7) applies.

15.7 Instability of a constrained adsorbate

In section 15.5 we discussed an instability of the (100) silicon surface to splitting into domains, which results from the logarithmically divergent interaction between steps on this surface. Such an interaction is also found in heteroepitaxy, if no misfit dislocations are present to relax the strain imposed by the substrate on the adsorbate. Therefore, an epitaxial (i.e. commensurate) adsorbate with a flat surface is also expected to be unstable.

In this section, the stability with respect to the formation of a truncated pyramid (Fig. 15.8) will be investigated. The following argument summarizes a calculation by Tersoff & LeGoues (1994).

Let h be the height of the pyramid, ℓ its average width, and θ its slope. The energy of the pyramid can be written as the sum of three terms.

i) The interaction between parallel steps of opposite signs, on opposite sides of the pyramids. The interaction energy per unit length between two steps is, on average, proportional to $-\ln \ell$, as in (15.6). Since the size of a terrace is of order ℓ, and the number of step pairs is h^2, the total energy of type (i) is (Tersoff & Tromp 1993)

$$W^{(i)} \approx -4Ch^2\ell \ln \ell .$$

C contains the elastic constants and is computed below.

ii) The interaction energy between parallel steps of the same sign, on the same side of the pyramid. The distance between these steps is proportional to $h \cot \theta$. The interaction energy per unit length between two steps is proportional, on average, to $\ln(h \cot \theta)$. Since the size of a terrace is of order ℓ and the number of step pairs is h^2, the total energy of type (ii) is

$$W^{(ii)} \approx 4Ch^2\ell \ln(h \cot \theta) .$$

iii) The interaction energy between steps of different orientations. This contribution does not modify the result qualitatively. The detailed calculation is just the integration of (15.7) over the pyramid, and the elastic energy of the pyramid turns out to be (Tersoff & LeGoues 1994):

$$W_{el} \approx -4Ch^2\ell \ln \frac{\ell e^{3/2}}{h \cot \theta} \qquad (15.11)$$

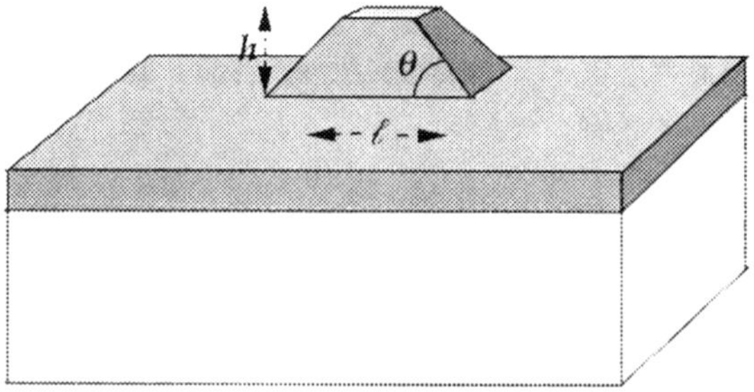

Fig. 15.8. An adsorbate island in the form of a truncated pyramid on a misfitting substrate.

which is essentially the sum of the contributions (i) and (ii). This formula is correct for $h \ll \ell$ and $0 < \theta < \pi/2$, and not too close to $\pi/2$. For $\theta = \pi/2$, the denominator inside the logarithm should be replaced by 1.

To the elastic energy (15.11), a positive surface energy ('capillarity') contribution equal to $4\gamma h\ell$ should be added, where γ is the step free energy. As in the preceding section, the logarithmic factor implies that the total energy

$$W \approx -4Ch^2\ell \ln \frac{\ell e^{3/2}}{h \cot\theta} + 4\gamma h\ell \tag{15.12}$$

is negative if ℓ is large enough, whatever the value of h. However, for small ℓ and h, the energy is positive. Therefore, there is an activation barrier for the birth of a pyramid. It will be determined for small C and for a given value of θ, not too close to $\pi/2$. It is convenient to take as variables $x = \ell/h$ and $y = h^2\ell$. With this choice, (15.12) reads

$$W \approx -4Cy \ln \frac{xe^{3/2}}{\cot\theta} + 4\gamma x^{1/3} y^{2/3} . \tag{15.13a}$$

The total energy must be minimized as a function of the aspect ratio $x = \ell/h$, at constant volume. The volume can be approximated by

$$V \approx h(\ell^2 + h^2 \cot^2\theta) = \left(xy + \frac{y}{x} \cot^2\theta \right) . \tag{15.13b}$$

Geometry imposes the constraint $x \geq \cot\theta$.

The requirement of constant volume yields y as a function of x, and the energy is then also a function of x only. The differentiation of (15.13a) with the condition given by the constancy of (15.13b), $V = V_0$, yields

$$W'(x) \approx 4CV_0 \frac{x^2 - \cot^2\theta}{(x^2 + \cot^2\theta)^2} \ln\left(\frac{xe^{3/2}}{\cot\theta}\right) - \frac{4CV_0}{x^2 + \cot^2\theta}$$
$$+ \frac{4}{3}\gamma V_0^{2/3} \frac{3\cot^2\theta - x^2}{(x^2 + \cot^2\theta)^{5/3}}$$

for $x \geq \cot\theta$. It is readily seen that, if C and θ are small enough, W' is positive at the boundary of the domain of existence of W, i.e. at $x = \cot\theta$. This is thus a minimum for the energy. The corresponding value of y is $y(\cot\theta) = h^3 \cot\theta$, the volume is $V = h^3 \cot^2\theta$ and the energy is

$$W \approx -6Ch^3 \cot\theta + 4\gamma h^2 \cot\theta = -6CV \tan\theta + 4\gamma V^{2/3} \tan^{1/3}\theta .$$

This energy is positive for small V, and negative for large V. The maximum is the activation energy. It corresponds to

$$V_0 = \left(\frac{4\gamma}{9C}\right)^3 \cot^2\theta$$

and the energy barrier is

$$W_0 \approx \frac{\gamma^3}{C^2}\cot\theta \,. \tag{15.14}$$

The constant C is easily deduced from (15.7) and it is equal to

$$C = \frac{1-\zeta^2}{\pi E}p_0^2 = \frac{1-\zeta}{2\pi\mu}p_0^2 = E\frac{1+\zeta}{1-\zeta}\left(\frac{\delta a}{a}\right)^2 .$$

Inserting this value into (15.14) shows that the activation energy is proportional to $(a/\delta a)^4$. Note that a complete calculation would imply a minimization also with respect to θ. The result would be a negative value of θ, i.e. an upside-down pyramid (a pit, in fact), and the formulae above would not be correct in that case. This implies that a pit is energetically favoured with respect to a hill, if the adsorbate is thick enough to allow for the depth of the pit; otherwise the hill appears first.

Problems

15.1. Calculate the strain created by an adsorbed droplet at the surface of a substrate, in the case of an isotropic surface tension.

Solution – The height of the adsorbate at $\mathbf{r}$ is $h(\mathbf{r}) = Rf(r/R)$, where the function f is the equation of a sphere of radius R and depends on the contact angle given by Young's formula of chapter 5. The displacement at ρ due to the effect of the adsorbate at $\mathbf{r}$ is proportional to $(\rho - \mathbf{r}) / \mid \rho - \mathbf{r} \mid^3$. Integrating the effects of the various points $\mathbf{r}$, the displacement at ρ is (Fig. 15.7), omitting a multiplicative factor which depends on the elastic constants,

$$u(\rho) = R\int_0^\infty r\mathrm{d}rf\left(\frac{r}{R}\right)\int_{-\pi}^{\pi}\mathrm{d}\theta\frac{\cos\varphi}{\mid \rho - \mathbf{r}\mid^2} .$$

The denominator is $\mid \rho - \mathbf{r} \mid^2 = \rho^2 + r^2 - 2r\rho\cos\theta = L^2$ and φ is defined by

$$\frac{\sin\varphi}{r} = \frac{\sin\theta}{L}$$

or

$$\cos\varphi = \frac{\rho - r\cos\theta}{\sqrt{\rho^2 + r^2 - 2r\rho\cos\theta}} .$$

Therefore

$$u(\rho) = R\int_0^\infty r\mathrm{d}rf\left(\frac{r}{R}\right)\int_{-\pi}^{\pi}\mathrm{d}\theta\frac{\rho - r\cos\theta}{\left(\rho^2+r^2-2r\rho\cos\theta\right)^{3/2}}$$

$$= R\int_0^\infty \frac{\mathrm{d}r}{r}f\left(\frac{r}{R}\right)g\left(\frac{r}{\rho}\right)$$

where g is a function of r/ρ. The change of variable $x = r/R$ yields

$$u(\rho) = R\int_0^\infty \frac{\mathrm{d}x}{x}f(x)g\left(\frac{Rx}{\rho}\right) = RG\left(\frac{\rho}{R}\right)$$

where G is a function which depends on f. The strain is

$$\frac{\mathrm{d}u}{\mathrm{d}\rho} = G'\left(\frac{\rho}{R}\right).$$

Near the droplet, $\mathrm{d}u/\mathrm{d}\rho = G'(1)$ which is independent of R as stated in the text.

15.2. Discuss the evolution of a system of monoatomic-height islands of a commensurate adsorbate (i.e. without misfit dislocations) taking elasticity into account. (The geometry is as in Fig. 5.1, but the adsorbate is commensurate in the sense of section 5.3). In section 8.5, it has been shown that, if only the surface tension is taken into account, big droplets become bigger and small droplets die. Is that still true if elasticity is taken into account?

Solution – A population of monoatomic-height islands evolves, in principle, toward a periodic structure made of identical clusters (Priester & Lannoo 1995). The case of three-dimensional clusters is more complicated (Shchukin *et al.* 1995) but the Lifshitz-Slyozov theory is certainly modified.

15.3. Calculate the elastic energy of a monoatomic-height, circular terrace of radius R, commensurate with a substrate of different chemical nature.

Solution – According to (15.3), the elastic energy is

$$W_{\mathrm{el}} = C\int_{\rho<R}\mathrm{d}^2\rho\int_{\rho'<R}\mathrm{d}^2\rho'\,|\rho'-\rho|^{-3}$$

where C is a constant. This energy is expected to contain a dominant term proportional to the terrace area, plus a correction of the form (15.6). Actually, the calculation can be performed exactly if the terrace is a circle of radius R, as will be seen. For given $\mathbf{r} = \rho' - \rho$, ρ lies in the portion Δ of the plane which is common to the circle Γ and to the circle obtained by a translation of Γ of a vector $\mathbf{r}$. The area of Δ is

$$\Phi(\alpha) = 2\alpha R^2 - 2R^2\sin\alpha\cos\alpha$$

where $\alpha = \arccos[r/(2R)]$ and varies from 0 to $\alpha_0 = \arccos[r/(2R)]$. The energy reads

$$W_{el}(R) = C\int_a^r \mathrm{d}^2 r r^{-3}\Phi(\alpha) = 2\pi CR\int_0^{\alpha_0}[\alpha - \sin\alpha\cos\alpha]\,\mathrm{d}\alpha$$

$$= 2\pi C R \left[\frac{\alpha_0}{\cos\alpha_0} \ln \frac{1+\sin\alpha_0}{1-\sin\alpha_0} + \sin\alpha_0 \right]$$

$$= 2\pi C R \left[\frac{\alpha_0}{\cos\alpha_0} \ln \frac{1+\sin\alpha_0}{1-\sin\alpha_0} + \sin\alpha_0 \right]$$

$$= 2\pi^2 C \frac{R^2}{a} - 4\pi C R \ln\left(\frac{R}{a}\right) .$$

The first term is proportional to the number of atoms in the terrace, and is just a correction to the chemical potential. The second term corresponds to (15.6). It modifies the line tension by a negative amount proportional to $\ln(R/a)$.

15.4. Take a two-dimensional array of magnetic dipoles perpendicular to the plane of the array, and coupled by ferromagnetic, nearest-neighbour exchange interactions stronger than the dipolar ones. Prove that they exhibit a striped superstructure analogous to Fig. 15.6. Compare with Bloch walls in a three-dimensional magnet.

Solution – The difference with the three-dimensional case is that, in two dimensions, the equilibrium domain size is independent of the sample size. See Gehring & Keskin (1993).

15.5. Instead of Fig. 15.6, one might consider the possibility of a checkerboard structure, with crossing steps. Discuss the relative stability of the striped and the checkerboard structures.

Solution – In the magnetic case, the solution is given by Gehring & Keskin (1993). Stripes are more stable, in contradiction with a previous statement by one of the authors of this book. The case of Si(100) is, according to Alerhand *et al.* (1988, 1990) more complicated.

16
General equations of an elastic solid

Do not imagine you can abdicate:
Before you reach the frontier you are caught;
Others have tried it and will try again
To finish what they did not begin.

W. H. Auden

As long as possible we have postponed a general study of elasticity and its partial differential equations. Here they are!

The elastic equilibrium of a solid is generally treated in the continuum approximation. The strain satisfies certain equations in the bulk, and other equations at the surface. This set of equations has an infinite number of solutions, and the correct one is that which minimizes a given thermodynamic potential or free energy. This minimization is not needed for a semi-infinite solid because the good solution in this case is the one which vanishes at infinity.

The power of continuous elasticity theory is limited. In particular it is not appropriate to investigate the surface relaxation, i.e. the change in the atomic distance near the surface. Nevertheless, the continuum approximation allows for spectacular predictions, for instance the Asaro-Tiller-Grinfeld instability, which is one of the major obstacles to layer-by-layer heteroepitaxial growth.

16.1 Memento of elasticity in a bulk solid

In this section, the theory of linear elasticity in a homogeneous solid away from the surface will be recalled.

In order to write the condition for mechanical equilibrium, one has to consider the forces acting on a volume δV of the solid (Fig. 16.1). There may be an external force $\delta \mathbf{f}^{\text{ext}}$, and there is a force produced by the part of the solid outside δV. Assuming short range interactions, this force is a sum of elementary components acting on the surface $\delta\Sigma$ of δV. The force $\mathbf{df}$ acting (from outside) on the surface element $\mathrm{d}\Sigma$ is proportional to $\mathrm{d}\Sigma$, but depends on the normal unit vector $\mathbf{n}$. A linear dependence is assumed, so that

$$\mathrm{d}f_\alpha = \sum_\gamma p_{\alpha\gamma} n_\gamma \mathrm{d}\Sigma \,. \tag{16.1}$$

This relation defines the *stress tensor* $p_{\alpha\gamma}(\mathbf{r})$ at the point $\mathbf{r}$. The indices α and γ designate the Cartesian coordinates. The stress is a symmetric tensor. Indeed, if p_{xy} and p_{yx} were different, the volume element δV would rotate (Fig. 16.1b). The force acting from the interior of the body is of course opposite to (16.1).

The total force exerted on the volume δV by the part of the solid outside δV (Fig. 16.1a) is the integral of (16.1) on the surface $\delta\Sigma$ of δV. This surface integral is easily transformed into a volume integral:

$$f_\alpha = \sum_\gamma \int_{\delta\Sigma} p_{\alpha\gamma} n_\gamma \mathrm{d}^2 r = \sum_\gamma \int_{\delta V} \partial_\gamma p_{\alpha\gamma} \mathrm{d}^3 r \,.$$

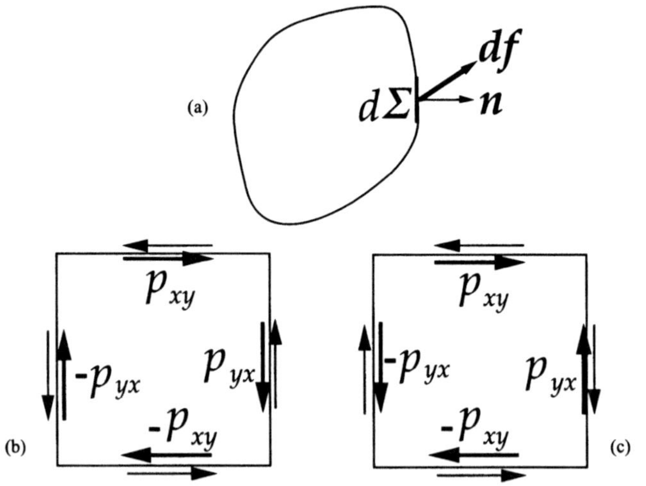

Fig. 16.1. a) Force on a fictitious surface element $\mathrm{d}\Sigma$ in a solid (cross section). b), c) Cross section of a fictitious prism inside a solid, through the xy plane. Thick arrows show the forces acting on the prism and resulting from the strain components p_{xy} and p_{yx}. The thin arrows show the forces exerted by the prism, which are opposite to the thick arrows because of the action-reaction principle. If p_{xy} and p_{yx} were different (b), the system would rotate until they become equal. The stress tensor should therefore be symmetric (c).

At equilibrium, this force must be opposite to the external force f^{ext}. Therefore the external force per unit volume is, at equilibrium:

$$\sum_{\gamma} \partial_\gamma p_{\alpha\gamma} = -\frac{\mathrm{d}f_\alpha^{\mathrm{ext}}}{\mathrm{d}V} \,. \tag{16.2a}$$

Gravity will generally be neglected, so that the external bulk force usually vanishes; thus, in a homogeneous medium at equilibrium

$$\sum_{\gamma} \partial_\gamma p_{\alpha\gamma} = 0 \,. \tag{16.2b}$$

However, in a heterogeneous medium, effective external forces may appear, and (16.2a) should apply, as will be seen later. In the presence of a uniform hydrostatic pressure, (16.2b) is still valid in the bulk.

Since the force $\mathbf{df}$ originates from interactions between the molecules of the solid, the stress tensor $p_{\alpha\gamma}(\mathbf{r})$ is a function of the local strain $\epsilon_{\alpha\gamma}(\mathbf{r})$. What is generally measured is this strain. To identify the strain (to be separated from rotations) one can characterize it by the variations of the scalar products $\mathbf{e}_\alpha \cdot \mathbf{e}_\gamma$, where the vector $\mathbf{e}_x$ is, in the unstrained system,

$$\mathbf{e}_x = (1, 0, 0) \tag{16.3}$$

and accordingly for y and z.

A *small* strain transforms $\mathbf{e}_x$ into $(1 + \partial_x u_x, \partial_x u_y, \partial_x u_z)$, and the scalar product $\mathbf{e}_x \cdot \mathbf{e}_y$ into $\partial_x u_y + \partial_y u_x$. The strain is therefore characterized by the following symmetric tensor called the *strain tensor*:

$$\epsilon_{\alpha\gamma} = \frac{1}{2}(\partial_\gamma u_\alpha + \partial_\alpha u_\gamma) \,. \tag{16.4}$$

Before giving the relation between stress and strain, it is necessary to say from which state the strain is defined. It is generally counted from the 'unconstrained' solid, i.e. with no stress and no pressure. Then, for weak strains, the following linear relation (*Hooke's law*) can be assumed:

$$p_{\alpha\gamma} = \sum_{\xi\zeta} \Omega_{\alpha\gamma}^{\xi\zeta} \epsilon_{\xi\zeta} \,. \tag{16.5}$$

It is now possible to compute the displacement inside a given volume filled with a homogeneous medium if the displacement is known on the surface. The equation to be solved is easily deduced from (16.2), (16.4) and (16.5), namely

$$\frac{1}{2} \sum_{\gamma} \sum_{\xi\zeta} \Omega_{\alpha\gamma}^{\xi\zeta} \partial_\gamma \left(\partial_\zeta u_\xi + \partial_\xi u_\zeta\right) = -\frac{\mathrm{d}f_\alpha^{\mathrm{ext}}}{\mathrm{d}V} \tag{16.6a}$$

or, in the absence of external forces:

$$\sum_{\gamma}\sum_{\xi\zeta}\Omega_{\alpha\gamma}^{\xi\zeta}\partial_{\gamma}\left(\partial_{\zeta}u_{\xi}+\partial_{\xi}u_{\zeta}\right)=0\,. \tag{16.6b}$$

The matrix elements $\Omega_{\alpha\gamma}^{\xi\zeta}$ are called the elastic constants.

The advantage in writing the equations in terms of the displacements, rather than the strains or the stresses, is that the components of the displacement are independent variables.

16.2 Elasticity with an interface

Consider a homogeneous solid bounded by an interface. The analysis of the preceding section, in which everything was continuous, has to be revised. Now indeed, there are discontinuities at the interface. For instance, the elastic coefficients $\Omega_{\alpha\gamma}^{\xi\zeta}$ are discontinuous. If both media are solids, the lattice constant is discontinuous, and the stress may also be discontinuous.

We now consider the forces on a small volume element which crosses the interface, and is assumed to have the shape of a cylinder of height h, whose top and bottom are parallel to the interface and have an area $\delta\Sigma$ (Fig. 16.2). There is a force from inside the material, which is given by (16.1), but now with a minus sign; and there is a force from outside which will be called external.

What is the nature of this external force? If the external medium is a fluid, its molecules exert a pressure, and therefore a force normal to the interface. If there are adsorbed atoms, they exert forces (Fig. 15.1a) whose resultant vanishes at equilibrium, but whose moment (force dipole) is responsible for the interactions studied in chapter 15. If the external medium is a solid, its atoms exert similar forces, with spectacular effects (Fig. 16.3). These forces will be called external, but, of course, if we write

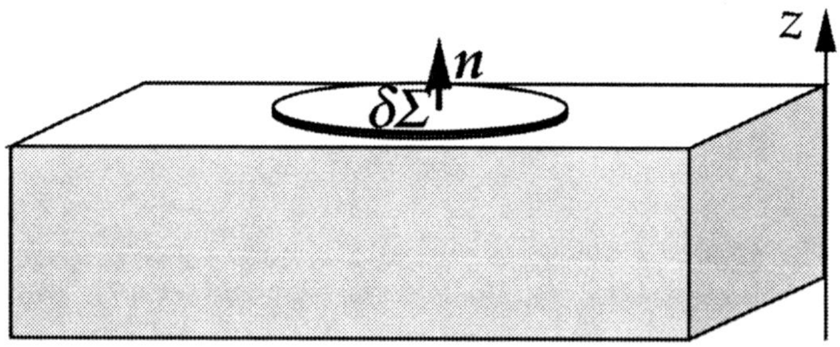

Fig. 16.2. The force acting on a thin cylinder crossing the surface of a solid body should vanish at equilibrium. Relation (16.8a) results, provided the force applied on the cross section of the cylinder by the plane vanishes (middle circle) or is equilibrated by internal forces.

(a)

(b)

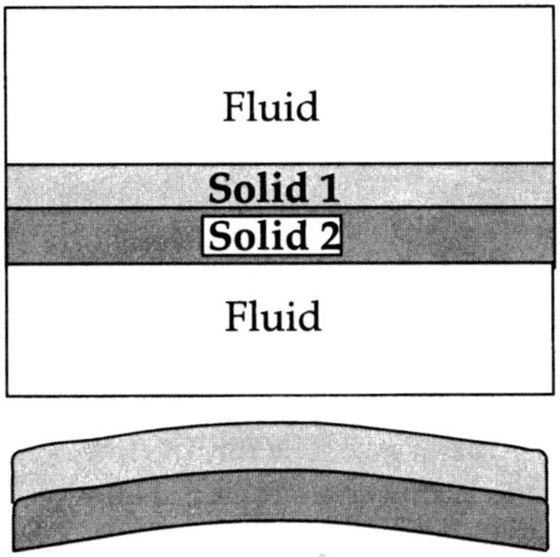

Fig. 16.3. A bilayer made of two different solids glued together. Equation (16.7) can only be applied inside each solid, provided effective external forces are introduced, taking the interaction of the other body into account. As it is well known, the resulting shape (b) depends on temperature and the system may be used as a thermometer.

the equilibrium condition on the other side of the interface, they will be regarded as internal, and previously internal forces will become external.

For the sake of simplicity, internal surface forces (the so-called capillarity forces, related to surface tension) will be neglected in this section. The external force acting on the interface will be assumed to be characterized by its density per unit area $\mathrm{d}F_\alpha^{\mathrm{ext}}/\mathrm{d}S$, assumed to be a continuous function. The force $\delta\mathbf{f}$ acting on $\delta\Sigma$ may be written as

$$\delta f_\alpha = -\sum_\gamma p_{\alpha\gamma} n_\gamma(\mathbf{r})\delta\Sigma + \frac{\mathrm{d}F_\alpha^{\mathrm{ext}}}{\mathrm{d}S}\delta\Sigma$$

where $\mathbf{n}(\mathbf{r})$ is the unit vector normal to the interface at the point $\mathbf{r}$. At equilibrium, the force vanishes and

$$\sum_\gamma p_{\alpha\gamma} n_\gamma(\mathbf{r}) = \frac{\mathrm{d}F_\alpha^{\mathrm{ext}}}{\mathrm{d}S} . \tag{16.7}$$

Inserting (16.5) and (16.4) into (16.7), one finds *on the interface*

$$\frac{1}{2}\sum_{\gamma\xi\zeta} \Omega_{\alpha\gamma}^{\xi\zeta} \left(\partial_\zeta u_\xi + \partial_\xi u_\zeta\right) n_\gamma = \frac{\mathrm{d}F_\alpha^{\mathrm{ext}}}{\mathrm{d}S} . \tag{16.8a}$$

For instance, in the case of a solid-fluid interface with a pressure P, then $\mathrm{d}F_\alpha^{\mathrm{ext}}/\mathrm{d}S = -Pn_\alpha$ and

$$\frac{1}{2}\sum_{\gamma\xi\zeta}\Omega_{\alpha\gamma}^{\xi\zeta}\left(\partial_\zeta u_\xi + \partial_\xi u_\zeta\right)n_\gamma = -Pn_\alpha\,. \tag{16.8b}$$

Thus, the force acts with different signs in (16.6) and in (16.8). The signs in (16.6a) and (16.8a) can be checked, assuming $n_x = n_y = 0$ and no shear. A positive pressure should produce a negative strain $\partial_z u_z < 0$, so that Ω_{zz}^{zz} should be positive in (16.8). On the other hand, in (16.6a), if the external force is gravity, and if the z axis is oriented upward, $\mathrm{d}f_z^{\mathrm{ext}}/\mathrm{d}V$ is negative, and the right-hand side is positive. But the strain should be more negative at lower heights, so that $\partial_z(\partial_z u_z) > 0$. One finds again that Ω_{zz}^{zz} is positive. Thus the opposite signs in (16.6a) and (16.8a) are indeed justified.

Force dipoles

As said in section 15.2 and illustrated by Fig. 15.1, external forces often form force dipoles, $m_{\alpha\gamma} = \sum_r r_\alpha F_\gamma^{\mathrm{ext}}$ $(\alpha, \gamma = x, y, z)$.

The case of a planar surface is particularly interesting. Let z be the direction normal to the surface. The strain field produced by an external force dipole of type m_{xx} or m_{yy} is calculated in appendix N, as well as the interaction between two such force dipoles. The strain field produced by an external force dipole of type m_{zz}, m_{zx} or m_{zy} is calculated in appendix O, as well as the interaction between two such force dipoles. The interaction between two identical force dipoles of any type is found to be repulsive.

In particular, a uniform distribution of tangential dipoles m_{xx} is equivalent to a surface tension. A uniform distribution of normal dipoles m_{zz} only produces surface relaxation.

A distribution of force dipoles on the surface is equivalent to a distribution of forces on the true surface, and an opposite distribution on a surface translated by a given vector $\mathbf{b}$. Thus, equation (16.6a) allows for the description of the effect of a distribution of force dipoles.

It will be seen in section 16.11 that, if capillary effects are taken into account, there are also internal forces localized at the interface, even for a clean surface without external forces. However, equations (16.8) are useful and are commonly written in textbooks (see e.g. Landau & Lifshitz (1959b), formula (2.8) and section 8 therein).

In order to be really able to use the key formulae (16.7) and (16.8), one needs to know the actual value of the external force per unit area, and the actual value of the elastic constants. We shall give simple examples in which these questions receive an answer, and we shall begin with the second question.

16.3 The isotropic solid

The equations of elasticity are difficult to solve, and a simplified model is often used, in which compressibility and resistance to shear are the same in all directions.

These conditions imply the following expression of the elastic constants defined in section 16.1:

$$\Omega_{\alpha\gamma}^{\xi\zeta} = \mu\left(\delta_{\alpha\xi}\delta_{\gamma\zeta} + \delta_{\alpha\zeta}\delta_{\gamma\xi}\right) + \lambda\delta_{\alpha\gamma}\delta_{\xi\zeta} \tag{16.9}$$

where λ and μ are called Lamé coefficients. The justification of expression (16.9) can be found in textbooks, and will be recalled in section 16.5.

Insertion of (16.9) into (16.6) and (16.8) yields

$$\sum_\gamma \partial_\gamma\left(\mu\partial_\alpha u_\gamma + \mu\partial_\gamma u_\alpha + \lambda\delta_{\alpha\gamma}\mathrm{div}\mathbf{u}\right) = 0 \qquad \text{(bulk)} \tag{16.10a}$$

$$\sum_\gamma \mu n_\gamma\left(\partial_\alpha u_\gamma + \partial_\gamma u_\alpha\right) + \lambda n_\alpha \mathrm{div}\mathbf{u} = \frac{\mathrm{d}F_\alpha^{\mathrm{ext}}}{\mathrm{d}S} \quad \text{(surface)}\,. \tag{16.10b}$$

An alternative form of (16.10a) is

$$\nabla(\mathrm{div}\mathbf{u}) + (1-2\zeta)(\nabla^2\mathbf{u}) = 0 \tag{16.11}$$

where

$$\zeta = \frac{\lambda}{2(\lambda+\mu)} \tag{16.12}$$

is called the Poisson ratio or coefficient. It is also usual to introduce the Young modulus

$$E = 2(1+\zeta)\mu = \frac{3\lambda+2\mu}{\lambda+\mu}\mu\,. \tag{16.13}$$

The following relations are easily derived from (16.12):

$$1-2\zeta = \frac{\mu}{\lambda+\mu} \qquad 1-\zeta = \frac{\lambda+2\mu}{2(\lambda+\mu)}\,.$$

Stability requires that the Poisson ratio ζ satisfies the condition $-1 \le \zeta \le 1/2$. In practice, the condition $0 < \zeta < 1/2$ is always satisfied (Landau & Lifshitz 1959b). Numerical values of the elastic constants are given in appendix P.

Hooke's law reads for an isotropic solid

$$p_{\alpha\gamma} = 2\mu\epsilon_{\alpha\gamma} + \lambda\delta_{\alpha\gamma}\sum_\xi \epsilon_{\xi\xi}\,.$$

Taking the divergence of (16.11) and observing that $\mathrm{div}\nabla\psi = \nabla^2\psi$ for any scalar function ψ, one obtains

$$\nabla^2(\mathrm{div}\mathbf{u}) = 0\,. \tag{16.14}$$

This can alternatively be written as

$$\nabla^2 \left(\sum_\alpha \epsilon_{\alpha\alpha} \right) = 0 . \tag{16.15}$$

Insertion of (16.5) into (16.15) yields

$$\nabla^2 \left(\sum_\alpha p_{\alpha\alpha} \right) = 0 . \tag{16.16}$$

In order to avoid mistakes in writing equations, it is useful to take note of the dimensions of the various quantities. They are listed in Table 16.1.

16.4 Homogeneous solid under uniform hydrostatic pressure

If a uniform hydrostatic pressure is present, equations (16.6b) and (16.8b) are consistent with a constant stress and a constant strain throughout the sample. Since there cannot be any other solution, it results that (16.8b) is valid inside the solid as well as on the surface for any vector **n**, namely

$$\frac{1}{2} \sum_{\xi\zeta} \Omega_{\alpha\gamma}^{\xi\zeta} \left(\partial_\zeta u_\xi + \partial_\xi u_\zeta \right) = -P \delta_{\alpha\gamma} \tag{16.17a}$$

or alternatively, by (16.4) and (16.5),

$$p_{\alpha\gamma} = -\delta_{\alpha\gamma} P . \tag{16.17b}$$

For an isotropic solid, (16.17a) reads

$$\partial_\alpha u_\gamma = -\frac{P}{3K} \delta_{\alpha\gamma} \tag{16.17c}$$

where the bulk modulus K is defined as

$$K = P \frac{V}{\delta V} = \lambda + \frac{2}{3} \mu \tag{16.17d}$$

Table 16.1. *Dimensions and units of the quantities which appear in elasticity.*

Quantity	Dimensions	Units
Stress $p_{\alpha\gamma}$	WL^{-3}	Pascal
Strain $\epsilon_{\alpha\gamma}$	Dimensionless	–
λ, μ, E, K, $\Omega_{\alpha\gamma}^{\xi\zeta}$	WL^{-3}	Pascal
Poisson ratio ζ	Dimensionless	–
Surface stress	WL^{-2}	Joule/m^2
Force dipole moment m	W	Joule

where the volume variation δV is caused by the pression P. Values of the bulk modulus for selected elements are displayed in Fig. 16.4. It should be emphasized again that the applicability of the above equations is very restricted. The set of two solid layers of Fig. 16.3 is under hydrostatic pressure but equations (16.17) do not apply because the system is heterogeneous. This problem will be addressed in section 16.6, in the case where one of the solids is infinitely thick.

16.5 Free energy

Let us again look at the system of Fig. 16.3. Assuming $d\mathbf{F}^{ext}/dS$ to be known, one can write the equations (16.10) in each of the solids. But how can $d\mathbf{F}^{ext}/dS$ be determined?

To solve a problem with two solids glued together, the general method consists of minimizing an appropriate 'thermodynamic potential' or 'generalized free energy'. If the external force is due to a hydrostatic pressure, the quantity to be minimized is the Gibbs free energy Φ. In vanishing pressure, Φ reduces to the Helmholtz free energy $\mathscr{F} = \Phi - PV$. At constant volume, the quantity to be minimized is $\mathscr{F}$, but the volume is practically never held fixed in surface physics. However, $\mathscr{F}$ is a nicer quantity to deal with, because it does not depend on external forces.

The generalized free energy contains a volume contribution $\mathscr{F}_V$ and a surface contribution $\mathscr{F}_S$. The former can be written, in an external potential, as

$$\mathscr{F}_V = \int_V \varphi\left(\{\epsilon(\mathbf{r})\}\right) d^3r - \int_V \sum_\alpha \frac{df_\alpha^{ext}(\mathbf{r})}{dV} u_\alpha(\mathbf{r}) d^3r \qquad (16.18)$$

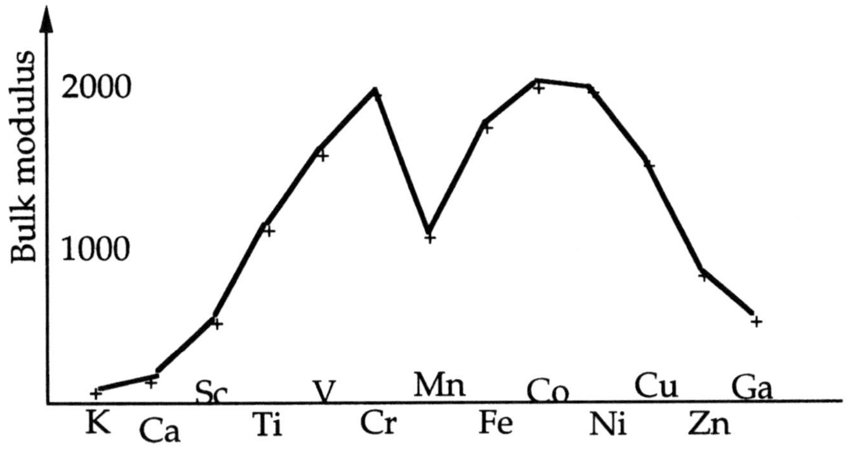

Fig. 16.4. The bulk modulus for selected elements. Units are kilobars (1600 kbar=1 eV/Å^2.)

where $\varphi(\{\epsilon(\mathbf{r})\})$ is an abbreviation for $\varphi(\epsilon_{xx}(\mathbf{r}), \epsilon_{xy}(\mathbf{r}), \epsilon_{xz}(\mathbf{r}), \epsilon_{yy}(\mathbf{r}) \ldots)$ and the ϵs are functions of the us through (16.4). The free energy density φ is determined by the fact that, for an infinite crystal at equilibrium, the displacement field u should minimize $\mathscr{F}_V$. Therefore,

$$\frac{\delta\varphi(\{\epsilon(\mathbf{r})\})}{\delta u_\alpha(\mathbf{r})} - \frac{\mathrm{d}f_\alpha^{\mathrm{ext}}(\mathbf{r})}{\mathrm{d}V} = 0\,.$$

Identifying this relation with (16.6a) yields, after a few pieces of variational calculus,

$$\varphi(\{\epsilon\}) = \frac{1}{2}\sum_{\alpha\gamma\xi\zeta} \Omega_{\alpha\gamma}^{\xi\zeta}\epsilon_{\alpha\gamma}\epsilon_{\xi\zeta}\,. \tag{16.19}$$

Since ϵ is a symmetric matrix according to (16.4), Ω may be chosen so as to satisfy the symmetry relations

$$\Omega_{\alpha\gamma}^{\xi\zeta} = \Omega_{\alpha\gamma}^{\zeta\xi} = \Omega_{\gamma\alpha}^{\xi\zeta}\,. \tag{16.20}$$

In the case of the isotropic solid, the quantity (16.19) should be invariant under rotation of the axes. The most general quadratic function which satisfies this condition is

$$\varphi(\{\epsilon\}) = \frac{\lambda}{2}\left(\sum_\alpha \epsilon_{\alpha\alpha}\right)^2 + \mu\sum_{\alpha\gamma}\epsilon_{\alpha\gamma}^2\,. \tag{16.21}$$

Comparison with (16.19) and use of the symmetry relations (16.20) yields (16.9).

Relations (16.19) and (16.21) hold for a homogeneous medium. What can be done for a heterogeneous medium (as e.g. in Fig. 16.3)?

The problem with equation (16.21) is that it holds only if the displacements and strains are counted from the unconstrained state, as noted at the end of section 16.1. This is generally not the best choice! For instance, in the case of Fig. 16.3, it is appropriate to use the same reference state for both solids. In such a reference state, one solid at least is generally constrained, and its strain with respect to the unconstrained state has a non-vanishing value $\epsilon_{\alpha\gamma}^0$. If $\epsilon_{\xi\zeta}$ is the strain counted from the common reference state, $\epsilon_{\xi\zeta}$ has to be replaced by $\epsilon_{\xi\zeta} + \epsilon_{\xi\zeta}^0$ in equations (16.19) and (16.21), which now read

$$\varphi(\{\epsilon\}) = \frac{1}{2}\sum_{\alpha\gamma\xi\zeta}\Omega_{\alpha\gamma}^{\xi\zeta}\left(\epsilon_{\alpha\gamma} + \epsilon_{\alpha\gamma}^0\right)\left(\epsilon_{\xi\zeta} + \epsilon_{\xi\zeta}^0\right) \tag{16.22a}$$

and

$$\varphi(\{\epsilon\}) = \frac{\lambda}{2}\left[\sum_\alpha(\epsilon_{\alpha\alpha} + \epsilon_{\alpha\alpha}^0)\right]^2 + \mu\sum_{\alpha\gamma}\left(\epsilon_{\alpha\gamma} + \epsilon_{\alpha\gamma}^0\right)^2\,. \tag{16.22b}$$

Thus, the free energy density now contains linear terms, which are similar to the second term of (16.18) except that the strain appears instead of the displacement. However, as will be seen later, the volume integral can be transformed into a surface integral where now the displacement appears. The effect of the solid 1 on the solid 2 is therefore equivalent to an external force acting on the surface of 2 (but not restricted to the 1-2 interface).

The explicit form of (16.21) and (16.22) in terms of the displacements, which are the independent variables, is

$$\varphi(\{u\}) = \frac{\lambda}{2}(\mathrm{div}\mathbf{u})^2 + \mu\left[(\partial_x u_x)^2 + (\partial_y u_y)^2 + (\partial_z u_z)^2\right] + \frac{\mu}{2}\left[(\partial_x u_y + \partial_y u_x)^2 + (\partial_x u_z + \partial_z u_x)^2 + (\partial_y u_z + \partial_z u_y)^2\right]. \tag{16.23}$$

To summarize, the free energy of a system of solids glued together (Fig. 16.3) in vanishing pressure, is the sum of the following two contributions.

i) For each homogeneous volume V, in the absence of *bulk* external force, the integral

$$\mathscr{F}_V = \int_V \varphi(\{u(\mathbf{r})\})\, \mathrm{d}^3 r \tag{16.24a}$$

where φ is given by (16.22a). Continuously varying impurity concentrations or other heterogeneities can be taken into account through variations of the elastic constants and of $\epsilon^0_{\alpha\gamma}$.

ii) For each interface or free surface Σ, add a contribution

$$\mathscr{F}_\Sigma = \int_\Sigma \sigma(\{\mathbf{n}(\mathbf{r}), \epsilon_1(\mathbf{r}), \epsilon_2(\mathbf{r})\})\, \mathrm{d}^2 r \tag{16.24b}$$

where $n(\mathbf{r})$, $\epsilon_1(\mathbf{r})$ and $\epsilon_2(\mathbf{r})$ are respectively the normal unit vector of the interface and the strains of both materials.

The global equilibrium state is obtained by minimizing the total free energy with respect to the strain or the displacement field, i.e. with respect to all $u_\alpha(\mathbf{r})$. This is a complicated problem of variational calculus. It is simpler, when possible, to use equations (16.6) and (16.10).

In the presence of an external pressure, the quantity which is to be minimized is, as before, the Gibbs free energy (see problem 16.1), which is obtained by adding a term PV to (16.24a).

16.6 The equilibrium free energy as a surface integral

If u satisfies (16.6) and (16.8), i.e. if the free energy has already been minimized with respect to the displacement inside the solid, then it is possible (and technically extremely useful) to transform the volume integral (16.24a) into a surface integral. This is most easily seen if a discrete form of (16.19) is used. Introducing the atom labels $\mathbf{R}$, the force $\mathbf{F}_R$ acting on them, their displacements $\mathbf{u}_R$ and a discrete 3×3 matrices $\tilde{\Omega}_{RR'}$, the discretized free energy is

$$\mathscr{F} = \frac{1}{2} \sum_{RR'} \sum_{\alpha\gamma} \tilde{\Omega}_{RR'}^{\alpha\gamma} \left(u_R^\alpha - u_{R'}^\alpha\right) \left(u_R^\gamma - u_{R'}^\gamma\right) - \sum_{R\alpha} F_R^\alpha u_R^\alpha . \tag{16.25}$$

The lattice coordinates $\mathbf{R}$ are not necessarily the positions of real atoms, but may be seen as a mathematical consequence of the discretization which appears in the Riemann theory of integration.

Minimization of (16.25) with respect to u_R^α yields

$$2 \sum_{R'} \sum_{\gamma} \tilde{\Omega}_{RR'}^{\alpha\gamma} \left(u_R^\gamma - u_{R'}^\gamma\right) = F_R^\alpha$$

where $\mathbf{F}_R$ is localized at the surface. Using this relation, the volume free energy becomes

$$\mathscr{F}_V = \sum_{RR'} \sum_{\alpha\gamma} \tilde{\Omega}_{RR'}^{\alpha\gamma} u_R^\alpha \left(u_R^\gamma - u_{R'}^\gamma\right) = \frac{1}{2} \sum_{R\alpha} F_R^\alpha u_R^\alpha$$

while the total free energy (16.25) is just the same with the opposite sign, namely

$$\mathscr{F} = -\frac{1}{2} \sum_{R} \mathbf{F}_R \cdot \mathbf{u}_R . \tag{16.26a}$$

Taking the continuum limit, this reads, in the case of surface forces:

$$\mathscr{F} = -\frac{1}{2} \int_\Sigma \mathrm{d}^2 r \sum_\alpha \frac{\mathrm{d}F_\alpha(\mathbf{r})}{\mathrm{d}S} u_\alpha(\mathbf{r}) . \tag{16.26b}$$

As seen in section 15.2, the forces are often grouped in sets $\mathbf{F}_R(\mathbf{r})$ of forces acting at points $\mathbf{r}$, such that $\sum_r \mathbf{F}_R(\mathbf{r}) = 0$. Then, (16.26a) reads

$$\begin{aligned} \mathscr{F} &= -\frac{1}{2} \sum_R \sum_r \mathbf{F}_R(\mathbf{r}) \cdot \mathbf{u}(\mathbf{r}) \\ &= -\frac{1}{2} \sum_R \sum_{r,\alpha} F_R^\alpha(\mathbf{r}) \left[u_\alpha(\mathbf{R}) + \sum_\gamma r_\gamma \partial_\gamma u_\alpha(\mathbf{R}) \right] \\ &= -\frac{1}{2} \sum_R \sum_{\alpha\gamma} m_{\gamma\alpha}(\mathbf{R}) \partial_\gamma u_\alpha(\mathbf{R}) . \end{aligned} \tag{16.26c}$$

If the force dipoles $m_{\gamma\alpha}$ are symmetric, this can be written as

$$\mathscr{F} = -\frac{1}{2} \int_\Sigma \mathrm{d}^2 r \sum_{\alpha\gamma} m_{\alpha\gamma}(\mathbf{r}) \epsilon_{\alpha\gamma}(\mathbf{r}) . \tag{16.26d}$$

We emphasize once more that, in this equation, the field $\mathbf{u}(\mathbf{r})$ should be solution of (16.10) for a given shape and given external forces. The relations (16.26) cannot be minimized with respect to the elastic displacement, but can be minimized with respect to the shape, i.e. with respect to plastic deformations. An example will be given in the next section. It turns out that formulae (16.26) are often much more convenient than (16.18). Examples will be given in the following section, in the appendices, and in the problems.

The methods presented in sections 16.5 and 16.6 might be used to determine the state of the bilayer of Fig. 16.3. One might postulate a given strain u_S at the interface, deduce the effective external force acting on the surface of both bodies, solve equations (16.10) in both solids, calculate the total free energy by (16.26) and minimize it with respect to u_S. We shall rather treat a simpler case, when one of the solids is infinite in the direction perpendicular to the surface. Its average strain is then zero if the external pressure is zero, and given by (16.17c) if there is a non-vanishing pressure.

16.7 Solid adsorbate in epitaxy with a semi-infinite crystal

Figs. 15.2, 15.7 and 16.5 give examples of this important special case. The adsorbate is forced to assume the lattice parameter a of the substrate, while it would like to have the interatomic distance $a - \delta a$. Such an adsorbate is called 'coherent' or 'commensurate' or simply 'epitaxial' or 'in epitaxy'.

The z axis will be chosen perpendicular to the interface. The sample is assumed to be confined in the volume defined by $-L/2 < x < L/2$ and $-L/2 < y < L/2$, where the limit $L = \infty$ will be taken. The substrate thickness is also assumed to be infinite.

If the surface of the adsorbate is flat, the strain is uniformly 0 in the substrate because an adsorbate of finite thickness cannot deform a substrate of infinite thickness. In the adsorbate, the strain (counted from the free adsorbate) is $\epsilon_{xx} = \epsilon_{yy} = \epsilon_0 = \delta a/a$ in the directions parallel to

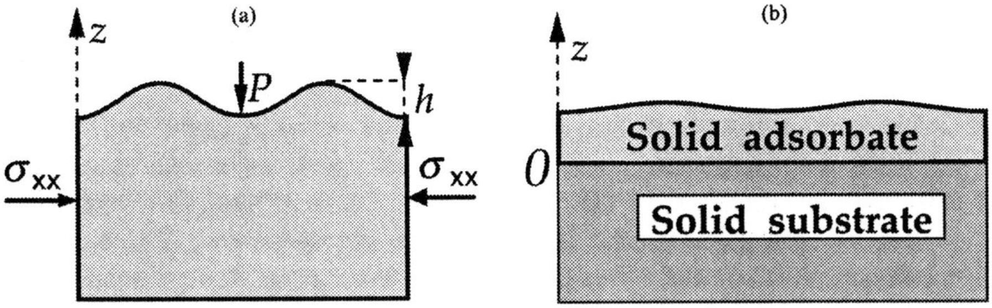

Fig. 16.5. A crystal subject to a uniaxial stress produced mechanically (a) or by an adsorbate (b).

the interface. One should not forget that there is also a strain ϵ^0_{zz} in the perpendicular direction. It will be shown below that $\epsilon^0_{zz} = -2\lambda\epsilon_0/(\lambda+2\mu)$.

The case of interest is when the surface is not exactly a plane, but has a weak modulation (Fig. 16.5) of thickness $\delta Z(x, y)$. The elastic free energy of the substrate, assumed to be an isotropic solid, is

$$\mathscr{F}_s = \frac{\lambda}{2}\int_s \mathrm{d}^3r \left[(\epsilon_{xx} + \epsilon_{yy} + \epsilon_{zz})^2\right] + \mu\int_s \mathrm{d}^3r \left[\sum_{\alpha,\gamma} \epsilon^2_{\alpha\gamma}\right] , \tag{16.27a}$$

where $\epsilon(x, y, z)$ is the local strain and the subscript s under the integral sign indicates an integration over the substrate. In the adsorbate, the free energy would also have the form (16.27a) if the strain were counted from the free adsorbate. However, in order to ensure the continuity of ϵ_{xx} and ϵ_{yy} at the interface, we prefer to count the strain from the constrained, homogeneous adsorbate, and therefore ϵ_{xx}, ϵ_{yy} and ϵ_{zz} should be replaced in (16.27a) by $\epsilon_{xx} + \epsilon_0$, $\epsilon_{yy} + \epsilon_0$, and $\epsilon_{zz} + \epsilon^0_{zz}$. Therefore, the free energy of the adsorbate reads, within an additive constant,

$$\mathscr{F}_a = \frac{\lambda}{2}\int_a \mathrm{d}^3r \left[\epsilon_{zz} + \epsilon_{yy} + \epsilon_{xx} + 2\epsilon_0 + \epsilon^0_{zz}\right]^2$$
$$+\mu\int_a \mathrm{d}^3r \left[\sum_{\alpha\gamma} \epsilon^2_{\alpha\gamma} + 2(\epsilon_{xx} + \epsilon_{yy})\epsilon_0 + 2\epsilon^0_{zz}\epsilon_{zz}\right] ,$$

where the subscript a under the integral sign indicates an integration over the adsorbate. For the sake of simplicity, the elastic constants will first be assumed to be the same in the substrate and the adsorbate. When expanding this expression, the terms linear in ϵ_{zz} can be made to vanish by choosing $\epsilon^0_{zz} = -2\lambda\epsilon_0/(\lambda+2\mu)$, as anticipated. Grouping terms linear in ϵ_{xx} and ϵ_{yy} and making use of the relation $2\mu(3\lambda+2\mu)/(\lambda+2\mu) = E/(1-\zeta)$, the previous expression reads, within an additive constant,

$$\mathscr{F}_a = \frac{\lambda}{2}\int_a \mathrm{d}^3r(\epsilon_{xx} + \epsilon_{yy} + \epsilon_{zz})^2 + \mu\int_a \mathrm{d}^3r \left[\sum_{\alpha,\gamma} \epsilon^2_{\alpha\gamma}\right]$$
$$+\frac{E\epsilon_0}{1-\zeta}\int_a \mathrm{d}^3r \left[\epsilon_{xx}(x, y, z) + \epsilon_{yy}(x, y, z)\right] . \tag{16.27b}$$

Since $\epsilon_{xx} = \partial u_x/\partial x$, the term linear in ϵ_{xx} can be integrated over x, and gives rise to an integral on the surface of the adsorbate, linear with respect to the local displacement. Such a term is the same as would result from an external force acting on the surface of the adsorbate. The term linear in ϵ_{yy} can be transformed and interpreted in the same way. Thus, one has to calculate the response of an elastic solid to external forces acting on its surface.

Such a response is not too difficult to calculate when the surface is a plane, and this is done in appendices N and O. However, as said above, the situation of interest is when the surface is not a plane. The case of a *weakly* undulating surface will now be investigated.

It will be assumed, and checked self-consistently at the end of the calculation, that the integral in the first two (quadratic) terms of (16.27b) can be replaced by an integral on $z < \bar{Z}$, where $z < \bar{Z}$ is the average height of the surface. Thus, the quadratic part of the free energy is that of a harmonic solid limited by a plane.

The last (linear) term of (16.27b) may be interpreted as resulting from a uniform density

$$\frac{\mathrm{d}m_{xx}}{\mathrm{d}v} = \frac{\mathrm{d}m_{yy}}{\mathrm{d}v} = -\frac{E\epsilon_0}{1-\zeta} \tag{16.28}$$

of external forces moment inside the adsorbate. Assuming $\epsilon_{xx}(x, y, z)$ and $\epsilon_{yy}(x, y, z)$ to be continuous, the linear part of the free energy can be rewritten as

$$\frac{E\epsilon_0}{1-\zeta}\int_{0<z<\bar{Z}} \mathrm{d}^3r\left[\epsilon_{xx}(x, y, z) + \epsilon_{yy}(x, y, z)\right]$$
$$+\frac{E\epsilon_0}{1-\zeta}\int \mathrm{d}x\mathrm{d}y\left[\epsilon_{xx}(x, y, 0) + \epsilon_{yy}(x, y, 0)\right]\delta Z(x, y)\,,$$

where $\delta Z(x, y) = Z(x, y) - \bar{Z}$. The first term of the above expression can be integrated once (ϵ_{xx} can be integrated over x, and ϵ_{yy} over y) and gives rise to an integral on the edge of the crystal. This integral is proportional to L and negligible for large L (and for an infinitely thick substrate) with respect to the second term, proportional to L^2, namely

$$\delta\mathscr{F}_a = \frac{E\epsilon_0}{1-\zeta}\int \mathrm{d}x\mathrm{d}y\left[\epsilon_{xx}(x, y, 0) + \epsilon_{yy}(x, y, 0)\right]\delta Z(x, y)\,. \tag{16.29a}$$

Thus the free energy is that of a harmonic solid limited by the plane $z = \bar{Z}$ and subject to a density of force dipoles per unit area acting in this plane and equal to

$$\frac{\mathrm{d}m_{xx}}{\mathrm{d}S} = \frac{\mathrm{d}m_{yy}}{\mathrm{d}S} = \frac{\mathrm{d}m}{\mathrm{d}S} = \frac{E}{1-\zeta}\frac{\delta a}{a}\delta Z\,. \tag{16.29b}$$

The strain resulting from these forces can be calculated, and this is done in appendices N and O. It is of the same order as δZ. When this response is inserted into (16.27b), the quadratic terms are seen to be of higher order and therefore negligible, consistently with the initial assumption.

In the above argument, we have assumed the elastic constants to be the same in the substrate and the adsorbate. Of course, it is generally not so. In formulae (16.27b) to (16.29b), the elastic constants E and ζ are those of the adsorbate. In the response functions, the substrate plays a part too, and if the adsorbate is thin, they contain only the elastic constants E and ζ of the substrate. The simplest case is when the adsorbate is thick–more precisely, thicker than the wavelength of all the Fourier components of the modulation function $\delta Z(x,y)$. Then, one should use the elastic constants of the adsorbate in the response functions. This is true for instance if there is a single Fourier component

$$\delta Z(x) = h\cos(qx)\,, \tag{16.30}$$

and if the thickness of the adsorbate is larger than $1/q$. This case will be addressed in the next section.

In the present section, continuum elasticity theory has been assumed to hold. In reality, h cannot be smaller than a single atomic distance. It is clear that continuum elasticity theory is not reliable in the vicinity of a step, and this can be confirmed with the help of integral equations (Hu 1979). Thus, an additional requirement for the validity of (16.29b) is that there are not too many steps. In the language of continuum elasticity, this means that the slope of the surface is small everywhere, or that only Fourier components of small wavevector q are present. Note that the strain produced by a sinusoidal external stress of wavevector q is proportional to q, as seen in appendix N. The quadratic terms in (16.27b) are therefore proportional to q^2 and negligible for small q.

To summarize this section, we have shown that, when an adsorbate is coherent with the substrate and its surface is not planar, it is subject to forces. The response to these forces can be approximately calculated analytically in certain simple cases. On the other hand, these forces tend to displace the atoms. Will this displacement smoothen the surface or increase the modulation and therefore make the planar surface unstable? The second possibility is the correct one as it will now be seen.

16.8 The Grinfeld instability

The last term of (16.27b) represents the effect of an external, anisotropic stress (or force dipole density). In the present section, we wish to investigate the consequences of such a stress on a *dislocation-free* solid. The stress can be produced by a substrate (Fig. 16.5) (as in the preceding section) or mechanically (Thiel *et al.* 1992). In contrast with the preceding section, it will be assumed that only the xx stress component is different from 0 and equal to p_0. Since we shall not go beyond linear response theory, the effect of a yy component, if present, can just be added.

If the external stress p_0 is produced by a substrate, then it is given, according to (16.29b), by

$$p_0 = -\frac{E}{1-\zeta}\epsilon_0 = \frac{E}{1-\zeta}\frac{\delta a}{a}\,.$$

To understand the effect of a uniaxial external stress on a solid, it is of interest to consider first the case of a liquid: if one squeezes a liquid slab between two plates, the liquid flows and the slab becomes narrower. The solid tries to do the same, i.e. atoms move and the device is unstable. However, since atomic motion in a solid is very difficult, a lot of things occur before the solid actually becomes narrower. In the dislocation-free solid, the easiest atomic motion is surface diffusion. One can guess that surface diffusion will lead to the formation of bubbles analogous to those of Fig. 5.1, because the stress is at least partially released at the top of the bubbles. Note that bubble formation in Stranski-Krastanov growth is generally a result of interface free energies balance (wetting), while the effect of interest here is elastic.

Considering bubbles would be too complicated, and we shall just make a linear analysis of the stability of the planar surface, as was done in chapter 10 in a different context (moving surface, and no elastic effects).

We are thus led to investigate the linear stability of the planar surface with respect to the perturbation (16.30). That is, we want to see whether the free energy variation due to (16.30) is positive or negative. The substrate thickness will be assumed infinite, so that the average atomic distance normal to z is fixed within complete layers. However, if some atomic layers are not complete, they can expand or shrink. The atomic layers of the adsorbate are happy to do so, since their lattice parameter thus gets closer to its natural value. Therefore we can expect them to split through a modulation of the surface (Fig. 16.5). This is the Grinfeld instability (Asaro & Tiller 1972, Grinfeld 1986, 1993).

Simplified argument

A simplified calculation will first be presented, introducing an average strain $\delta\epsilon$ (with respect to the flat surface $h = 0$) instead of the complete strain field. The free energy per unit area contains three contributions.

i) The capillary energy (due to chemical bonds which are broken when forming the surface) is given in section 2.1 for a non-singular surface. For the sake of simplicity, the surface stiffness $\tilde{\sigma}$ will be assumed isotropic and thus equal to the surface tension σ. It follows from equation (2.4) that the average capillary energy per unit area is increased by the modulation by a quantity

$$d\mathscr{F}_{cap}/d\mathscr{A} = \sigma h^2 q^2/2 .$$

ii) The energy gained due to the relaxation in the undulating region is proportional to the height h of this region, to the average strain ϵ, and to the external stress p_0:

$$d\mathscr{F}_{relax}/d\mathscr{A} \approx -hp_0\epsilon .$$

iii) This energy gain is partially compensated by the elastic energy paid to the inhomogeneity of the strain. This energy is mainly concentrated below the undulating region. Its three-dimensional density is proportional to ϵ^2 through an elastic constant C. In order to obtain the energy per unit area in the xy plane, one should multiply by the depth of the strained region. This depth is of order $1/q$. This is seen in appendix N, but the reason is essentially that the solution of the linear, partial derivative equation (16.6b) in a semi-infinite medium has the form $\mathbf{u}_0 \exp(iqx + qz)$, because this form allows compensation of $\partial^2/\partial x^2 = -q^2$ by $\partial^2/\partial z^2 = q^2$. The elastic energy per unit area is then

$$\mathrm{d}\mathscr{F}_{\mathrm{el}}/\mathrm{d}\mathscr{A} \approx C\epsilon^2/(2q)\,.$$

Minimizing the sum of contributions (ii) and (iii) with respect to ϵ yields

$$\epsilon \approx hqp_0/C$$

so that the variation of the total free energy per unit area resulting from the modulation is the sum of the three contributions (i), (ii) and (iii), namely

$$\mathrm{d}\mathscr{F}/\mathrm{d}\mathscr{A} = \sigma h^2 q^2/2 - h^2 p_0^2\, q/(2C)\,.$$

If q is small enough, the positive term due to surface tension is unable to compensate the negative term, so that increasing the modulation amplitude h lowers the free energy. There is an instability.

Detailed calculation

From the above argument, the surface tension has been seen to be negligible at long wavelengths. Thus, it will first be neglected in the forthcoming detailed calculation.

As seen in the preceding section, the adsorbed layer with its modulation is equivalent to a surface density of force dipoles given by (16.29b) and (16.30), namely

$$\frac{\mathrm{d}m_{\alpha\gamma}}{\mathrm{d}S} = hp_0\delta_{\alpha x}\delta_{\alpha\gamma}\cos(qx)\,, \tag{16.31}$$

The interaction energy might be deduced from (15.3) or (15.4), but it is simpler to calculate directly the elastic response to the sinusoidal field (16.31), since this calculation is an intermediate step in the derivation of (15.3) or (15.4) in appendix N. Formula (N.17) yields the variation of the free energy due to the surface modulation (16.31) as

$$\Delta\mathscr{F} = -\frac{1-\zeta^2}{2E}\mathscr{A}h^2p_0^2\,|q|\,. \tag{16.32}$$

The minus sign reveals that the modulated state has a lower free energy than the flat surface. Thus, there is an instability, provided there is no positive contribution to the free energy. Indeed, there is one, which has been neglected so far: the surface tension! Assuming an isotropic surface tension σ, the surface free energy, calculated in chapter 2, is

$$\mathscr{F}_{\text{surf}} = L\frac{\sigma}{2}\int_0^L \left(\frac{\mathrm{d}z}{\mathrm{d}x}\right)^2 = \frac{\mathscr{A}}{2}\sigma h^2 q^2 \,. \tag{16.33}$$

It is positive, but proportional to q^2, while the negative elastic contribution (16.32) is proportional to $|q|$. It then follows that the planar surface is unstable with respect to modulations of long enough wavelength. It may be of interest to note that the role of the surface tension here is somewhat similar to its role in the Mullins-Sekerka instability of chapter 10: it tries in vain to stabilize the planar interface.

The above treatment, valid in the absence of hydrostatic pressure, can easily be extended if there is a pressure P, so that the external stress components are given by (16.17b) except $p_{xx} = p_0 - P$. The Helmholtz free energy $\mathscr{F}$ should be replaced by the Gibbs free energy Φ and the displacement should be counted from the equilibrium state under hydrostatic pressure with $p_0 = 0$. The above formulae are still valid, with $p_0 = p_{xx} + P$.

Apart from the surface tension, another stabilizing effect has been neglected in the above treatment, namely gravity. It does stabilize the planar surface if $| \, p_{xx} + P \, |$ is not too large (Nozières 1991).

Below the roughening transition, the above discussion is no longer valid. Indeed, the variation of the surface tension due to a modulation of amplitude h is in that case

$$\gamma q \int_0^{1/q} |z'(x)| \mathrm{d}x = \gamma q h \,,$$

where γ is the line tension of the steps. For small h, this contribution dominates that of (16.32). In fact, an instability does show up, but an activation barrier has to be overcome (Tersoff & LeGoues 1994). This is indeed the problem which has been treated in section 15.7.

In the case of a stepped surface, the Grinfeld instability causes step bunching (Duport *et al.* 1995a,b).

Non-linear investigations of the Grinfeld instability have been given by Nozières (1993) and Orr *et al.* (1992) while the structure of an isolated bubble has been investigated by Johnson & Freund (1996).

Experimentally observed instabilities of coherent multilayers can often be explained by the Grinfeld instability (Pidduck *et al.* 1992, Ponchet *et al.* 1993). A remarkable feature is the fairly regular distribution of bubbles (in fact, pyramids) in heteroepitaxial layers (Leonard *et al.* 1993, Moison *et al.* 1994, Noetzel *et al.* 1994, Shchukin *et al.* 1995).

16.9 Dynamics of the Grinfeld instability

How is the amplitude $h(t)$ of the modulation (16.30) expected to change with the time t ? This problem will be addressed assuming the evolution to take place through adatom diffusion on the surface, so that equations (8.10) and (8.11) hold. The problem is then to express the free energy per atom μ on the surface, described by (16.29a). In the absence of elasticity, relation (8.5) holds, and it yields $\delta\mu(x) \sim q^2\delta z(x)$, with a positive coefficient. When elasticity is included, we will just assume that $\delta\mu(x) = \alpha(q)\delta z(x)$. Then, $\alpha(q)$ may be deduced from the excess free energy associated with a sinusoidal surface modulation of amplitude h and wavevector q, which for an area $\mathscr{A}$ is

$$\delta\mathscr{F} = \frac{1}{a}\int \delta\mu(x)\delta z(x) = \frac{1}{a}\int \alpha(q)\delta z^2(x) = \frac{1}{2a}h^2\alpha(q)\mathscr{A} .$$

From (16.32) it follows that $\alpha(q) \sim q$ for small q, with a negative coefficient. Combining with (8.10) and (8.11) gives

$$(\partial/\partial t)\delta z \sim q^3\delta z$$

and therefore the modulation (16.30) increases as $\exp(\mathrm{Const} \times t/q^3)$. The detailed derivation has been given by Srolovitz (1989) and by Spencer *et al.* (1991).

16.10 Surface stress and surface tension: Shuttleworth relation

Consider now a solid bounded by a clean surface, for instance in vanishing pressure. Usually, the surface atoms do not like to have the same lattice parameter as the bulk atoms. In the direction normal to the surface, they do adjust their distance, and this is called surface relaxation. In the direction parallel to the surface, they cannot relax completely their 'frustration'. The consequence being that, in the cylinder of Fig. 16.2 the force applied by the outer atoms on the interior atoms is different at the surface and in the bulk. Therefore, there is a linear distribution of forces acting at the intersection of the cylinder with the surface. The force acting on a line element $d\ell$ orthogonal to the unit vector $\mathbf{n}$ on the surface can be written as

$$df_\alpha = \sum_{\gamma=x,y} \pi_{\alpha\gamma} n_\gamma d\ell , \qquad (16.34)$$

where the tensor $\pi_{\alpha\gamma}$ is called surface stress and has the dimension of an energy per unit area. The microscopic origin of the surface stress is illustrated by a simple model in appendix M.

The surface stress has been ignored in (16.6a). This is often correct, at least to a good approximation. Indeed, a qualitatively correct picture of the surface stress is obtained by imagining the surface as a thin coating of a different material with a different natural lattice parameter (Fig. 16.6). Since the coating is very thin, it does not affect very strongly the rest of the solid, which is generally thick.

For instance, on the infinite, planar surface of Fig. 16.2, a force resulting from the surface stress is exerted by the outer part of the cylinder onto its inner part, but it is automatically cancelled by an opposite force from the inner atoms. On the other hand, if the surface stress were taken into account in the calculation of the interaction between two adatoms, formula (15.3) would not be modified (see problem 16.2). In the bilayer of Fig. 16.3, the surface stress at the solid-fluid interfaces should in principle be included, but it is not expected to be dramatically strong. The bending is not due to the surface stress.

On the other hand, the surface stress is present in the general theory of section 16.5. As an example, the surface free energy (16.24b) will be written for a planar solid, uniformly stretched in the x direction by a small amount ϵ_{xx}. For a surface area $\mathscr{A}$, (16.24b) reads

$$\mathscr{F}_\Sigma = \mathscr{A}\sigma(\epsilon_{xx}) = (\mathscr{A}_0 + \delta\mathscr{A})\left[\sigma(0) + \frac{\partial\sigma}{\partial\epsilon_{xx}}\epsilon_{xx}\right],$$

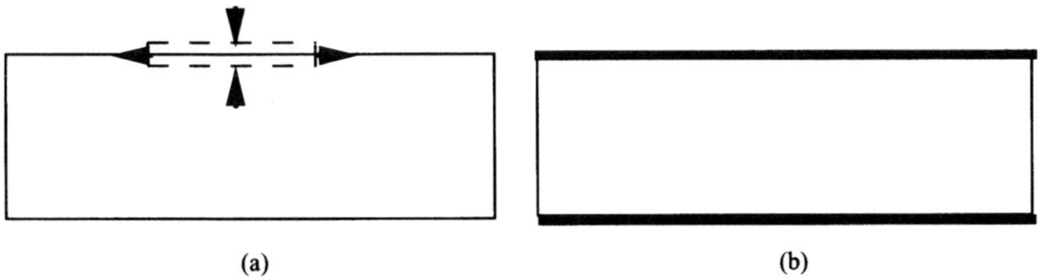

Fig. 16.6. a) Cross section of the cylinder of Fig. 16.2. In reality, there are forces (horizontal arrows) acting on the side of the cylinder, which do not vanish with its height. Their intensity per unit length is called the surface stress. For a planar surface, the surface tension does not invalidate the relation (16.6a) because it is equilibrated by internal forces having the same origin. In particular, p_{zz} inside the solid equilibrates the external pressure (vertical arrows). b) One may imagine the surface stress to be produce by a coating of atomic thickness of a material with a different interatomic distance. The bending of Fig. 16.3 does not take place because all sides are coated. The modulation of Fig. 16.5 does not take place because the coating thickness is fixed. But, in principle, if the sample is very thin, its length is modified by the surface stress.

where $\delta\mathscr{A} = \mathscr{A}_0\epsilon_{xx}$ is the area increase. The variation of the surface free energy corresponding to a stretching is therefore

$$\delta\mathscr{F}_\Sigma = \delta\mathscr{A}\left(\sigma + \frac{\partial\sigma}{\partial\epsilon_{xx}}\epsilon_{xx}\right) .$$

This can be identified to the mechanical work accomplished by the surface stress, which is $\pi_{xx}\delta\mathscr{A}$. Therefore,

$$\pi_{xx} = \sigma + \frac{\partial\sigma}{\partial\epsilon_{xx}} . \tag{16.35}$$

This formula, called the Shuttleworth relation, allows for a comparison between a solid and a liquid. In a liquid, one can modify the surface area by an amount $\mathscr{A}_0\delta\epsilon_{xx}$. This is not accompanied by any microscopic change of the atomic architecture, so that $\partial\sigma/\partial\epsilon_{xx} = 0$. The bulk contribution to the work also vanishes in a liquid, so that the force per unit length necessary to increase the area can be measured. If we call it π_{xx}, then

$$\pi_{xx} = \sigma . \tag{16.36}$$

Thus, (16.35) can be considered as the extension of (16.36) to the case of a solid. Since π_{xx} is a *tensile* force per unit length, equation (16.36) justifies, in the case of a liquid, the expression 'surface tension' given to the free energy per unit area. Note that, in this section, σ is the free energy per unit area. This is the usual convention, which contrasts with the remainder of this book where, for the sake of simplicity, σ designates the free energy per surface site. For an incompressible solid both definitions are equivalent. For a compressible solid body, they are not.

The surface stress can be experimentally determined by measuring the external force necessary to stretch a sample by a certain amount. This force should equilibrate the surface stress plus (in the case of a solid) a bulk force which can be calculated (see problem 16.2).

16.11 Force dipoles, adatoms and steps

As said in section (15.2), adatoms exert force dipoles on the surface. Force dipoles are characterized by their moments. The moment of a dipole constituted by two forces $-\mathbf{f}$ and $\mathbf{f}$ acting at points $\mathbf{R}$ and $\mathbf{R}+\mathbf{b}$, has six components given by (15.2), namely

$$m_{\alpha\gamma} = b_\alpha f_\gamma \tag{16.37}$$

where b is the atomic distance in the surface. In the case of a single adatom on a high symmetry surface, symmetry imposes that only m_{zz} and $m_{xx} = m_{yy} = m$ are different from zero. Moreover, the real force dipole exerted by the adatom may be replaced by an effective force dipole of moments $m'_{zz} = 0$ and $m'_{xx} = m'_{yy} = m - m_{zz}\zeta/(1-\zeta)$. This results from formula (15.4) and is demonstrated in the appendix O. The effective dipole produces the same strain field as the real one.

The forces exerted by an adatom on a surface can be calculated within a model, e.g. the pair potential model described in appendix M. The forces exerted by adatoms can be inserted into (16.6a) or (16.10b) as 'external forces' which replace the adatom.

The case of a step (Fig. 15.4) is more complicated. We are going to argue that (16.6) or (16.10) can still be used, but the meaning of $\mathrm{d}F_\alpha^{\mathrm{ext}}/\mathrm{d}S$ has to be more carefully defined. Indeed, it seems impossible to define an 'external' force.

Far from the surface, the equilibrium equations (16.6b) or (16.10a) have been obtained by writing that the sress p is related to the strain ϵ by Hooke's law (16.5). Assuming Hooke's law to hold near the surface as well, one would obtain (16.6a) or (16.10b) with $\mathrm{d}F_\alpha^{\mathrm{ext}}/\mathrm{d}S = 0$, in the absence of true external forces. The solution would obviously be $\mathbf{u} = 0$.

However, as will be seen a few lines below, this result is not correct. A step does induce a strain, and this strain is not localized, but decays as $1/r^2$ at a distance r from the step.

This means that, when a step is present, there are 'extra forces' $\mathrm{d}F_\alpha^{\mathrm{ext}}/\mathrm{d}S$ which should be inserted into (16.6) and (16.10). The superscript 'ext' now means 'extra' rather than 'external'. The extra forces are not determined by the elastic constants, but can be in principle determined from a microscopic model.

For definiteness, we shall assume that the energy is the sum of pair potentials $\mathscr{V}(|\mathbf{r}-\mathbf{r}'|)$ and three-body potentials $\mathscr{V}_3(|\mathbf{r}-\mathbf{r}'|, |\mathbf{r}-\mathbf{r}''|, |\mathbf{r}'-\mathbf{r}''|)$. Pair interactions alone would not be adequate for diamond and semiconductors. These potentials will be assumed to be short-ranged, i.e., they vanish if the distance $|\mathbf{r}-\mathbf{r}'|$, etc., is larger than some value r_1. Four-body potentials, etc., could easily be included.

Within such a model, it is straightforward to calculate the forces acting on each atom in the zero-strain state. These forces are easily seen to be different from zero near a step, so that the zero-strain state, solution of (16.6) with $\mathrm{d}F_\alpha^{\mathrm{ext}}/\mathrm{d}S = 0$, is not an equilibrium state. This proves the existence of extra forces (Fig. 16.7).

As a matter of fact, in the state where the displacement of all atoms (with respect to the periodic crystal) vanishes, the forces do not vanish even on a smooth surface, or on a surface far from a step. However, a vanishing force (and therefore mechanical equilibrium) is obtained if the atomic distances are suitably modified within a microscopic thickness (i.e. a few atomic distances) near the surface. This is the *surface relaxation.* More precisely, the relaxation decays exponentially when the distance from the surface increases. Within continuum elasticity theory, this surface relaxation can

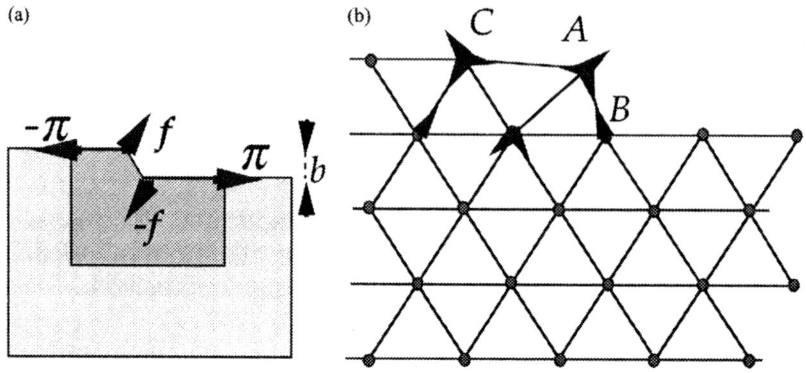

Fig. 16.7. a) Local forces (f, $-f$) exerted on a surface by a step. The forces applied to the dark area are these two forces and the forces π and $-\pi$ resulting from the surface stress. The absence of rotation yields condition (16.38) for the local forces dipole. b) An example. See problem 16.4.

just be ignored, because the strain near the surface should be defined as the strain at a small, but macroscopic distance from the surface. With this definition, the strain does vanish at equilibrium in the absence of steps, so that the extra force $\mathrm{d}F_\alpha^{\text{ext}}/\mathrm{d}S$ vanishes in (16.6a) and (16.10b).

If a step is present, the extra forces contain at least a non-vanishing part related to the surface stress $\tilde{\pi}$. Indeed, if one considers the forces acting on a volume element (Fig. 16.7) containing a length y of a step, the two forces $\delta F_1 = \tilde{\pi} y$ and $\delta F_2 = -\tilde{\pi} y$ acting on both sides of the step have a non-vanishing moment density (per unit length)

$$(\mathrm{d}/\mathrm{d}y) m_{zx} = \tilde{\pi} c \tag{16.38}$$

where c is the step height. Since $\tilde{\pi}$ is not related to the elastic constants (e.g. E and ζ), the forces related to this non-vanishing moment are extra forces in the sense defined above. At mechanical equilibrium, the extra moment has to be compensated by an opposite dipole moment resulting, through Hooke's law, from an appropriate strain.

Beside the m_{zx} component resulting from the surface stress, we expect the extra force moment to have components m_{zz}, m_{xx} and m_{xz}.

From the above argument, one concludes that a step parallel to the y direction produces an extra force dipole line density $\mathrm{d}m_{zx}/\mathrm{d}y$. Dipole moment density components $\mathrm{d}m_{xx}/\mathrm{d}y$, $\mathrm{d}m_{zz}/\mathrm{d}y$ and $\mathrm{d}m_{xz}/\mathrm{d}y$ are also expected to be present. However, it will now be seen that the long range effect of a step can actually be described by a single parameter (in addition to the surface stress).

Firstly, as said in the beginning of this section, the component $\mathrm{d}m_{zz}/\mathrm{d}y$ has the same effect as a density of dipole moment $\mathrm{d}m'_{xx}/\mathrm{d}y$ equal to

$-[\zeta/(1-\zeta)]\mathrm{d}m_{zz}/\mathrm{d}y$. The zz component of the dipole moment density can therefore be ignored, provided the xx component is modified.

Similarly, it is shown in appendix O that a force dipole density $\mathrm{d}m_{zx}/\mathrm{d}y$ is equivalent to a force dipole density $\mathrm{d}m_{xz}/\mathrm{d}y$ with opposite sign and same absolute value. The zx component of the dipole moment density can therefore be assumed to vanish, provided the xz component is modified.

We now have only two components xx and xz as generally assumed (Marchenko & Parshin 1980, Andreev & Kosevich 1981, Nozières 1991).

Thus, the effect of a step is given by the solution of (16.6), or (16.10) for an isotropic solid, with extra force dipole moments $\mathrm{d}m_{xx}/\mathrm{d}y$, and $\mathrm{d}m_{xz}/\mathrm{d}y$. These equations must, in principle, be solved in the presence of a step. In other words, the step has to be taken into account both through the extra force and through the appropriate boundary condition. Fortunately, it can be argued, for instance using perturbation theory, that the response to the extra force far from the step is *not* modified by the step, and is therefore well approximated by the response of a solid with a smooth surface. To our knowledge, no careful proof has been published and, since the argument would be pretty long, we decided not to give it here either.

Another point which would deserve a more detailed discussion is the localized nature of the extra forces. In the zero-strain state, we have seen that they are localized near the step. They are generally assumed to be localized at mechanical equilibrium too (Marchenko & Parshin 1980, Andreev & Kosevich 1981, Nozières 1991).

From the above results and appendix N, it follows that the strain produced by a step on a clean surface decays as $1/r^2$ at long distance r from the step.

The elastic interaction per unit length between two straight, parallel steps at long distance r can then be deduced from (16.26d) if the strain produced by each step is known. Since this strain has been seen to be proportional to $1/r^2$, the same rule applies for the interaction, namely

$$\frac{\mathrm{d}W_{\mathrm{int}}(r)}{\mathrm{d}y} = \frac{\mathrm{Const}}{r^2} . \tag{16.39}$$

From this chapter and the previous one, elasticity is seen to introduce serious complications, mainly related to the long range interactions between adatoms, steps and other defects. An additional difficulty comes from the fact that elasticity is a very old science, so that the bibliographical effort cannot be limited to the second half of the twentieth century. A typical example is the exact calculation of the stress in an adsorbate having the shape of a bar whose cross-section is an arc of a circle. The solution of this cylindric version of Fig. 15.8 (a real computational *tour de force*) is contained in a paper by Ling (1948) which contains references to earlier articles published in the German and Japanese literatures.

Problems

16.1. Write the Gibbs free energy of a homogeneous elastic medium in the presence of a hydrostatic pressure P, as a function of a small strain counted from the equilibrium state.

Solution – The Gibbs free energy is $\Phi\left(\{\epsilon\}\right) = \mathcal{F}\left(\{\epsilon\}\right) + PV$.

16.2. Calculate the work necessary to stretch an elastic parallelepiped by an amount $\delta\epsilon_{xx}$.

Solution – Assume all $\epsilon_{\alpha\gamma} = 0$ except ϵ_{xx} (this implies choosing appropriate values for $p_{\alpha\gamma}$). Letting $K = \lambda + 2\mu$, the Gibbs free energy reads

$$\Phi = \frac{1}{2} L_x L_y L_z \; K\epsilon_{xx}^2 - L_x L_y L_z p_{xx}\epsilon_{xx} - L_x L_y \pi_{xx}\epsilon_{xx} \, .$$

It must be minimized with respect to ϵ_{xx}, which yields

$$\epsilon_{xx} = \frac{1}{K} p_{xx} + \frac{1}{L_z K}\pi_{xx}$$

and thus

$$\Phi' = -\frac{1}{2K} L_x L_y L_z (p_{xx} + \frac{1}{L_z}\pi_{xx})^2 \, .$$

If $p_{xx} = 0$, define $\Phi_0 = \Phi'(p_{xx} = 0)$, and $\epsilon_{xx}^{(0)} = \epsilon_{xx}(p_{xx} = 0) = \pi_{xx}/(KL_z)$. Then, $\delta\epsilon_{xx} = \epsilon_{xx} - \epsilon_{xx}^{(0)} = 1/K p_{xx}$ and

$$\delta W = \Phi' - \Phi_0 = -\frac{1}{2} L_x L_y L_z \; K(\delta\epsilon_{xx})^2 - L_x L_y \pi_{xx}\delta\epsilon_{xx} \, .$$

16.3. Show that the interaction between two point defects vanishes to lowest order in an infinite isotropic solid.

Solution – Each point defect is equivalent to a set of external forces $\mathbf{F}_R$ acting on the neighbouring atoms at $\mathbf{R}$. Let the forces corresponding to the two defects be $\mathbf{F}_R$ and $\mathbf{F}'_R$, and the displacements at the point $\mathbf{R}$ due the two defects be $\mathbf{u}_R$ and $\mathbf{u}'_R$. The free energy is given by (16.26), and the part of it which corresponds to the interaction between both defects is

$$\mathcal{F} = -\frac{1}{2}\sum_{R\alpha} F_R^\alpha u_R'^\alpha - \frac{1}{2}\sum_{R\alpha} F_R'^\alpha u_R^\alpha \, .$$

Let the origin be chosen at the site of the 'primed' impurity. Then, as seen

$$u_R'^\alpha = C\frac{R_\alpha}{R^3}U(R) \qquad (1)$$

where C is a constant. Similarly, if the other impurity is at $\mathbf{R}_0$, the force at $\mathbf{R} = \mathbf{R}_0 + \mathbf{a}$ is

$$F^\alpha_{\mathbf{R}_0+\mathbf{a}} = a_\alpha \varphi(a)$$

where φ is a function of $|\mathbf{a}| = a$ only. All these formulae are approximate since, for instance, the force due to one impurity is slightly modified by the other. This point will be discussed later. The above formulae yield

$$\mathscr{F} = -C \sum_{R\alpha} a_\alpha \varphi(a) \left(R^0_\alpha + a_\alpha\right) \frac{1}{R^3} \,. \tag{2}$$

At long distance:

$$\begin{aligned}
\mathscr{F} &\simeq -\frac{C}{R_0^3} \sum_{a\alpha} a_\alpha \varphi(a) \left(R^0_\alpha + a_\alpha\right) \left(1 - \frac{3}{R_0^2} \mathbf{R}_0 \cdot \mathbf{a}\right) \\
&\simeq -\frac{C}{R_0^3} \sum_{a\alpha} a_\alpha \varphi(a) R^0_\alpha - \frac{C}{R_0^3} \sum_{a\alpha} a_\alpha^2 \varphi(a) + \frac{3}{R_0^5} \sum_{a\alpha\gamma} \varphi(a) a_\alpha a_\gamma R^0_\alpha R^0_\gamma \,.
\end{aligned}$$

The first term vanishes when summing over $\mathbf{a}$.

$$\mathscr{F} \simeq -\frac{C}{R_0^3} \sum_{a\alpha} a_\alpha^2 \varphi(a) + \frac{3C}{R_0^5} \sum_{a\alpha} \varphi(a) a_\alpha^2 (R^0_\alpha)^2 \,.$$

Now, in 3 dimensions and in spherical or cubic symmetry, one has

$$\sum_a a_\alpha^2 \varphi(a) = \frac{1}{3} \sum_a a^2 \varphi(a)$$

and

$$\begin{aligned}
\mathscr{F} &\simeq -\frac{C}{R_0^3} \sum_a a^2 \varphi(a) + \frac{C}{R_0^5} \sum_{a\alpha} \varphi(a) a^2 (R^0_\alpha)^2 \\
&\simeq -\frac{C}{R_0^3} \sum_a a^2 \varphi(a) + \frac{C}{R_0^5} \sum_a \varphi(a) a^2 R_0^2 = 0 \,.
\end{aligned}$$

As said above, this calculation is only approximate. Formulae (1) and (2) should be corrected and the correction is proportional to the local strain which is proportional to R_0^{-3}. The detailed calculation (Hardy & Bullough 1966, Siems 1968) should take into account the atomic nature of matter and yields an interaction proportional to R^{-5}. The interaction is anisotropic, attractive or repulsive according to the direction, and vanishes on the average.

16.4. Consider a triangular lattice (Fig. 16.7b) in which the energy is a sum of harmonic pair interactions between nearest neighbours defined as follows. i) The minimum energy of a pair is obtained for the same value a of the distance except if both atoms have an odd number of neighbours. ii) The force constant is the same for all pairs where at least one atom has 6 neighbours, and much weaker for other pairs. a) What is the surface stress far from the step? b) Show that the extra forces defined in section 16.11 are localized.

Solution. The surface stress vanishes far from the step. Only interactions between surface atoms are different from bulk interactions and can contribute to extra forces as defined in section 16.11. All surface bonds except AB and AC in Fig. 16.7 have the same length as in the bulk, so that the corresponding force is weak. Consequently the extra forces act only on atoms A, B, C.

17
Technology, crystal growth and surface science

I ask a lot of these "dumb" questions:
"Is a cathode plus or minus?
Is an an-ion this way, or that way? "

R. Feynman *(Surely you are joking, Mr Feynman!)*

While the first chapter of this book dwelt on the confirmation of the fundamental ideas of the philosophers of the fifth century BC, this last chapter will be devoted to the twentieth century and to the transformation of our materialistic everyday life brought about by electronics.

The early developments employed electronic tubes which owed nothing to surface physics and crystal growth. Then came the transistor and the realm of silicon, of which huge, dislocation-free crystals can be grown from the melt. In this end of the twentieth century, electromagnetic waves play an increasingly important part. Their production and detection require more and more complex semiconducting materials, often grown by molecular beam epitaxy.

17.1 Introduction

The modernity of surface science lies for a large part in the fact that new experimental techniques now make surfaces accessible to investigation. But the present interest in surface physics is also much motivated by the technological importance of crystals grown with a well-controlled composition and a high degree of perfection, and this depends very much on the state of the crystal surface during growth.

The purpose of this chapter is to outline the technological importance of materials with a controlled purity and morphology.

We shall mainly discuss semiconductors. It is not very easy to find simple articles or books explaining semiconductor electronics to the ignorant, but the book by Parker (1994) accomplishes this function. Since we want to talk to a reader who has no knowledge of electronic technology, we think it useful to give a short description of the whole history of electronic technology throughout the twentieth century, even though solid state devices appeared only in the middle of this century.

17.2 The first half of the twentieth century: the age of the radio

From 1900 to 1940, radio penetrated our world and deeply changed the habits of human beings. The most spectacular transformation probably took place in the military domain, and did not concern only the soldiers themselves. During the Second World War, the radio took an essential place in the life of civilians as well, and for instance the French who were already born at that time and are still alive remember listening to the BBC which kept them in contact with free France.

The essential parts of their radio sets were electronic tubes, which ensured amplification of the tiny signal captured by the antenna, where the appropriate wavelength was selected by a tunable resonating circuit. These electronic tubes, mainly triodes, were big, fairly expensive and operated at high temperature, so that their lifetime was not very long. In the second half of the twentieth century, the triodes were replaced by transistors.

17.3 The third quarter of the twentieth century: the age of transistors

From 1948 to 1952, as the growth of silicon crystals of controlled purity became possible, the first transistors were fabricated. They motivated the Nobel prize awarded in 1956 to Bardeen, Brattain and Schockley. A few years later, semiconductors devices replaced electronic tubes for most of their functions: rectifiers, amplifiers, oscillators. They were cheaper, smaller, and more robust.

The virtue of semiconductors was their tunable conductivity, which could be obtained by doping them with impurities. If you put impurities into a metal, you modify the conductivity, but you still have a conductor. If you put impurities into an ionic crystal, it is still an insulator. Semiconductors are so near to be metals, that they do become metals when they are doped. They are actually much more than metals: a metal (Fig. 17.1a) conducts through its electrons and its holes, which are both in the same band. Semiconductors can conduct either through electrons (in the normally empty conduction band, Fig. 17.1b) or through holes (in the

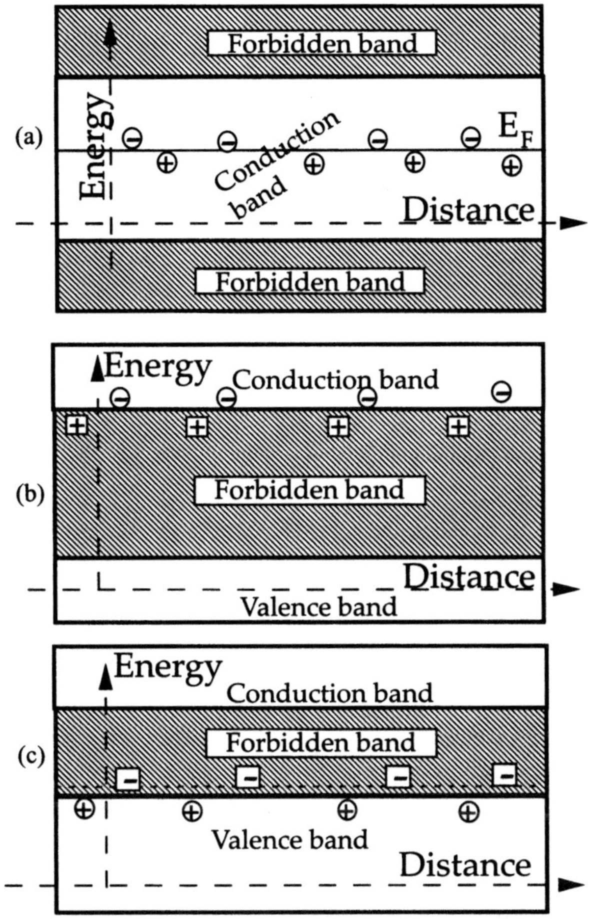

Fig. 17.1. a) The conduction band of a metal. At finite temperature, both electrons (− charges above the Fermi level) and holes (+ charges below E_F) are present and contribute to electrical conduction as 'charge carriers'. b) In an n-type semiconductor, the only carriers are electrons in the normally empty 'conduction band' (circles with - signs). The squares correspond to impurities which have lost an electron. c) In a p-type semiconductor, the only carriers are holes in the normally full 'valence band' (circles with + signs). The squares correspond to impurities which have captured an electron. The dotted line denotes the energy of electrons bound to these acceptor impurities.

normally full valence band, Fig. 17.1c). This is marvellous, because if you now make a junction of two semiconductors of these two different types, you do get a current flowing, but in one direction only (Fig. 17.2). Indeed, for one of the two possible current directions, both electrons and holes flow to the junction, where they recombine, so that there is a current. For the other current direction, both electrons and holes flow away from the junction, and after a short time there are no electrons nor holes available at the junction, so there is no current. Thus, the junction is a rectifying diode.

A semiconductor which conducts through electrons (Fig. 17.1b) is called n-type (n for negative) and the impurities are called donors (of electrons). A semiconductor which conducts through holes (Fig. 17.1c) is called p-type (p for positive) and the impurities are called acceptors. The diode of Fig. 17.2 is a p-n junction. There are many types of diodes, and their function is not necessarily to be a rectifier, as we shall see. A diode, quite generally, is an object with two poles, with a non-linear response to an applied voltage.

Similarly, there are many types of transistors. Their common property is to have three poles which can be connected to electric wires. The most common transistor, at least in the early stage of semiconductor electronics, is an n-p-n or p-n-p sequence (Fig. 17.3a). The sandwiched part, called *base*, is very thin, say 1 μm. One wire is connected to the base and the other two are connected to the two lateral parts, called emitter and collector. The electric current intensity I_C through the collector is a highly non-linear function (Fig. 17.3b) of the current i_B through the base. In the almost linear, intermediate region of the curve, a very weak variation of i_B causes a very large variation of I_C, so that the transistor can be used as an intensity amplifier. The other two parts of the curve $I_C = f(i_B)$ correspond respectively to a vanishingly small value of I_C

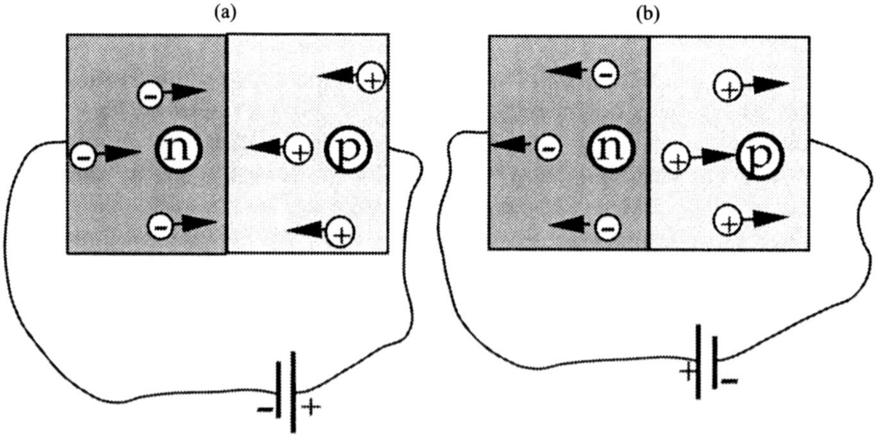

Fig. 17.2. A p-n junction. a) Conducting connection to generator. Both types of carriers flow to the junction where they annihilate. b) The current cannot flow in the other direction because both carriers flow away from the junction and thus there are no carriers available.

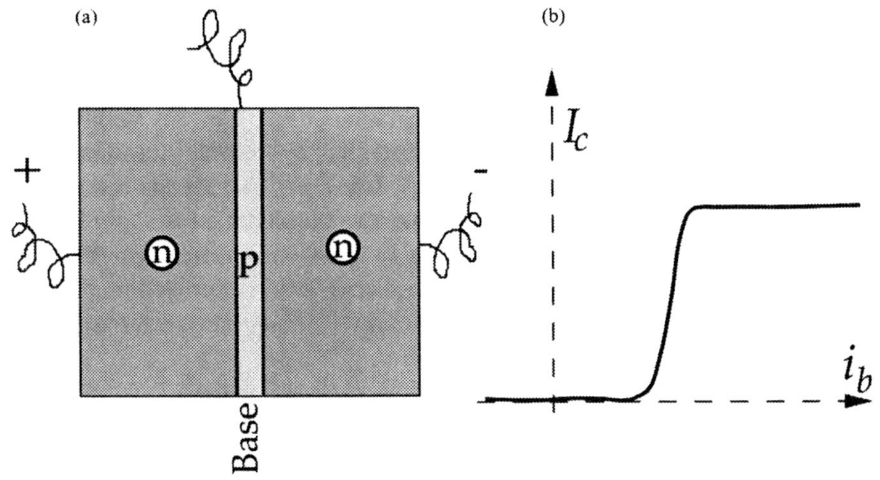

Fig. 17.3. a) An n-p-n transistor. b) The collector current as a function of the base current.

and to a practically constant, non-vanishing value. These parts allow the transistor to be used as a two-level system in a logic circuit.

17.4 The last quarter of the twentieth century: the age of chips

The controlled concentration of useful dopants and the absence of undesirable impurities were remarkable achievements, even though they would not fulfill the requirements of contemporary technology. The interfaces were not smooth and the linear size, although about 50 times as small as those of an electronic tube, was still macroscopic. This was not inconvenient to making radio sets, but this was a serious hindrance to making high performance computers, for instance, which should have very large memories. The memories of a computer can use magnetic materials rather than semiconductors, but electronic devices are still necessary to use the memory, i.e. to read the information. Thus, in order to store much information in a necessarily restricted space, one needs very small transistors, and more generally very small semiconductor devices.

There are other reasons to look for small sizes: one is the need to get results in reasonably short times. One of the authors of this book owns an old computer with an old program to play chess. At levels 1, 2 and 3, he generally wins. At level 4, the computer needs several hours to play a single move, and becomes so hot that it seems preferable to switch it off. Indeed, another reason to require small sizes is the reduction of energy consumption and heating–the major cause of breakdown in old electronic tubes.

This size reduction is of course limited by the atomic size, and more generally by the atomic nature of matter. The most widely used material in the end of the twentieth century is still doped silicon, and the thickness should therefore be sufficient to ensure a reasonably uniform density of dopants in the other two directions. Moreover, the techniques used for doping are diffusion at high temperature, or implantation by accelerators: they do not provide a very good control of the concentration. Molecular beam epitaxy would be better, but it is only usual in technology for III-V compounds as GaAs. MBE growth of silicon, often mentioned in this book, is only done in research laboratories, and 95% of the semiconductor technology uses silicon.

As a result of size reduction, electronic devices are grouped in *chips*. A chip is a rigid set of thousands of transistors, connections, etc. gathered on a silicon plate, whose dimensions vary from a few millimetres to a few centimetres. Such chips are present on our credit cards, protected and hidden by a patch of about 0.4 cm^2. They store our confidential code and are able to check that the number keyed in by the user is the same as the one which is stored. One can thus say that they are little, special-purpose computers.

It is necessary to control the composition in three directions during growth. In MBE, the composition in the direction perpendicular to the surface is controlled by switching the beam on and off so as to deposit the required number of atomic layers. The composition in the directions parallel to the surface is determined by methods which are called lithographic, because they use coatings and corrosion, as the art of lithography has being doing for centuries. The realization of a chip is a very complicated operation which lasts one or two months and requires much care. However, since many identical chips are fabricated at the same time, chips are cheap.

17.5 MOSFETS and memories

As we said above, there are many types of transistors. Some of them use MOS (Metal-Oxide-Semiconductor) devices. In a MOS device, one of the electrodes is surrounded by an insulator (silica) so that the current through it is zero. The applied voltage does exert an action through the resulting electric field. Fig. 17.4a shows a MOS diode, which is not at all a rectifier (as the diode of Fig. 17.2), but a kind of capacitor.

Fig. 17.4b shows a Metal-Oxide-Semiconductor Field Effect Transistor or MOSFET. It can be used as a switch–an important function in a computer or any other electronic device. Like any transistor, it has 3 poles, now called source (instead of emitter), gate (instead of base) and

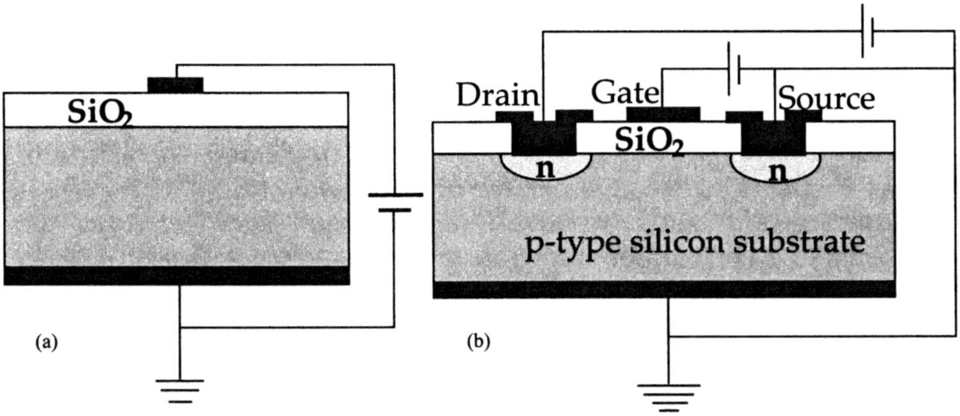

Fig. 17.4. a) A MOS diode. b) A MOS transistor or MOSFET. It is conducting or not, according to the voltage applied to the metal.

drain (instead of collector). The drain current is controlled by the gate, but now the current through the gate is stricly zero, due to the oxide layer, and the voltage at the gate generates an electric field which allows or forbids the electric current in the semiconductor.

Another MOS device is the memory sketched in Fig. 17.5. It is a dead memory, i.e. it works even in the absence of a current. The most obvious difference with Fig. 17.4 is the presence of a conducting 'floating grid' embedded in the insulator (SiO_2) and not connected. The floating grid is able to store electrons when a small positive voltage is applied to the drain and a strong positive voltage to the metallic command. The presence or absence of electrons is equivalent to a bit. Our credit cards contain such memories, which turn out to be safer than magnetic memories used for instance in floppy disks.

Computers also contain memories which work only when an electric current is present. The conception of such a memory, using objects with

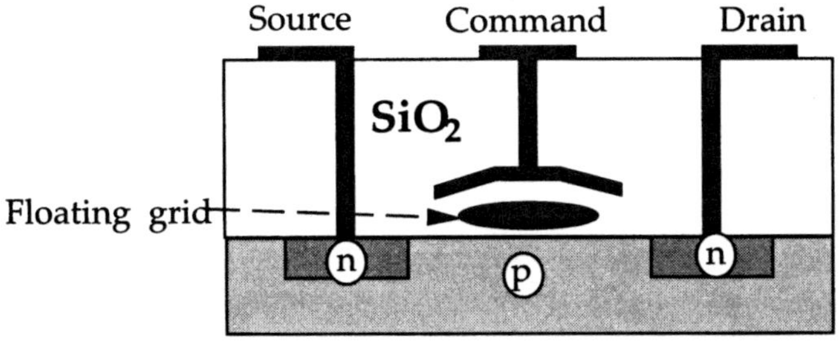

Fig. 17.5. A dead memory using MOS devices.

non-linear response as in Fig. 17.3, is just a logical operation. On the contrary, the conception of the dead memory of Fig. 17.5 requires precise physical calculations taking into account the tunnel effect through the insulator.

The oxide layer of a MOS junction is grown by surface oxidation of silicon. Oxidation of the first layers makes no problem, but the oxygen has to diffuse through the oxide to form the deeper oxide layers, and the random nature of the diffusion process unavoidably generates a rough Si/SiO_2 interface (see problem 17.1). It turns out that the oxide layer may not be thinner than about 4 nm. For smaller thicknesses, leak currents occur. Replacing the oxide by an insulator grown for instance by MBE might improve the performances. Another possibility is to replace the oxide by grafted polymers. Their drawback is that they cannot be heated at more than 450 °C in the subsequent steps.

17.6 From electronics to optics

Certain functions accomplished until 1990 by electrons are now accomplished by photons, and the role of optics will probably increase in the near future (these lines are written in 1997). Indeed, for the transport of information, photons have great advantages: they travel very fast and do not interact much with appropriately chosen materials, thus avoiding energy dissipation. Of course, electronics will always be necessary for many uses ..., especially to produce photons.

Light can be produced, in particular, by Light Emitting Diodes (LED). Even if these objects are not used as rectifiers, they can still be called diodes because they are objects with two poles. As a familiar application, LEDs are commonly used as 'on/off' indicators. If an instrument is on, there is a red, yellow or green light. LEDs are diodes in which the recombination of electrons and holes at the p-n junction is accompanied by the emission of photons which ensure energy conservation. However, energy conservation can also be achieved by phonons. Therefore, not any p-n junction is a LED.

In order to discuss whether a semiconductor can or cannot emit light by electron-hole recombination, one should go beyond the simple-minded idea of characterizing a semiconductor just by the fact that the gap between the valence and the conduction band is narrow. The band gap can indeed be direct or indirect (Fig. 17.6). In the former case, the gap is between electron states of the same wavevector k (often $k = 0$). In the latter case, it is not so. Only materials with a direct band gap are appropriate to emit photons. The best direct band gap material presently known (in 1997) is GaAs. Note that an appropriate Al concentration transforms it into an indirect band gap material (Fig. 17.6).

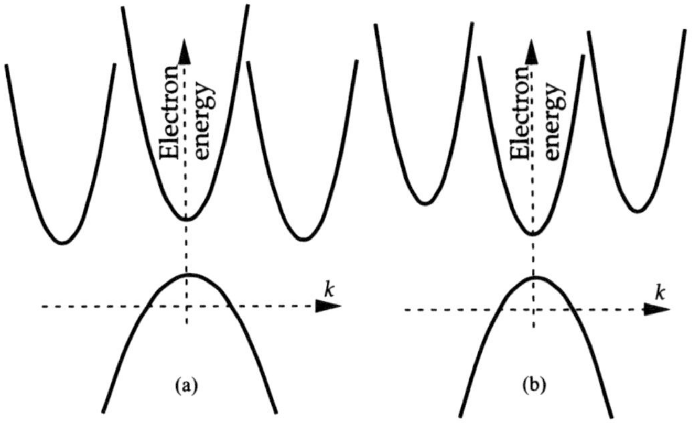

Fig. 17.6. a) Indirect band gap. b) Direct band gap. Both cases can be realized in GaAlAs for different Al concentrations.

The importance of binary semiconductors is mainly due to their adequacy for light production and light control. For other uses, silicon is more appropriate. GaAs is a 'III-V' semiconductor, i.e. it combines an element of the third column of the Mendeleiev table with an element of the fifth column. Other direct band gap semiconductors are II-VI semiconductors, e.g. ZnSe or ZnTe. Unfortunately they do not easily accept being doped at the user's will.

17.7 Semiconductor lasers

Another device able to produce light is the laser. There are many types of lasers, very big ones and very small ones. The small ones are made with semiconductors. Their merit is to provide a very thin light beam with a reasonably high power. When writing this book, we have used a laser printer, an object able to produce a high temperature in a small area (0.01 mm^2 if the letters are 0.1 mm thick) chosen by the word processor. When we listen the music contained in a compact disc, we use a laser beam which can read the information contained in an even smaller area. This information consists of small bumps, about 0.1 μm high. To give an idea of their size parallel to the surface, we shall just say that a compact disc is able to store 10^{10} bits per cm^2 or 100 bits per μm^2. A 'bit' is a variable which can take two values, e.g. 'yes' and 'no', or + and −, or 0 and 1. This gives an idea of how thin the laser beam should be in order to read each bit, for instance by being absorbed or reflected in a particular direction.

What is a laser? i) The word means Light Amplification by the Stimulated Emission of Radiation. A laser is actually a light emitter rather than an amplifier, but it contains a device which amplifies the emission of a particular wavelength. ii) A particular device is the laser LED (Fig.

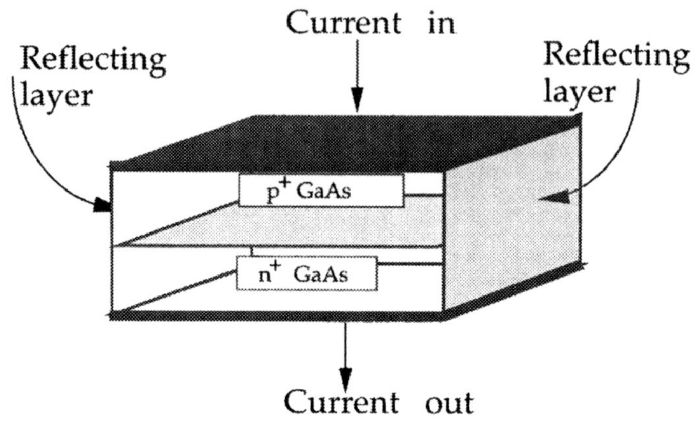

Fig. 17.7. A laser diode. Recombination at the n-p junction generates light, the reflecting layers select a particular wavelength which forms standing waves which stimulate the emission of light of exactly the same wavelength with a very high intensity.

17.7) which is just a LED, two ends of which are able to reflect partly the light emitted by the junction in the direction perpendicular to the mirrors. The reflected light comes back to the junction and stimulates emission of identical light, i.e. with the same wavelength and the same direction. iii) What is stimulated emission of light? Quantum theory of light emission shows that, if the number of electrons in an excited state is larger than in the ground state, light emission is more intense in the presence of an incoming photon, than it would spontaneously be. The stimulated photons have exactly the same frequency as the incoming one, and they are coherent with it. Thus, the laser is an amplifier, but only for photons of a particular wavevector: those which form stationary waves between the two mirrors. The junction produces other photons by spontaneous emission, but they are not amplified and can be neglected.

The number of bits stored in a compact disc (100 per μm^2 as said above) is amazing. The light beam must obviously have a strong angular opening 2θ, since the wavelength λ and the linear resolution d are related to θ by the Young-Fresnel relation $\lambda = 2d \sin\theta$. This is essentially Heisenberg's uncertainty relation, and therefore cannot be violated. Thus, a lens is needed, and a sophisticated signal processing. It is also convenient to put the laser, and the lens, very close to the disc. Finally, it is preferable to reduce the wavelength. Presently, red light is used. When physicists find an appropriate blue laser, the allowed information density will be multiplied by 10 and it will be possible to have Beethoven's nine symphonies (together with his concertos) on just one disc.

17.8 Quantum wells

Quantum wells are layers of very small thickness (say 10 to 20 atomic layers) of a semiconductor embedded in another semiconductor, e.g. GaAs in GaAlAs. Assuming the system to be infinite in the directions parallel to the layers, the electronic states are, of course, characterized by a wavevector $\mathbf{k} = (k_x, k_y, k_z)$, where z is the direction perpendicular to the layer. For given k_x and k_y, e.g. $k_x = k_y = 0$, the third component k_z is quantized, and only the lowest value is allowed at room temperature.

The main interest of quantum wells is their adjustable gap, which depends on the size. Another interest is to provide a strong interaction between matter and light, thus allowing the control of electromagnetic waves which should carry information.

The realization of quantum wells requires a microscopic type of crystal growth, i.e. MBE (Fig. 17.8).

Quantum wells are two-dimensional, electron-confining systems. It would be of interest to realize, one-dimensional electron-confining systems (quantum rods) or 'zero-dimensional' electron-confining systems (quantum dots). 'Zero-dimensional' means here a microscopic length in all directions. In a quantum dot, only the lowest values of the three components of $\mathbf{k}$ would be populated.

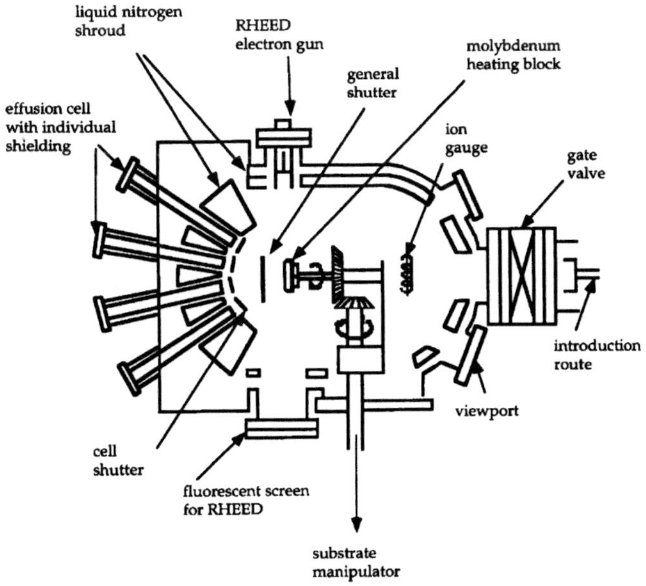

Fig. 17.8. Apparatus for Molecular Beam Epitaxy seen from above (horizontal cross section). An essential component is hidden in the (invisible) lower part, namely the pump which ensures ultra-high vacuum.

Problems

17.1. Discuss qualitatively the roughness of a Si/SiO_2 interface, grown by diffusion of oxygen through the oxide.

17.2. Does a laser provide a light with a well-defined wavelength or wavevector?

Answer – Only the wavelength is well-defined, not the direction.

17.3. Check that, if a single electron is present in a quantum dot of thickness 10 atomic layers in all three directions, then only its ground state is populated at room temperature.

Appendix A
From the discrete Gaussian model to the two-dimensional Coulomb gas

(After Chui & Weeks, 1976.)

Let i, j be the different columns of Fig. 1.8, and h_i the height of the i–th column. The *discrete Gaussian model* is defined by its partition function

$$Z_{\mathrm{dG}} = \sum_{\{h_i\}} \exp\left\{-\beta J \sum_{i,j} (h_i - h_j)^2\right\}.$$

In the SOS model, as a matter of fact, $(h_i - h_j)^2$ is replaced by $|h_i - h_j|$. However, the two models are expected to possess analogous properties, or, in the phase transition jargon, to belong to the same universality class. The discrete Gaussian model has the non-minor advantage of being allowed to exploit the following transformation:

$$Z_{\mathrm{dG}} = \int \mathscr{D}u \exp\left\{-\beta J \sum_{i,j} (u_i - u_j)^2\right\} \prod_i \sum_{m_i=-\infty}^{\infty} \delta(u_i - m_i)\,.$$

This peculiar transformation has the only purpose of allowing application of the following formula–a special case of Poisson's summation formula–whose proof is recalled at the end of this appendix:

$$\sum_{m_i=-\infty}^{\infty} \delta(u_i - m_i) = \sum_{n=-\infty}^{\infty} \mathrm{e}^{2i\pi n u_i} \tag{A.1}$$

where m, n are integers.

The two preceding formulae permit the following manipulations of the partition function:

$$Z_{\mathrm{dG}} = \sum_{\{n\}} \int \mathscr{D}u \exp\left\{-\beta J \sum_{i,j} (u_i - u_j)^2 + 2i\pi \sum_i n_i u_i\right\} \tag{A.2}$$

$$= \sum_{\{n\}} \int \mathscr{D}u \exp\left\{ \sum_q [-\beta JK^2(\mathbf{q})|u_q|^2 + 2i\pi n_{-q} u_q] \right\}$$

$$= \sum_{\{n\}} \int \mathscr{D}u \exp\left\{ \sum_q [-\beta JK^2(\mathbf{q})|u_q - \frac{i\pi n_q}{\beta JK^2(\mathbf{q})}|^2 - \pi^2 T \frac{n_{-q} n_q}{JK^2(\mathbf{q})}] \right\}$$

where

$$K^2(\mathbf{q}) = 4\sin^2(\frac{1}{2}q_x) + 4\sin^2(\frac{1}{2}q_y)\,. \tag{A.3}$$

Performing the elementary integration over the continuous variable u_q, (A.2) factors,

$$Z_{\text{dG}} = Z_{\text{h}} Z_{\text{Cg}} \tag{A.4}$$

where

$$Z_{\text{h}} = \prod_q \frac{T}{2JK^2(\mathbf{q})}$$

is the partition function of a system of classical harmonic oscillators, and

$$Z_{\text{Cg}} = \sum_{\{n\}} \exp\left\{-\beta\tilde{\mathscr{H}}\right\},$$

where

$$\tilde{\mathscr{H}} = \frac{T^2\pi^2}{J} \sum_q \frac{n_q n_{-q}}{K^2(\mathbf{q})} \tag{A.5}$$

is the partition function of the so-called *Coulomb gas*.

Before going on writing equations, let us see where we are. For small q, $K^2(\mathbf{q}) \sim q^2$. Formula (A.5) describes thus essentially a pair particle interaction $U(r)$ whose Fourier transform behaves basically as $V(q) \sim 1/q^2$ at small q. The Fourier transform of the equation $q^2 V(\mathbf{q}) = 1$ is $\nabla^2 V(\mathbf{r}) = \delta(\mathbf{r})$, within a factor of 2π. This is Poisson's equation.

Some theorists decided in the 1960s to call such a $V(\mathbf{r})$ 'Coulomb potential' irrespective of the space dimension. An applied physicist would undoubtly give this name to the interaction between two electric charges, which always goes like $1/r$, even if the charges are constrained to move in a plane.

As a matter of fact, the Fourier transform of $1/q^2$ in two dimensions is not $1/r$, but rather $\ln r$. The simplest way to see this is to consider the electrostatic interaction $L\tilde{V}(r)$ between two very long bars of length L, parallel to the x axis at mutual distance r. According to Gauss's theorem, $\tilde{V}(r)$ verifies the equation

$$\left(\frac{\partial^2}{\partial x^2} + \frac{\partial^2}{\partial y^2}\right) \tilde{V}(x, y) = 0$$

for $\mathbf{r} = (x, y) \neq 0$. This is the two-dimensional Poisson's equation. On the other hand, $\tilde{V}(\mathbf{r})$ is straightforwardly obtained by integrating the three-dimensional potential over the bars, and this just gives $\ln r$.

As a consequence, (A.5) describes a system of particles interacting with a potential proportional to $\ln r$ at long distance. All qualitative properties of the system are dominated by this divergence at long distance, and the short distance behaviour is trifling. After computing the numerical prefactor C, equation (A.5) may be replaced by

$$\tilde{\mathscr{H}} = C \sum_{i,j} n_i n_j \ln r_{ij}$$

where the n_is are again the positive or negative integers defined by (A.1).

Finding the actual value of C will force us to pass through some more equations. First of all, note that (A.5) implies

$$n|_{q=0} = 0 \tag{A.6}$$

since $K(q = 0)$ vanishes according to (A.3). On the other hand, in direct space (A.5) reads

$$\tilde{\mathscr{H}} = \sum_{i,j} \tilde{V}(\mathbf{r}_{ij}) n_i n_j \tag{A.7}$$

where

$$\tilde{V}(\mathbf{r}_{ij}) = \frac{T^2\pi^2}{JN} \sum_{\mathbf{q}\neq 0} \frac{\exp(-i\mathbf{q}\cdot\mathbf{r})}{K^2(\mathbf{q})} . \tag{A.8}$$

This expression is meaningful only for a finite volume, while we would prefer a formula also applicable at an infinite volume. To this end, note that (A.6) can be also written $\sum_i n_i = 0$, expressing thus the 'electrical neutrality' of the system, which allows rewriting (A.7) in the form

$$\tilde{\mathscr{H}} = \sum_{i,j} \left\{\tilde{V}(\mathbf{r}_{ij}) - \tilde{V}(0)\right\} n_i n_j = \sum_{i,j} V(\mathbf{r}_{ij}) n_i n_j \tag{A.9}$$

where $V(\mathbf{r}_{ij}) = \tilde{V}(\mathbf{r}_{ij}) - \tilde{V}(0)$ is finite (in fact, zero) for $\mathbf{r} = 0$. At large r, we can replace K^2 by q^2 and the integration over Brillouin's zone by an integration over a sphere. Equation (A.8) becomes then

$$\tilde{V}(\mathbf{r}_{ij}) \simeq \frac{T^2\pi}{2J} \int_0^\pi \frac{1 - J_0(qr)}{q} \mathrm{d}q \simeq \frac{T^2\pi}{2J} \int_{1/r}^\pi \frac{\mathrm{d}q}{q} \simeq \frac{T^2\pi}{2J} \ln r .$$

We can now specify C and write:

$$\tilde{\mathscr{H}} = \frac{\pi T^2}{2J} \sum_{i,j} n_i n_j \ln r_{ij} . \tag{A.10}$$

We close this appendix by proving (A.1). Introducing the small positive quantity ϵ, we have

$$\begin{aligned}\sum_{n=-\infty}^{\infty} \exp\{-2i\pi nu - |n|\epsilon\} &= 2\mathrm{Re}\sum_{n=0}^{\infty} \exp\{-n(2i\pi u+\epsilon)\} - 1\\ &= \frac{2-2\cos(2\pi u)\mathrm{e}^{-\epsilon}}{1-2\cos(2\pi u)\mathrm{e}^{-\epsilon}+\mathrm{e}^{-2\epsilon}} - 1\\ &= \frac{1-\mathrm{e}^{-2\epsilon}}{(1-\mathrm{e}^{-\epsilon})^2+4\sin^2(\pi u)\mathrm{e}^{-\epsilon}}\\ &\simeq \frac{2\epsilon}{\epsilon^2+4\sin^2(\pi u)}\,.\end{aligned}$$

In the $\epsilon \to 0$ limit, this vanishes unless u is an integer. Relation (A.1) follows with the correct coefficient from the integral

$$\int_{-\frac{1}{2}\pi}^{\frac{1}{2}\pi} \frac{2\epsilon}{\epsilon^2+4\sin^2(\pi u)} \simeq \int_{-\infty}^{\infty} \frac{2\epsilon}{\epsilon^2+4\pi^2u^2} = 1\,.$$

The derivation of (A.9) or (A.10) is not very complicated, but it is quite abstract. There seems to be no remedy for that–we do not know of any simple interpretation of the variables n_i in connection with surface roughness, as is the case for the XY model of magnetism (Kosterlitz & Thouless, 1973).

Formula (A.5) shows that the quantity J/T^2 plays the role of a dielectric constant. The renormalization of the latter will be the subject of the next appendix.

Appendix B
The renormalization group applied to the two-dimensional Coulomb gas

In appendix A we have shown the equivalence of the SOS model with a two-dimensional Coulomb gas. The roughening transition of the SOS model thus corresponds to the conductor-insulator transition of the Coulomb gas. We describe now the theoretical approach to this transition following A. P. Young (1978).

A conductor-insulator transition manifests itself through the dielectric constant ϵ, which becomes infinite in the conducting phase. In fact, the electrostatic force $qq'/\epsilon r$ between two charges q and $-q'$ at distance r must vanish due to screening. It is really the force which varies as $1/r$ since the Coulomb energy in two dimensions–as we defined it in appendix A–varies as $\ln r$.

Actually the force only vanishes at infinite distance, and it is necessary to introduce a distance-dependent dielectric constant $\epsilon(r)$, defined by the requirement that the force $f(r)$ between two charges q and $-q'$ at mutual distance r is (with an appropriate choice of the unit of charge, which introduces a factor 2π):

$$f(r) = \frac{2\pi qq'}{\epsilon(r)r} \,. \tag{B.1}$$

The interaction energy between the two charges is thus:

$$W(r) = \int_a^r \frac{2\pi qq' \mathrm{d}r'}{\epsilon(r')r'} = qq' \int_{\ln a}^{\ln r} \frac{2\pi}{\epsilon(r')} \mathrm{d}(\ln r') = 2\pi U(r) \ln\left(\frac{r}{a}\right) . \tag{B.2}$$

For future use we introduced a function $U(r)$ which equals qq' in the absence of screening ($\epsilon(r) = 1$). We also introduced a short scale cutoff a, of the order of the interatomic distance 1 (in our units).

Such equations translate the following physical situation: the interaction between two charges q and $-q'$ would be equal to (B.1) and (B.2) with $\epsilon(r) = 1$ if no other charge was present. In fact, there are other charges between q and $-q'$: we can think of these extra charges as making up dipoles, i.e. very close *pairs* of charges with opposite signs. We are thus implying that no *isolated* charge q'' is present between q and $-q'$. This is reasonable if the sign of q'' is opposite to either that of q or that of $-q'$, because then q'' could form a dipole with the opposite charge. The only problematic case is when the three charges have the same sign. But this occurrence is quite unlikely in the insulating phase, while in the conducting phase it simply means that it is useless to consider the very distant charges q and $-q'$, for ϵ is already almost infinite for the closer pairs q and q'' or q'' and $-q'$.

In three-dimensional electrostatics, we know that the effect qualitatively described above is translated by the equation:

$$\epsilon = 1 + 4\pi\chi \,. \tag{B.3}$$

It is not difficult to convince oneself that this equation also holds in our two-dimensional electrostatics, if one notes that our choice of the unit of charge transforms the factor 4π into $4\pi^2$.

The dielectric susceptibility χ, for a system of dipoles of fixed moment qr' is equal to $\beta q^2 r'^2 v$, where v is the dipole number density. In our problem we have dipoles of any size. Assuming that $|q| = 1$ for all charges, (B.3) must be written in the form:

$$\epsilon(r) = \epsilon(a) + 4\pi^2\beta \int_a^r r'^2 v(r')\mathrm{d}r' \tag{B.4}$$

where $v(r)\mathrm{d}r$ is the product of the area element $2\pi r \mathrm{d}r$ times Boltzmann's weight $\exp\{-\beta W(r)\}$. It is useful to multiply (B.4) by an overall constant factor representing the charge chemical potential, which depends on the coupling constant J of appendix A. Using now the function $U(r)$ defined by (B.2), we define a parameter y_0 through the relation:

$$\epsilon(r) = \epsilon(a) + \pi^3 y_0^2 \beta \int_a^r (r'/a)^{3-2\pi\beta U(r')}\mathrm{d}(r'/a) \,. \tag{B.5}$$

Letting

$$y^2(r) = y^2(a)(r/a)^{4-2\pi\beta U(r)} \tag{B.6}$$

with $y(a) = y_0$, equation (B.5) takes the simpler form

$$\epsilon(r) = \epsilon(a) + \pi^3\beta \int_{\ln a}^{\ln r} y(r')\mathrm{d}\ln(r'/a) \,. \tag{B.7}$$

Using again the definition (B.2), we see that (B.6) may be written as:

$$y^2(r) = y^2(a)\exp\left(4\ln(r/a) - \beta\int_{\ln a}^{\ln r} \frac{2\pi}{\epsilon(r')}\mathrm{d}(\ln r')\right) \,. \tag{B.8}$$

We recall that our purpose is to find $\epsilon(\infty)$, which must take a finite value in the insulating and an infinite value in the conducting phase. We have reduced the problem to the solution of the system of integral equations (B.7) and (B.8). A system of differential equations is however easier to handle. If we differentiate both equations with respect to $\ln r$, we obtain from (B.7):

$$
\begin{aligned}
2y(r)\frac{\mathrm{d}y(r)}{\mathrm{d}\ln r} &= y^2(a)\exp\left\{4\ln(r/a) - \int_{\ln a}^{\ln r}\frac{2\pi\beta}{\epsilon(r')}\mathrm{d}(\ln r')\right\}\left(4 - \frac{2\pi\beta}{\epsilon(r)}\right) \\
&= 2\left(2 - \frac{\pi\beta}{\epsilon(r)}\right)y^2(r)\,.
\end{aligned}
$$

Simplifying and combining with the derivative of (B.8), we find

$$
\frac{\mathrm{d}y(r)}{\mathrm{d}\ln r} = \left[2 - \frac{\pi\beta}{\epsilon(r)}\right]y(r) \tag{B.9}
$$

$$
\frac{\mathrm{d}\epsilon(r)}{\mathrm{d}\ln r} = 4\pi^3 y^2(r)\,. \tag{B.10}
$$

These equations were obtained first by Kosterlitz (1974) in a somewhat different context and language. Our notation corresponds essentially to that of Itzykson & Drouffe (1989), the only difference being the definition of the dielectric constant: our $\beta/\epsilon(r)$ is in their case replaced by the notation $\beta(r)$.

We are now going to show that the transition takes place at $\epsilon(\infty) = \frac{1}{2}\beta\pi$.

- Consider a value of r making $2 - \pi\beta/\epsilon(r)$ positive. Then, according to (B.9) $y(r)$ increases, and $\epsilon(r)$ diverges with r. This is the conducting phase.
- If r is such that $2 - \pi\beta/\epsilon(r)$ is negative, $y(r)$ decreases. Moreover, if $y(r)$ is initially quite small, $\epsilon(r)$ never reaches the value $\frac{1}{2}\pi\beta$, and stays thus finite.

Appendix C
Entropic interaction between steps or other linear defects

An interesting type of interaction between steps is of entropic nature: steps cannot cross each other, so that the entropy of a step is reduced by its neighbours. This interaction is obviously repulsive since each step tries to have as much space as possible, just like somebody in the underground. The interaction free energy per unit step length will be found to be proportional to $1/\ell^2$, just as the elastic and electrostatic interactions. The length ℓ is the average distance between steps.

We shall give an approximate derivation, because we would like to avoid considering many steps. For this purpose, the two steps neighbouring a given step will be replaced by straight lines parallel to the y axis (Fig. C.1). The position of the step at ordinate y will be assumed unique and called x_y. It satisfies the condition $-\ell/2 < x_y < \ell/2$. Due to the discrete nature of matter, x and y may be assumed to be integers if the length unit is the

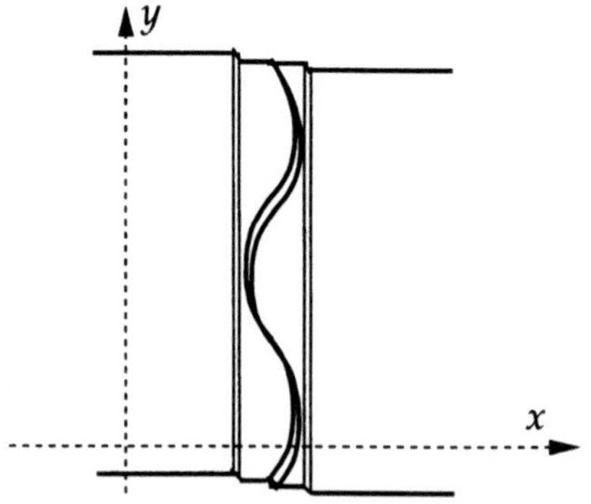

Fig. C.1. A fluctuating step between two other steps of same sign, approximated as straight. The fluctuating step cannot cross its neighbours, therefore its entropy is reduced with respect to that of an isolated step, by a factor which increases when the average distance between steps decreases.

atomic distance. In order to avoid uninteresting algebraic complications, the energy will be assumed to have the form

$$W\left(x_1, x_2, \ldots, x_L\right) = \sum_{n=1}^{L} w(x_n - x_{n-1}) \tag{C.1}$$

where L is the sample size in the y direction and periodic boundary conditions are assumed, $x_0 = x_L$. The elementary energy is an even function, $w(x) = w(-x)$. The free energy per step is equal to $-k_B T = -1/\beta$ multiplied by the logarithm of

$$Z = \prod_{n=1}^{L} \{\exp[-\beta w(x_n - x_{n-1})]\} \ .$$

The partition function Z can be evaluated by a standard method of statistical mechanics, the transfer matrix method. The trick is to consider each factor as the element of a $(\ell - 1) \times (\ell - 1)$ matrix:

$$\exp[-\beta w(p-q)] = \left\langle p \left| \hat{\Theta} \right| q \right\rangle \ , \tag{C.2}$$

which makes sense because x_n and x_{n-1}, or p and q, are integers. In terms of the eigenvalues θ_m, the partition function reads

$$Z = \mathrm{Tr}\ \hat{\Theta}^L = \sum_{m=1}^{\ell-1} \theta_m^L \ . \tag{C.3}$$

Our task is now to calculate the eigenvalues θ_m. This is easy, because the eigenvectors of $\hat{\Theta}$ have the form

$$\varphi_p = u \exp(ikp) + v \exp(-ikp) \tag{C.4}$$

where k is a real quantity which can take only a discrete set of values. Again, we take the simplest possible case, which corresponds to

$$\begin{cases} w(0) = \gamma_0 \\ w(1) = w(-1) = \gamma_0 + W_0 \\ w(p) = \infty \quad (|p| > 1) \end{cases}$$

where the step energy per atom γ_0 and the kink energy W_0 are constants. The eigenvector (C.4) satisfies the eigenvalue equation

$$\sum_q \langle p|\hat{\Theta}|q\rangle \varphi_q = \theta_p \varphi_p$$

for $-\ell + 3/2 \leq p \leq \ell - 3/2$, if

$$\theta_p = \exp[-\beta w(0)] + 2\exp[-\beta w(1)]\cos k \ . \tag{C.5}$$

The eigenvalue equation is also satisfied for $p = \pm(\ell - 1/2)$, if

$$\varphi_{\pm(\ell+1/2)} = u \exp\left[\pm ik(\ell + 1/2)\right] + v \exp\left[\mp ik(\ell + 1/2)\right] = 0\,, \qquad \text{(C.6)}$$

or

$$u \exp\left[ik(\ell + 1/2)\right] + v \exp\left[-ik(\ell + 1/2)\right] = u \exp\left[-ik(\ell + 1/2)\right] + v \exp\left[ik(\ell + 1/2)\right] = 0 \qquad \text{(C.7)}$$

which implies

$$u\{\exp\left[ik(\ell + 1/2)\right] - \exp\left[-ik(\ell + 1/2)\right]\} = v\{\exp\left[ik(\ell + 1/2)\right] - \exp\left[-ik(\ell + 1/2)\right]\}\,.$$

There are two families of solutions. The first one corresponds to

$$\exp\left[ik(\ell + 1/2)\right] - \exp\left[-ik(\ell + 1/2)\right] = 0\,.$$

Then, equation (C.7) yields $u = -v$.

The second family of solutions corresponds to $u = v$. Insertion into (C.7) yields

$$\exp\left[ik(\ell + 1/2)\right] + \exp\left[-ik(\ell + 1/2)\right] = 0\,.$$

To summarize, all acceptable values of k are integer multiples of $\pi/(2\ell+1)$:

$$k = \frac{\pi m}{2\ell + 1}\,. \qquad \text{(C.8)}$$

However, the value $m = 0$ is excluded because (C.6) would then imply $\varphi_p = 0$ for all ps.

We are interested in large systems. For large L, (C.3) reduces to

$$Z = (\theta_1)^L \qquad \text{(C.9)}$$

where θ_1 is the largest eigenvalue. According to (C.8) and (C.5), it corresponds to $m = 1$. Insertion of (C.5) into (C.9) yields

$$Z = \left\{\exp\left[-\beta w(0)\right] + 2\exp\left[-\beta w(1)\right]\cos\frac{\pi}{2\ell + 1}\right\}^L\,.$$

For large ℓ, this may be replaced by

$$Z(\ell) = \left\{\exp\left[-\beta w(0)\right] + 2\exp\left[-\beta w(1)\right] - \frac{\pi^2}{4\ell^2}\exp\left[-\beta w(1)\right]\right\}^L = Z(\infty)\exp\left(-\frac{CL}{\ell^2}\right) \qquad \text{(C.10)}$$

where

$$C = \frac{\pi^2}{4}\frac{\exp\left[-\beta w(1)\right]}{\exp\left[-\beta w(0)\right] + 2\exp\left[-\beta w(1)\right]}\,.$$

The free energy per step is

$$f(\ell) = -k_B T \ln Z(\ell) = f(\infty) + \frac{CL}{\ell^2} k_B T \,. \tag{C.11}$$

The second term is the interaction free energy, and it is seen to be proportional to $1/\ell^2$, as claimed before. The free energy per unit area is

$$\phi(\ell) = \frac{f(\ell)}{L\ell} = \frac{A}{\ell} + \frac{C}{\ell^3} k_B T \,. \tag{C.12}$$

The constant C can be related to the line stiffness $\tilde{\gamma}$. This can be done, for instance by making $\ell = \infty$ and calculating $\langle (x_n - x_0)^2 \rangle$. The result is

$$\langle (x_n - x_0)^2 \rangle = \frac{CL}{2\pi^2} \,.$$

On the other hand, this should be equal to $k_B T L/\tilde{\gamma}$ (see problem 2.3). It follows that $C = \pi^2 k_B T/(8\tilde{\gamma})$. However, the above calculation is only approximate since each step has been assumed to fluctuate between two straight, fixed steps.

The correct result (Haldane & Villain 1981) is

$$C = \frac{\pi^2 k_B T}{6\tilde{\gamma}} \,. \tag{C.13}$$

This result is twice as large as that obtained by Jayaprakash *et al.* (1983) because of a factor 1/2 in the second term of equations (3) and (13) of these authors. This factor has been corrected by Rolley *et al.* (1995). Equation (C.13) is also in agreement with Yamamoto *et al.* (1994).

The reader is invited to extend this result to the following cases:

i) When $w(x_y - x_{y'})$ is infinite only for $|\, y - y' \,|$ larger than some value $M > 1$. In that case, the relation (C.8) is replaced by a much more complicated one.
ii) When the interaction $w(x_y - x_{y'})$ between neighbouring rows is replaced by the interactions $w_s(x_y - x_{y+s})$ between rows at distance s, which vanish only for $|\, s \,|$ larger than some value $r > 1$. In that case, the transfer matrix Θ is an $r(\ell - 1) \times r(\ell - 1)$ matrix.

Another exercise is the application of the transfer matrix method to the system of steps without making approximations. This is possible because the steps can be formally considered as the trajectories of non-interacting fermions in space-time. The fermion method is the trick used by Pokrovski & Talapov (1979), but a simpler derivation of (C.11) has been given by De Gennes (1968) and by Villain (1980).

Appendix D
Wulff's theorem finally proved

We will give in this appendix a proof of Wulff's construction in three dimensions. The first proof of this kind was given by Dinghas (1944) for fully polyhedral (faceted) crystals. We will follow a more recent and general paper by E.J. Taylor (1974).

Let the crystal surface $S(x, y, z)$ be a (piecewise differentiable) function defining a bounded region containing a given volume V.

Let $\mathbf{n}$ be any vector of unit length in the three-dimensional Euclidean space. Consider a continuous function $\sigma(\mathbf{n})$ such that the following integral over S

$$F(S) = \int_S \sigma(\mathbf{n}_r)\mathrm{d}\mathcal{S} \tag{D.1}$$

gives the *surface free energy* of S. Here $\mathbf{n}_r$ is the normal to the surface S at the point $\mathbf{r}$; it may be not defined at most along a countable number of curves (the crystal edges).

Let W be the Wulff shape, that is, the inner envelope of the planes defined by the Wulff construction seen in section 3.2. Let us call ∂W its surface.

We have seen in section 3.2 that the Wulff shape ∂W has the following property. Let M be a point on ∂W, and let $\mathbf{r}$ be the vector $\mathbf{OM}$. Define

$$\hat{w}(\mathbf{n}) = \max_{\mathbf{r}}(\mathbf{n} \cdot \mathbf{r})\,. \tag{D.2}$$

We have shown (see equation 3.5) that, if the crystal surface is an analytic function, $\hat{w}(\mathbf{n})$ coincides (within a scale factor) with the surface tension,

$$\hat{w}(\mathbf{n}) = \sigma(\mathbf{n})\,. \tag{D.3}$$

What happens when the crystal surface contains sharp edges? Consider Fig. D.1; let Σ be the polar plot of the surface free energy. The shaded

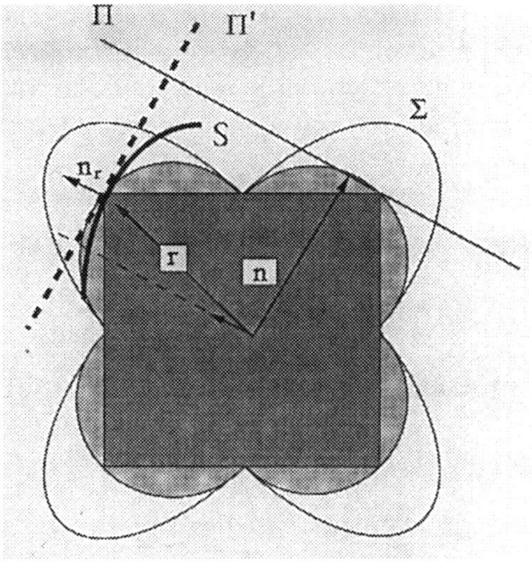

Fig. D.1. Wulff's construction for a crystal shape with sharp edges. The shaded square is the shape obtained from the surface free energy curve Σ. The region in light grey is delimited by the graph of $\hat{w}(\mathbf{n})$ defined in (D.2).

square is the shape derived by Wulff's construction. Choose now a direction $\mathbf{n}$, and construct $\hat{w}(\mathbf{n})$ according to (D.2): its plot yields the curve bounding the lightly shaded region. It is clear that for any direction $\mathbf{n}$ which defines a plane belonging to the Wulff shape, the equality (D.3) holds. When the chosen direction does not appear in the Wulff shape, as it occurs at an edge (cf. the plane Π in Fig. D.1), $\hat{w}$ is less than–at most equal to–the surface free energy, so that the following always holds:

$$\hat{w}(\mathbf{n}) \leq \sigma(\mathbf{n}) . \tag{D.4}$$

Furthermore, if one takes any surface S differentiable in $\mathbf{r}$ (bold line in Fig. D.1), the value of the free energy corresponding to the direction $\mathbf{n}_r$ of the plane Π' tangent to S in $\mathbf{r}$ (dashed line in Fig. D.1) is *strictly* greater than $\hat{w}(\mathbf{n}_r)$, so that (D.4) holds as a strict inequality in this case.

We shall now prove the following theorem (Taylor, 1974):

Wulff's theorem – Let σ be a continuous function. The crystal surface ∂W found according to Wulff's construction and bounding a given volume V, uniquely minimizes the integral of σ, compared to any other piecewise differentiable surface enclosing the same volume.

First of all, some other definitions. We call a crystal a finite region of space bounding a volume V. Taking as the origin O any point in the interior of this region, we call P the set of the vectors $\mathbf{OH}$ joining the origin to point H on the crystal surface. The 'increment' $P + W_h$ of P is defined by adding to each element (vector) of P each element of W_h, the latter being the crystal homothetical to W with homothety factor h. For convenience, we also define a function $\tilde{V}_3(P)$ which measures the volume of P (the index 3 reminds us that we are working in three dimensions). By hypothesis, $\tilde{V}_3(P) = V$.

If $\mathrm{d}\mathscr{V}$ and $\mathrm{d}\mathscr{S}$ are respectively the volume and the surface element, it is a direct application of Gauss's theorem to show that

$$V_3(P) = \int_P \mathrm{d}\mathscr{V} = \int_{\partial P} \mathbf{r}\cdot\mathbf{n}\mathrm{d}\mathscr{S}\,.$$

To prove Wulff's theorem, we need to know the value of limit

$$\lim_{h\to 0}\left\{\tilde{V}_3(P+W_h) - \tilde{V}_3(P)\right\}/h\,.$$

Since, by definition, each element $\mathbf{r}$ in $P + W_h$ can be written as the sum of one element of P, $\mathbf{r}_P$, and one of W_h, $\mathbf{r}_{W_h} = h\mathbf{r}_W$, we find:

$$\begin{aligned}\int_{\partial(P+W_h)} \mathbf{r}\cdot\mathbf{n}\,\mathrm{d}\mathscr{S} &= \int_{\partial(P+W_h)} \mathbf{r}_P\cdot\mathbf{n}\,\mathrm{d}\mathscr{S} + \int_{\partial(P+W_h)} \mathbf{r}_{W_h}\cdot\mathbf{n}\,\mathrm{d}\mathscr{S}\\ &= \int_{\partial(P+W_h)} \mathbf{r}_P\cdot\mathbf{n}\,\mathrm{d}\mathscr{S} + h\int_{\partial(P+W_h)} \mathbf{r}_W\cdot\mathbf{n}\,\mathrm{d}\mathscr{S}\end{aligned}$$

so that, letting $h \to 0$,

$$\int_{\partial(P+W_h)} \mathbf{r}\cdot\mathbf{n}\,\mathrm{d}\mathscr{S} - \int_{\partial P} \mathbf{r}\cdot\mathbf{n}\,\mathrm{d}\mathscr{S} = h\int_{\partial P} \mathbf{r}_W\cdot\mathbf{n}\,\mathrm{d}\mathscr{S}\,. \qquad \text{(D.5)}$$

Using (D.2) and (D.4), equation (D.5) yields

$$\begin{aligned}\lim_{h\to 0}[\tilde{V}_3(P+W_h) - \tilde{V}_3(P)]/h &= \int_{\partial P} \hat{w}(\mathbf{n})\mathrm{d}\mathscr{S}\\ &\le \int_{\partial P} \sigma(\mathbf{n})\mathrm{d}\mathscr{S}\,.\end{aligned} \qquad \text{(D.6)}$$

The proof now requires the Brunn-Minkowsky inequality (Dinghas 1944):

$$\tilde{V}_3(P+W_h) \ge [\tilde{V}_3(P)^{\frac{1}{3}} + \tilde{V}_3(W_h)^{\frac{1}{3}}]^3\,, \qquad \text{(D.7)}$$

where the equality holds when P has the form $P = W_{h_0}$.

Note that geometrical similarity implies $\tilde{V}_3(W_h) = h^3\tilde{V}_3(W)$. Using this property together with (D.7), we obtain

$$\lim_{h\to 0}\left\{\tilde{V}_3(P+W_h) - \tilde{V}_3(P)\right\}/h = 3\tilde{V}_3(P)^{\frac{2}{3}}\tilde{V}_3(W)^{\frac{1}{3}} = 3V \qquad \text{(D.8)}$$

where we used the assumption $\tilde{V}_3(P) = \tilde{V}_3(W) = V$.

We now approach the conclusion; the first equality in (D.6) holds independently of P, provided it encloses the volume V. In particular, we can choose P to be the Wulff shape W: (D.6), (D.7) and (D.8) become then equalities. We have therefore the result:

$$\int_{\partial W} \sigma(\mathbf{n}) \mathrm{d}\mathscr{S} \leq \int_{\partial P} \sigma(\mathbf{n}) \mathrm{d}\mathscr{S} \qquad \text{(D.9)}$$

and the Wulff shape W minimizes the integral of σ.

Remarks.

1) If the crystal shape P possesses tangent planes which are not also tangent to W, then according to (D.4) and (D.5) the inequality in formula (D.9) is strict.
2) On the other hand, if P has the same tangent planes as W, but it is not homothetical to it, then the Brunn-Minkowsky inequality (D.6) is strict, and the surface energy of W is strictly less than that of P.

We conclude that W uniquely minimizes the integral of σ among all piecewise differentiable surfaces enclosing the volume V.

The problem of minimizing the surface free energy with any given boundary conditions is not a simple one. Some useful observations may be found in Herring (1951) and Taylor (1974).

Appendix E
Proof of Frank's theorem

We will give in this appendix a proof of Frank's theorem in three dimensions.

Let the crystal surface $S(x, y, z; t)$ be the function defining the surface of a growing crystal at time t.

Let $\mathbf{n}$ be the normal unit vector to the surface. Our goal is the determination of the trajectory $\Gamma(\mathbf{n})$ of a point on S, $M(\mathbf{n}_0; t)$, such that the surface normal in M has a fixed value $\mathbf{n}_0$ (Fig. E.1). We choose a coordinate system such that locally the surface profile is a function $z = z(x, y; t)$ at each time t. Then, the coordinates of M at time t are x_M, y_M and $z(x_M, y_M; t)$. The point follows its trajectory with a velocity

$$\tau = (\dot{x}, \dot{y}, \dot{z}), \tag{E.1}$$

where

$$\dot{x} = \frac{\mathrm{d}x_M}{\mathrm{d}t}, \quad \dot{y} = \frac{\mathrm{d}y_M}{\mathrm{d}t}, \quad \dot{z} = \frac{\partial z}{\partial x_M}\dot{x} + \frac{\partial z}{\partial y_M}\dot{y} + \frac{\partial z}{\partial t}. \tag{E.2}$$

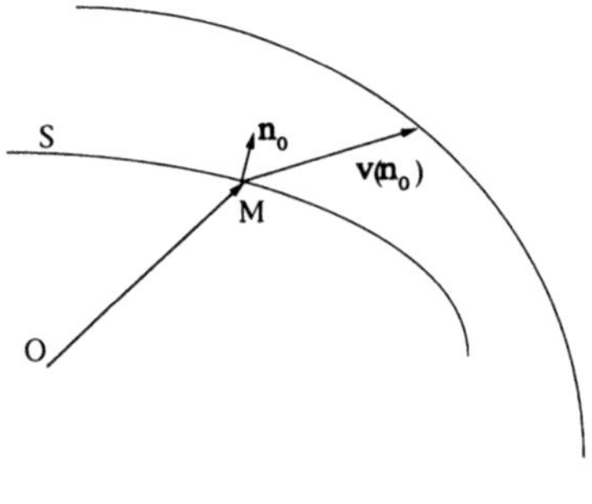

Fig. E.1. Within Frank's model it is possible to determine the trajectory Γ of the point $M(\mathbf{n})$, when the normal $\mathbf{n}$ is held fixed.

On the other hand, in this coordinate system the normal reads

$$\mathbf{n} = \frac{1}{\sqrt{1+z_x^2+z_y^2}}(-z_x, -z_y, 1) \, . \tag{E.3}$$

If $\mathbf{n}$ is kept fixed during the motion of M along the curve $\Gamma(\mathbf{n})$, the partial derivatives z_x and z_y are constants of the motion, and thus one finds

$$\frac{\mathrm{d}z_x}{\mathrm{d}t} = z_{xx}\dot{x} + z_{yx}\dot{y} + z_{tx} = 0 \tag{E.4a}$$

$$\frac{\mathrm{d}z_y}{\mathrm{d}t} = z_{xy}\dot{x} + z_{yy}\dot{y} + z_{ty} = 0 \tag{E.4b}$$

where $z_{\alpha\gamma}$ stands for the partial derivatives of $z(x, y; t)$ with respect to x, y and t.

Let v be the crystal growth velocity. If $(x, y, z)(t)$ is a point on the surface at time t, then the point $(x + n_x v\delta t, y + n_y v\delta t, z(x, y; t) + n_z v\delta t)$ is on the surface at time $t + \delta t$, due to the definition of v. The z component of the distance between these two points can thus be written in two ways; either $n_z v\delta t$, or, using (E.2), $(\partial z/\partial x)n_x v\delta t + (\partial z/\partial y)n_y v\delta t + (\partial z/\partial t)\delta t$. Since both ways must give the same result, we can write

$$n_z = \frac{\partial z}{\partial x}n_x + \frac{\partial z}{\partial y}n_y + \frac{1}{v}\frac{\partial z}{\partial t} \, .$$

From (E.3) now follows

$$\frac{1}{\sqrt{1+z_x^2+z_y^2}} = -\frac{z_x^2+z_y^2}{\sqrt{1+z_x^2+z_y^2}} + \frac{1}{v}z_t \, ,$$

and thus (see Fig. E.2)

$$v = \frac{z_t}{\sqrt{1+z_x^2+z_y^2}} \, . \tag{E.5}$$

In Frank's model that we are discussing here, the growth velocity v is a function of the surface normal only. According to (E.3) and (E.5), this implies that z_t is a function of z_x and z_y only. Differentiating this function with respect to x and y, we find:

$$z_{xt} = \frac{\partial z_t}{\partial z_x}z_{xx} + \frac{\partial z_t}{\partial z_y}z_{xy}$$

$$z_{yt} = \frac{\partial z_t}{\partial z_x}z_{yx} + \frac{\partial z_t}{\partial z_y}z_{yy} \, .$$

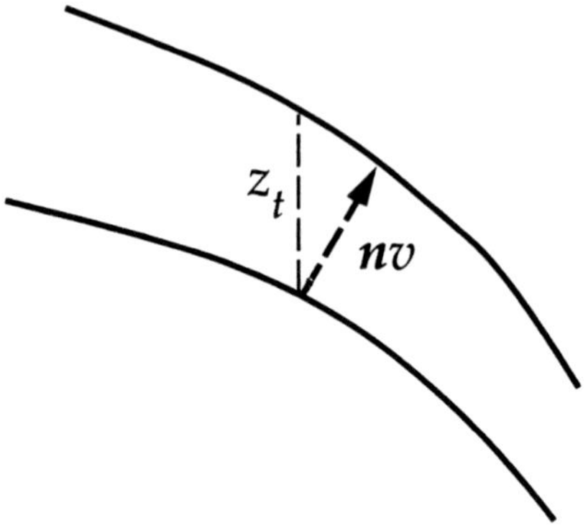

Fig. E.2.

Inserting these two expressions into (E.4), we obtain:

$$z_{xx}\left(\dot{x}+\frac{\partial z_t}{\partial z_x}\right)+z_{xy}\left(\dot{y}+\frac{\partial z_t}{\partial z_y}\right)=0$$

$$z_{xy}\left(\dot{x}+\frac{\partial z_t}{\partial z_x}\right)+z_{yy}\left(\dot{y}+\frac{\partial z_t}{\partial z_y}\right)=0\,,$$

and thus

$$\dot{x}+\frac{\partial z_t}{\partial z_x}=0 \tag{E.6a}$$

$$\dot{y}+\frac{\partial z_t}{\partial z_y}=0\,. \tag{E.6b}$$

Let us consider now any variation $\delta\mathbf{n}$ of $\mathbf{n}$. It is associated to the variations δz_x, δz_y, and

$$\delta z_t=\delta z_x\frac{\partial z_t}{\partial z_x}+\delta z_y\frac{\partial z_t}{\partial z_y}\,.$$

Equations (E.6) imply the following:

$$\left(\dot{x}+\frac{\partial z_t}{\partial z_x}\right)\delta z_x+\left(\dot{y}+\frac{\partial z_t}{\partial z_y}\right)\delta z_y=0\,,$$

which can also be written as

$$(\dot{x},\dot{y},\dot{z})\cdot(\delta z_x,\delta z_y,0)+\delta z_t=0$$

and in an alternative way yet:

$$\tau\cdot\delta\left(\mathbf{n}\sqrt{1+z_x^2+z_y^2}\right)-\delta z_t=0\,. \tag{E.7}$$

On the other hand, relation (E.2) implies

$$\tau \cdot \mathbf{n}\sqrt{1 + z_x^2 + z_y^2} - z_t = 0 \,. \tag{E.8}$$

Combining (E.7) and (E.8), we see that for any variation of $\mathbf{n}$, we have

$$\tau \cdot \delta \frac{\mathbf{n}\sqrt{1 + z_x^2 + z_y^2}}{z_t} = \frac{\delta z_t}{z_t} - z_t \delta \frac{1}{z_t} = 0 \,. \tag{E.9}$$

The vector $\mathbf{n}\sqrt{1 + z_x^2 + z_y^2}/z_t$ coincides with $\mathbf{n}/v(\mathbf{n})$ according to (E.5). Its infinitesimal variations lie in a plane which only depends on $\mathbf{n}$. Therefore, (E.9) shows that the vector τ, which is tangent to the trajectory $\Gamma(\mathbf{n})$, is orthogonal to a plane Π which only depends on $\mathbf{n}$, and is thus the same for all points of Γ. In other words, Γ is a straight line. It follows from (E.9) that the plane Π is tangent to the polar diagram of the reluctance $1/v(\mathbf{n})$: this is just Frank's theorem.

Problem – Show that the kinematic Wulff construction is a special case of the Frank construction (Fig. E.3).

Let A be a point on the curve defined by $\mathbf{r} = \mathbf{n}v(\mathbf{n})$, and B its inverse on the curve $\mathbf{r} = \mathbf{n}/v(\mathbf{n})$. Let D be the normal in A to the line OA, and M the point where D cuts its envelope. According to the kinematic Wulff construction, if the surface moves homothetically to itself, M should span this surface. If this is true, Frank's theorem shows that OM must be orthogonal to the tangent BT in B to the curve $\mathbf{r} = \mathbf{n}/v(\mathbf{n})$. We will now verify that it is actually so.

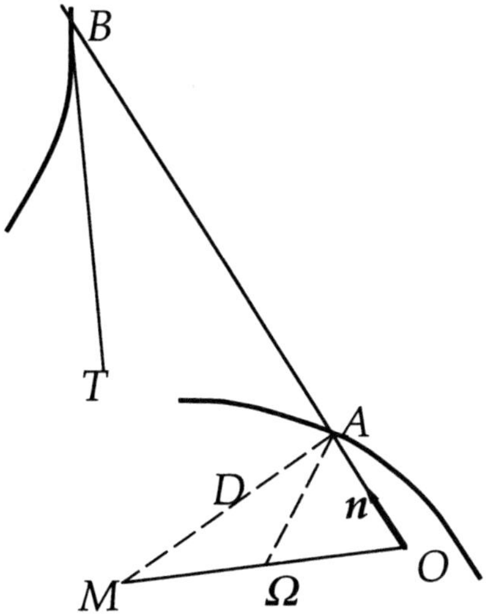

Fig. E.3. Kinematic Wulff's construction as a special case of Frank's construction.

Let A' be a point neighbouring A on the curve $\mathbf{r} = \mathbf{n}v(\mathbf{n})$. The intersection of the normals in A and A' to OA and OA', respectively, goes to M when A goes to A'. When A' is infinitely close to A, both A and A' lie on the circle whose diameter is OM. Let Ω be the centre of this circle. The straight line AA' is orthogonal to ΩA, so that

$$\widehat{AA'B} = \frac{1}{2}\pi - \widehat{\Omega AO} = \frac{1}{2}\pi - \widehat{MOA}\,.$$

Since inversion is a conformal transformation, we also have $\widehat{A'AB} = \widehat{TBA}$. It follows

$$\widehat{TBA} = \frac{1}{2}\pi - \widehat{MOA}\,.$$

This is just the property we were seeking to prove.

Appendix F
Step flow with a Schwoebel effect

We discuss in this appendix step flow in the presence of a barrier at descending steps. We start from equation (6.1) where we make the quasi-static approximation $\dot{\rho} = 0$. The solution has again the form (6.5) with the same value (6.6) of κ. The step velocity will now be calculated. Inserting (6.5) into (6.17a), one obtains

$$\lambda\kappa \exp(\kappa\ell/2) - \mu\kappa \exp(-\kappa\ell/2) = \frac{\Delta'\rho_0}{D} - \frac{F\Delta'}{D}\tau_v - \frac{\lambda\Delta'}{D}\exp(\kappa\ell/2) - \frac{\mu\Delta'}{D}\exp(-\kappa\ell/2) \; . \tag{F.1}$$

Inserting (6.5) into (6.17b), one obtains

$$-\lambda\kappa \exp(-\kappa\ell/2) + \mu\kappa \exp(\kappa\ell/2) = \frac{\Delta''\rho_0}{D} - \frac{F\Delta''}{D}\tau_v - \frac{\lambda\Delta''}{D}\exp(-\kappa\ell/2) - \frac{\mu\Delta''}{D}\exp(\kappa\ell/2) \; . \tag{F.2}$$

Equations (F.1) and (F.2) determine λ and μ. They can be written

$$\lambda\,(1 + \kappa D/\Delta')\exp(\kappa\ell/2) + \mu\,(1 - \kappa D/\Delta')\exp(-\kappa\ell/2) = \rho_0 - F\tau_v$$

and

$$\lambda\,(1 - \kappa D/\Delta'')\exp(-\kappa\ell/2) + \mu\,(1 + \kappa D/\Delta'')\exp(\kappa\ell/2) = \rho_0 - F\tau_v \; .$$

Letting $D/\Delta' = d'$ and $D/\Delta'' = d''$, it follows:

$$\lambda = \frac{1}{2}(\rho_0 - F\tau_v)\frac{(1 + \kappa d'')\exp(\kappa\ell/2) - (1 - \kappa d')\exp(-\kappa\ell/2)}{(1 + \kappa^2 d' d'')\sinh(\kappa\ell) + \kappa\,(d' + d'')\cosh(\kappa\ell)}$$

and

$$\mu = \frac{1}{2}(\rho_0 - F\tau_v)\frac{(1 + \kappa d')\exp(\kappa\ell/2) - (1 - \kappa d'')\exp(-\kappa\ell/2)}{(1 + \kappa^2 d' d'')\sinh(\kappa\ell) + \kappa\,(d' + d'')\cosh(\kappa\ell)} \; .$$

Consider two steps at positions $-\ell/2$ and $\ell/2$. The velocity of the lower step (at $\ell/2$) is the sum of a contribution due to the terrace in front of it, and of the term

$$v' = -D\rho'(\ell/2) = -\lambda\kappa D \exp(\kappa\ell/2) + \mu\kappa D \exp(-\kappa\ell/2)$$

$$= \kappa D\,(\rho_0 - F\tau_0)\,\frac{1 - \cosh(\kappa\ell) - \kappa d'' \sinh(\kappa\ell)}{(1 + \kappa^2 d' d'')\sinh(\kappa\ell) + \kappa\,(d' + d'')\cosh(\kappa\ell)}$$

which is (6.19a).

Similarly, the velocity of the higher step (at $-\ell/2$) is the sum of a contribution due to the terrace behind it and of the term

$$v'' = -D\rho'(\ell/2) = \lambda\kappa D \exp(-\kappa\ell/2) - \mu\kappa D \exp(\kappa\ell/2)$$

$$= \kappa D\,(\rho_0 - F\tau_0)\,\frac{1 - \cosh(\kappa\ell) - \kappa d' \sinh(\kappa\ell)}{(1 + \kappa^2 d' d'')\sinh(\kappa\ell) + \kappa\,(d' + d'')\cosh(\kappa\ell)}$$

which is (6.19b).

When κD is small with respect to Δ' and Δ'', these formulae reduce to

$$v' = v'' = v/2 = (\rho_0 - F\tau_v)\,\frac{2 - \exp(-\kappa\ell) - \exp(\kappa\ell)}{\exp(\kappa\ell) - \exp(-\kappa\ell)}\,\kappa D$$

$$= -(\rho_0 - F\tau_v)\,\frac{4\sinh^2(\kappa\ell/2)}{2\sinh(\kappa\ell)}\,\kappa D$$

$$= -(\rho_0 - F\tau_v)\,\frac{\sinh(\kappa\ell/2)}{\cosh(\kappa\ell/2)}\,\kappa D\,.$$

This coincides with (6.9) if $\ell' = \ell$.

A very strong Schwoebel effect implies $\kappa d' \gg 1$, while usually sticking is easy from the lower side, so that $\kappa d'' \ll 1$. Then the part v' of the velocity which corresponds to adatoms which go down the step, is very small, as one would expect.

In the second part of this appendix, the equations (6.21) and (6.22) will be solved. Assuming $F = 0$, we linearize them as

$$D\rho'' - \delta\rho/\tau_v - \Delta\sigma_0\delta\rho - \Delta\rho_0\delta\sigma = \rho_0/\tau_v \tag{F.3a}$$

$$\Lambda\sigma'' - \Delta\sigma_0\delta\rho - \Delta\rho_0\delta\sigma = 0\,. \tag{F.3b}$$

The solution has the form

$$\delta\rho(x) = A + B\cosh(\kappa x) + C\cosh(\kappa' x) \tag{F.4a}$$

$$\delta\sigma(x) = A' + B'\cosh(\kappa x) + C'\cosh(\kappa' x) \tag{F.4b}$$

where κ and κ' are roots of the equation

$$\left(D\kappa^2 - 1/\tau_v - \Delta\sigma_0\right)\left(\Lambda\kappa^2 - \Delta\rho_0\right) - \Delta^2\rho_0\sigma_0 = 0\,.$$

In the limit where τ_v is very large, the values of the roots to first order in $1/\tau_v$ are, respectively

$$\kappa^2 = \frac{1}{2D\tau_v} - \frac{1}{2D\tau_v}\frac{\Lambda\sigma_0 - D\rho_0}{\Lambda\sigma_0 + D\rho_0} = \frac{1}{\tau_v}\frac{\rho_0}{\Lambda\sigma_0 + D\rho_0}$$

and

$$\kappa'^2 = \frac{\Delta\sigma_0}{D} + \frac{\Delta\rho_0}{\Lambda} + \frac{1}{D\tau_v}\frac{\Lambda\sigma_0}{\Lambda\sigma_0 + D\rho_0} .$$

The constants A, B, C, A', B', C' are obtained by inserting (F.4) into (F.3) and into the boundary conditions at steps $\delta\rho = \delta\sigma = 0$ (without the Schwoebel effect). The result is that C and C' are zero at order 0 in $1/\tau_v$. To this order, without the Schwoebel effect and for $F = 0$, one finds

$$\begin{cases} \delta\rho(x) = -\rho_0 + \dfrac{\rho_0}{\cosh(\kappa\ell/2)}\cosh(\kappa x) \\ \delta\sigma(x) = \sigma_0 - \dfrac{\sigma_0}{\cosh(\kappa\ell/2)}\cosh(\kappa x) . \end{cases}$$

Appendix G

Dispersion relations for the fluctuations of a train of steps

In this appendix we give the derivation of the dispersion relation in the case of a train of completely *asynchronized* steps. We shall denote by x the direction parallel to the steps, by z_m the average position of the m-th step, and by ζ_m its instantaneous profile. The step profile is thus a function $\zeta_m = \zeta_m(x, t)$. It is convenient to write the adatom density as $u(z) = \rho(z) - F\tau$. After the quasistationary approximation $\dot{\rho} = 0$ has been made, the 'diffusion field' u obeys the following equation

$$\nabla^2 u - \frac{u}{x_s^2} = 0 . \tag{G.1}$$

On both sides of the step we have the boundary conditions

$$D[\frac{\partial u}{\partial n}]_+ = \Delta''(u + \tau F - \Gamma\kappa_m) \tag{G.2a}$$

$$D[\frac{\partial u}{\partial n}]_- = -\Delta'(u + \tau F - \Gamma\kappa_m) , \tag{G.2b}$$

where the '+' and '−' signs refer to the lower and upper terraces respectively, $\partial u/\partial n$ stands for the normal derivative at the step edge, with the normal pointing in the positive z direction, Δ'' and Δ' are the sticking coefficients from the lower and the upper terraces respectively, and κ_m is the step curvature, counted positive for a convex step profile.

Once the concentration gradients on both sides of the step are known we can evaluate the growth velocity by means of the requirement of mass conservation at the step:

$$v_n = D\Omega \left\{ [\frac{\partial u}{\partial n}]_+ - [\frac{\partial u}{\partial n}]_- \right\} . \tag{G.3}$$

The set of equations from (G.1) to (G.3) completely describes the step dynamics.

This set admits a straight-step solution $\zeta_m(x,t) = z_m$ moving at a constant velocity v_0 along the z-direction. The zeroth-order solution is a train of equidistant steps separated by the distance ℓ. The general solution of the associated diffusion field is given by

$$u_0(z) = A\cosh(z/x_s) + B\sinh(z/x_s)\,.$$

The two integration constants A and B are easily obtained by making use of the two boundary conditions equations (G.2). The result is

$$A = -\tau\Delta F\frac{\Delta'\Delta''\sinh(\ell/x_s) + D/x_s[\Delta''\cosh(\ell/x_s) + \Delta']}{[(D/x_s)^2 + \Delta'\Delta'']\sinh(\ell/x_s) + D/x_s(\Delta' + \Delta'')\cosh(\ell/x_s)}$$

$$B = \tau\Delta F\frac{D\Delta''/x_s\sinh(\ell/x_s) + \Delta'\Delta''[\cosh(\ell/x_s) - 1]}{[(D/x_s)^2 + \Delta'\Delta'']\sinh(\ell/x_s) + D/x_s(\Delta' + \Delta'')\cosh(\ell/x_s)}\,.$$

Making use of (G.3) and letting $d' = D/\Delta'$, $d'' = D/\Delta''$, we obtain the step flow velocity

$$v_0 = \Omega x_s\Delta F\frac{2[\cosh(\ell/x_s) - 1] + (d' + d'')/x_s\sinh(\ell/x_s)}{[1 + d'd''/x_s^2]\sinh(\ell/x_s) + (d' + d'')/x_s\cosh(\ell/x_s)}\,.$$

The linear stability of the straight-step solution is investigated by assuming a step profile of the form

$$\zeta_m(x,t) = A_m e^{iqx+\omega t} + \text{c.c.}$$

where A_m stands for the complex amplitude of the deviation of the m-th step from being straight, ω is the growth (or damping) rate of the perturbation and q is its wavenumber. Any modification in the shape of the step reverberates on the adatom density through the boundary conditions (G.2). The response of the adatom density to step perturbations may be written as

$$u_{1m} = [\alpha_m\sinh(\Lambda_q z) + \beta_m\cosh(\Lambda_q z)]e^{iqx+\omega t} + \text{c.c.}\,, \qquad \text{(G.4)}$$

where $\Lambda_q = \sqrt{q^2 + \kappa^2}$, and $\kappa = 1/x_s$. Equation (G.4) gives the diffusion field on the terrace just in front of of the step (i.e., on its lower side). The constants α_m and β_m are determined by equations (G.2). Care should be taken when using these boundary conditions. Equation (G.2a) should be satisfied at $z = \zeta_m$, taking the position of the m-th step as the origin ($z_m = 0$). Since u_0 (the straight step solution) is of order zero in the perturbation, terms like $\zeta_m(\partial u_0/\partial z)_{z=\zeta_m}$ must be taken into account. On the other hand, the constants α_m and β_m must be evaluated on the terrace between the step under scrutiny at $z = \zeta_m$ and its $m+1$ neighbour at $z = z_{m+1} + \zeta_{m+1}(x,t) = \ell + \zeta_{m+1}$. This means that the adatom density is determined by using the second condition (G.2b) at $z = \ell + \zeta_{m+1}$.

Taking care of these points, the calculation is straightforward and yields the following expressions

$$\alpha_m = -(\mathscr{D}x_s)^{-1}\{\mathscr{B}A_{m+1} + 1/\Delta''\mathscr{A}A_m[D\Lambda_q \sinh(\Lambda_q\ell) + \Delta'\cosh(\Lambda_q\ell)]\}$$
$$\beta_m = [D\alpha_m\Lambda_q + \kappa\mathscr{A}A_m]/\Delta''$$

where $\mathscr{A}$, $\mathscr{B}$ and $\mathscr{D}$ are given by

$$\mathscr{A} = \Delta''\Gamma x_s q^2 - D\tau\Delta F\Big\{(1+\frac{d'}{d''})\sinh(\kappa\ell) + \kappa d'\cosh(\kappa\ell) + \frac{1}{\kappa d}[\cosh(\kappa\ell)-1] + \kappa d''\Big\}/\,[x_s\Psi(0)]\ ,$$
$$\mathscr{B} = -\Delta'\Gamma x_s q^2 + D\tau\Delta F\Big[\frac{1}{\kappa d'} - \kappa d' - \left(\frac{1}{\kappa d'} + \kappa d''\right)\cosh(\kappa\ell) - \left(1+\frac{d''}{d'}\right)\sinh(\kappa\ell)\Big]/\,[x_s\Psi(0)]\ ,$$
$$\mathscr{D} = D\Lambda_q\left(1+\frac{d'}{d''}\right)\cosh(\Lambda_q\ell) + D\left(d' + \Lambda_q^2/d''\right)\sinh(\Lambda_q\ell)\,,$$

where $\Psi(0) \equiv \Psi(q=0)$ and

$$\Psi(q) = \left(1+\Lambda_q^2 d'd''\right)\sinh(\kappa\ell) + \Lambda_q(d'+d'')\cosh(\kappa\ell)\,.$$

Note that for the determination of u_m both the amplitudes A_m and A_{m+1} are needed. In order to find the $\omega(q)$ dispersion relation, we make use of equation (G.3). For that purpose, we need to evaluate the normal derivative of u at $z = \delta\zeta_m - 0^+$. This can be done by solving the diffusion field $u_{1(m-1)}$ in the domain $[\zeta_{m-1} - \ell < z < \zeta_m]$. It is easy to realize that the field u_{m-1} is given by an expression analogous to that of u_m where in α_m and β_m we replace the amplitudes of deformation A_m and A_{m+1} by A_{m-1} and A_m respectively. Having determined the expression of the field on both sides of the step we are in the right position to use equation (G.3). The calculation involves tedious algebra, and the result can be finally written in the following form

$$\begin{aligned}\omega A_m = &-q^2\Gamma D\Omega\Lambda_q \\ &\times\frac{[2\cosh(\Lambda_q\ell) + \Lambda_q(d'+d'')\sinh(\Lambda_q\ell)]A_m - A_{m-1} - A_{m+1}}{\Lambda_q(d'+d'')\cosh(\Lambda_q\ell) + (1+d'd''\Lambda_q^2)\sinh(\Lambda_q\ell)} \\ &+\Omega\Delta F\Big\{(d'-d'')\{\Lambda_q(d'+d'')[x_s\Lambda_q\sinh(\Lambda_q\ell)\sinh(\kappa\ell)] \\ &-\cosh(\Lambda_q\ell)\cosh(\kappa\ell)] + \sinh(\Lambda_q\ell)[1-\cosh(\kappa\ell)]x_s^2q^2\}A_m \\ &+\Lambda_q\cosh(\kappa\ell)[d'^2A_{m+1} - d''^2A_{m-1}] - \Lambda_q[x_s(d'+d'')\sinh(\kappa\ell) \\ &+(d'd''+x_s^2)\cosh(\kappa\ell) - x_s^2](A_{m-1} - A_{m+1})\Big\}/\,[x_s\Psi(q)\Psi(0)]\ .\end{aligned} \tag{G.5}$$

Equation (G.5) gives the relation between the dynamics of a step and that of its neighbours, when each step is moving in an asynchronized way with respect to the others. The general solution of equation (G.5) is given by $A_m = Ce^{im\phi}$ (for periodic boundary conditions) where ϕ is an arbitrary real number representing the phase shift between two consecutive steps, and C is a real constant. Inserting this expression of A_m into (G.5) we obtain the dispersion relation $\omega(q)$ parametrized by the phase ϕ. Making explicit the complex character of $\omega = \omega_r + i\omega_i$, we have

$$\omega_r = -q^2\Gamma D\Omega\Lambda_q \frac{2[\cosh(\Lambda_q\ell) - \cos\phi] + \Lambda_q(d'+d'')\sinh(\Lambda_q\ell)}{\Lambda_q(d'+d'')\cosh(\Lambda_q\ell) + (1+d'd''\Lambda_q^2)\sinh(\Lambda_q\ell)}$$
$$+\frac{\Omega\Delta F(d'-d'')}{x_s\Psi(q)\Psi(0)}\Big\{\Lambda_q(d'+d'')[x_s\Lambda_q\sinh(\Lambda_q\ell)\sinh(\kappa\ell)$$
$$-\cosh(\Lambda_q\ell)\cosh(\kappa\ell) + \cos\phi] + \sinh(\Lambda_q\ell)\,[\cosh(\kappa\ell) - 1]\,x_s^2q^2\Big\} \tag{G.6a}$$

$$\omega_i = \frac{\Lambda_q\Delta F\sin\phi}{x_s\Psi(q)\Psi(0)}\Big[2x_s(d'+d'')\sinh(\kappa\ell) + 2(d'd''+x_s^2))\cosh(\kappa\ell)$$
$$+d'^2 + d''^2 - 2x_s^2\Big]\,. \tag{G.6b}$$

It can be checked that for a synchronized train ($\phi = 0$) we recover the result of Bales & Zangwill (1990a).

For fluctuations in phase opposition ($\phi = \pi$) the dispersion relation takes the form

$$\omega = -q^2 f(q) + \Omega(d'-d'')\Delta F g(q)\,,$$

where

$$f(q) = \Gamma D\Omega\Lambda_q \frac{2[\cosh(\Lambda_q\ell) + 1] + \Lambda_q(d'+d'')\sinh(\Lambda_q\ell)}{\Lambda_q(d'+d'')\cosh(\Lambda_q\ell) + (1+d'd''\Lambda_q^2)\sinh(\Lambda_q\ell)}\,,$$

and

$$g(q) = \Big\{\Lambda_q(d'+d'')\,[x_s\Lambda_q\sinh(\Lambda_q\ell)\sinh(\kappa\ell) - \cosh(\Lambda_q\ell))\cosh(\kappa\ell) - 1]$$
$$+\sinh(\Lambda_q\ell)\,[\cosh(\kappa\ell) - 1]\,x_s^2q^2\Big\}/\,[x_s\Psi(q)\Psi(0)]\,.$$

Note that the imaginary part of ω only vanishes for $\phi = 0$ or $\phi = \pi$, i.e. when the steps fluctuate in phase or in phase opposition, respectively. The non-vanishing imaginary part means that propagative effects enter in the way by which perturbations from a step are carried to another one.

Appendix H
Adatom diffusion length ℓ_s and nucleation

In chapter 11, we have assumed that trimers, tetramers and larger clusters could not dissociate or diffuse. In this appendix, we will account for both cluster dissociation and motion. The rate equation for an n-mer (an island containing n adatoms) is assumed to take the following form, for $n \leq i^*$:

$$\dot{\rho}_n = D\rho_1\rho_{n-1} - \frac{\rho_n}{\tau_n} - D\rho_1\rho_n + \frac{\rho_{n+1}}{\tau_{n+1}} - \frac{D_n}{\ell_s^2}\rho_n \,. \tag{H.1}$$

This equation is a generalization of (11.18a). The meaning of the various terms on the right-hand side should be clear. The quantity τ_n is the average lifetime of an n-mer, and D_n its diffusion constant. As in section 11.8, coalescence of mobile islands is neglected. Note that (H.1) holds also for $n > i^*$, if one lets $D_n = 1/\tau_n = 1/\tau_{n+1} = 0$ for such ns. The major flaw is the error made in neglecting coalescence. This approximation has been justified in chapter 11 for $i^* = 2$. We will not justify it here for the general case.

The coefficients of the 'adatom capture terms' $\rho_1\rho_n$ in (H.1) are equal to the adatom diffusion constant D only in order of magnitude. Indeed, it is customary in the literature to write these terms in the form $\Delta_n\rho_1\rho_n$, where Δ_n/D is a 'capture coefficient', which is an increasing function of n. It is also a function of the coverage, in general. It can be shown, however, that the *average* value of the capture coefficient only depends on the coverage. A detailed proof can be found in Bartelt *et al.* (1995). Formula (H.1) is thus valid for the determination of an average quantity such as ℓ_s. It is not adequate for describing the distribution of island sizes, i.e. the probability of finding a cluster of given size n.

We consider average values of the cluster densities. The average is done over space and over a time longer than the deposition time of one layer, or the period of diffraction oscillations. We can thus assume that a stationary regime has been reached, where the time derivatives $\dot{\rho}_n$ vanish. A set of algebraic equations results, which reads:

$$
\begin{aligned}
&D\rho_1^2 - \frac{\rho_2}{\tau_2} - D\rho_1\rho_2 + \frac{\rho_3}{\tau_3} - \frac{D_2}{\ell_s^2}\rho_2 = 0\,, \\
&D\rho_1\rho_2 - \frac{\rho_3}{\tau_3} - D\rho_1\rho_3 + \frac{\rho_4}{\tau_4} - \frac{D_3}{\ell_s^2}\rho_3 = 0\,, \\
&D\rho_1\rho_3 - \frac{\rho_4}{\tau_4} - D\rho_1\rho_4 + \frac{\rho_5}{\tau_5} - \frac{D_4}{\ell_s^2}\rho_4 = 0\,, \\
&\qquad\qquad \ldots \\
&D\rho_1\rho_{i^*} - \frac{\rho_{i^*}}{\tau_{i^*}} - D\rho_1\rho_{i^*} - \frac{D_{i^*}}{\ell_s^2}\rho_{i^*} = 0\,.
\end{aligned}
\tag{H.2}
$$

We must add to this set the equations (11.1) and (11.4)–which are still valid–and an equation which generalizes (11.7). This last equation is obtained by writing that the nucleation rate of stable islands coincides with the formation rate of $i^* + 1$-mers, namely

$$
\frac{1}{\tau_{\mathrm{nuc}}} \approx D\rho_1\rho_{i^*}\,. \tag{H.3}
$$

Combining (H.3) with (11.1), we are able of getting rid of τ_{nuc}. Moreover, ρ_1 can be replaced by $F\ell_s^2/D$ (equation 11.4), which yields

$$
\begin{aligned}
&-\frac{\rho_2}{\tau_2} - F\ell_s^2\rho_2 + \frac{\rho_3}{\tau_3} - \frac{D_2}{\ell_s^2}\rho_2 = -\frac{F^2\ell_s^4}{D}\,, \\
&F\ell_s^2\rho_2 - \frac{\rho_3}{\tau_3} - F\ell_s^2\rho_3 + \frac{\rho_4}{\tau_4} - \frac{D_3}{\ell_s^2}\rho_3 = 0\,, \\
&\qquad\qquad \ldots \\
&F\ell_s^2\rho_{i^*} - \frac{\rho_{i^*}}{\tau_{i^*}} - F\ell_s^2\rho_{i^*} - \frac{D_{i^*}}{\ell_s^2}\rho_{i^*} = 0\,, \\
&F\ell_s^2\rho_{i^*} = \frac{F}{\ell_s^2}\,.
\end{aligned}
\tag{H.4}
$$

The last equation follows from (H.3) and (11.4). This set of i^* linear equations in the $i^* - 1$ densities $\rho_2, \ldots, \rho_{i^*}$, has a solution if its determinant vanishes. Letting

$$
\mathcal{Y}_i(\ell_s) = F\ell_s^2 + \frac{D_i}{\ell_s^2} + \frac{1}{\tau_i}\,,
$$

this condition reads

$$\begin{vmatrix} -\mathcal{Y}_2 & 1/\tau_3 & 0 & \cdots & 0 & F^2\ell_s^4/D \\ F\ell_s^2 & -\mathcal{Y}_3 & 1/\tau_4 & \cdots & 0 & 0 \\ 0 & F\ell_s^2 & -\mathcal{Y}_4 & \ddots & 0 & 0 \\ \vdots & \vdots & \ddots & \ddots & 1/\tau_{i^*} & \vdots \\ 0 & 0 & 0 & \ddots & -\mathcal{Y}_{i^*} & 0 \\ 0 & 0 & 0 & 0 & F\ell_s^2 & -F/\ell_s^2 \end{vmatrix} = 0 . \tag{H.5}$$

This equation gives ℓ_s. The adatom difusion length ℓ_s is thus given by an *algebraic equation*, of which the power law functions as (11.9), (11.23) and (11.25) are special cases.

The comparison between (11.9), which holds at very low temperature, and (11.23) or (11.25), which hold at slightly higher temperatures, shows that ℓ_s increases with T *faster* than an Arrhenius exponential $\exp(-W/k_B T)$: the apparent activation energy W increases with temperature. This agrees with experiments and computer simulations, as we discussed in chapter 11.

Appendix I
The Edwards-Wilkinson model

This appendix contains elementary, approximate algebraic calculations to make explicit formula (12.9) for the height-height correlation function $G(\mathbf{r},t) = \left\langle \left[z(\mathbf{r}',t) - z(\mathbf{r}'+\mathbf{r},t)\right]^2 \right\rangle$.

Height fluctuations for $d_c^\ell < d < d_c^u$

The case $r < \xi(t)$ is treated in the text. The opposite case $r > \xi(t)$ will now be addressed by similar methods. Replacing the squared sine by 1/2 in (12.9), one obtains

$$G(\mathbf{r},t) \simeq F \int q^{d'-1} \mathrm{d}q \frac{1 - \exp\left(-2(\nu q^2 + K q^4)t\right)}{\nu q^2 + K q^4} . \tag{I.1}$$

If $\nu \neq 0$:

$$G(\mathbf{r},t) \approx F \int_{1/\sqrt{\nu t}}^{1} q^{d'-1} \frac{\mathrm{d}q}{\nu q^2} .$$

Integration yields, for $\nu \neq 0$ and $d < 2 + 1$:

$$G(\mathbf{r},t) \approx \tau_0 \delta f^2 \, (\nu t)^{(3-d)/2} / \nu \tag{I.2a}$$

and for $\nu \neq 0$, $d = 3$:

$$G(\mathbf{r},t) \approx \frac{1}{2\nu} \tau_0 \delta f^2 \ln(\nu t) . \tag{I.2b}$$

If $\nu = 0$, $d < 4 + 1$:

$$G(\mathbf{r},t) \simeq \tau_0 \delta f^2 \int_{(Kt)^{-1/4}}^{1} q^{d-2} \frac{\mathrm{d}q}{K q^4} \simeq \tau_0 \delta f^2 \, (Kt)^{(5-d)/4} / K . \tag{I.3}$$

The case $r < \xi$, $d \leq 2+1$ for small v

The restriction 'small v' means

$$v < \sqrt{K/t} \tag{I.4}$$

It is useful to split (12.9) into two terms, which can both be written in an approximate but simpler form. The first term was neglected up to now.

$$G(\mathbf{r}) \simeq \frac{\tau_0 \delta f^2}{2} \int_0^{1/r} q^{d'-1} q^2 r^2 \frac{1-\exp\left[-2(vq^2+Kq^4)t\right]}{vq^2+Kq^4} \mathrm{d}q$$
$$+\tau_0 \delta f^2 \int_{1/r}^{1} q^{d'-1} \frac{\mathrm{d}q}{vq^2+Kq^4} \,. \tag{I.5}$$

In the second term on the right-hand side, v can be neglected provided $vr^2 < K$. Now, this condition is fulfilled, as follows from (I.1) and (I.4), and from the assumption $r < \xi$. Therefore one can write

$$G(\mathbf{r},t) \approx \frac{1}{2} r^2 \tau_0 \delta f^2 \int_0^{\sqrt{v/K}} q^{d'+1} \frac{1-\exp\left(-2vq^2t\right)}{vq^2} \mathrm{d}q$$
$$+\frac{1}{2} r^2 \tau_0 \delta f^2 \int_{\sqrt{v/K}}^{1/r} q^{d'+1} \frac{1-\exp\left(-2Kq^4t\right)}{Kq^4} \mathrm{d}q$$
$$+\tau_0 \delta f^2 \int_{1/r}^{1} q^{d'-1} \frac{\mathrm{d}q}{Kq^4} \,.$$

The first term on the right-hand side is small if $K > v$. It will thus be neglected. In the second term, the fraction can be replaced by 1 if $q < (2Kt)^{-1/4}$. This quantity is equal to $1/\xi$ according to (I.1). It is therefore smaller than $1/r$ because $r < \xi$. But, $1/\xi$ is larger than $\sqrt{Kv}$ according to (I.4). The preceding equation becomes then:

$$G(\mathbf{r},t) \approx \frac{1}{2} r^2 \tau_0 \delta f^2 \int_{1/\xi}^{1/r} q^{d'+1} \frac{\mathrm{d}q}{Kq^4} + \tau_0 \delta f^2 \int_{1/r}^{1} q^{d'-1} \frac{\mathrm{d}q}{Kq^4} \,.$$

For $d < 2+1$ this gives:

$$G(\mathbf{r},t) \approx \frac{r^2}{2K} \tau_0 \delta f^2 \xi^{3-d} \tag{I.6a}$$

and, for $d = 3$:

$$G(\mathbf{r},t) \approx \frac{r^2}{2K} \tau_0 \delta f^2 \ln(\xi/r) \,. \tag{I.6b}$$

Equations (I.6) are mathematically sound consequences of the model (12.6). However, one has to be sure that the right-hand side of (I.6) is, for $r = 1$, not larger than one interatomic distance. This would be physically absurd, and it would point to the inconsistency and unacceptability of the model defined by (12.6) in $d \leq 3$ for $v = 0$. For $v \neq 0$, the condition (I.4) implies $\xi < \sqrt{K/v}$. In this case, (I.6b) and the model (12.6) are only acceptable if the right-hand side is small for $r = 1$ and $d = 3$, so that

$$\frac{1}{4K}\tau_0 \delta f^2 \ln(K/v) < 1. \qquad \text{(I.7)}$$

If this condition is not satisfied, the model (12.6) becomes unacceptable at long times t.

Appendix J

Calculation of the coefficients of (13.1) for a stepped surface

The derivation of (13.4) in chapter 13 is purely phenomenological. It would be nice to calculate the coefficients λ and ν within a microscopic model. In this appendix, this problem is partly treated in the easier case of a stepped surface growing by MBE. Actually, only λ will be calculated, or rather, the two coefficients λ_x and λ_y, which will be found to have different signs.

As shown schematically by Fig. 5.4b, any freshly landed adatom diffuses until it reaches a step or evaporates. The evaporation probability depends mainly on the density of steps, which is $|\nabla z| = 1/\ell$, where ℓ is the local distance between steps and z is the coordinate perpendicular to terraces. If ℓ is small, adatoms have no time to evaporate, and the growth rate is just

$$\dot{z}(\mathbf{r},t) = f(\mathbf{r},t) + \text{diffusion terms.} \tag{J.1a}$$

For larger values of ℓ, adatoms can evaporate, and the growth rate is given by equation (6.9), where, however, fluctuations of the beam intensity F and of the evaporation rate $1/\tau_v$ should be taken into account. The modified equation (6.9) may be written as

$$\dot{z}(\mathbf{r},t) = f(\mathbf{r},t)\varphi(1/\ell^2) + \text{diffusion terms} \tag{J.1b}$$

where

$$\varphi(1/\ell^2) = \frac{2}{\ell}\sqrt{D_s\tau_v}\tanh\left(\frac{\ell}{2\sqrt{D_s\tau_v}}\right) \tag{J.2}$$

where D_s is the adatom diffusion constant.

Expanding φ in (J.1b) with respect to the average slope $(\bar{z}_x, 0)$ of the surface, and neglecting diffusion, one obtains

$$\dot{z}(\mathbf{r},t) = f(\mathbf{r},t)\varphi(z_x^2 + z_y^2) = f(\mathbf{r},t)\varphi_0 + f(\mathbf{r},t)\left(z_x^2 - \bar{z}_x^2 + z_y^2\right)\varphi_0'$$
$$+\frac{1}{2}f(\mathbf{r},t)\left(z_x^2 - \bar{z}_x^2 + z_y^2\right)^2\varphi_0'' + \dots .$$

In this formula, y is the average step direction, $\bar{z}_x = 1/\ell$ and φ_0, φ_0' and φ_0'' designate the values of φ and its derivatives for $z_x = \bar{z}_x$ and $z_y = 0$. Introducing the coordinate Z perpendicular to the average direction of the surface, and neglecting fluctuations of f everywhere except the first term, the above equation becomes

$$\dot{Z}(\mathbf{r}, t) = f(\mathbf{r}, t)\varphi_0 + cZ_x + \frac{1}{2}\lambda_x Z_x^2 + \frac{1}{2}\lambda_y Z_y^2 + \dots$$

where

$$\lambda_x = 2F(\varphi_0' + 2\bar{z}_x^2 \varphi_0''), \qquad \text{(J.3a)}$$

$$\lambda_y = 2F\varphi_0', \qquad \text{(J.3b)}$$

and $c = 2F\bar{z}_x$.

In the equation for $\dot{Z}$, the constant part of the first term on the right-hand side can be eliminated by a translation $Z \to Z - F\varphi_0 t$, and the second term by the Galilean transformation $x \to x - ct$. Finally, we get

$$\dot{Z}(\mathbf{r}, t) = \delta f(\mathbf{r}, t)\varphi_0 + \frac{1}{2}\lambda_x Z_x^2 + \frac{1}{2}\lambda_y Z_y^2. \qquad \text{(J.4)}$$

Because of evaporation, the beam intensity fluctuation is renormalized by a factor φ_0. This factor has been generally omitted in chapter 13.

It can be checked from (J.3) and (J.2) that λ_x and λ_y have opposite signs.

If there is no evaporation, φ_0 vanishes identically, and it follows from (J.3) that $\lambda_x = \lambda_y = 0$, in agreement with the observation of chapter 14. The growth equation is therefore linear and reads

$$\dot{Z}(\mathbf{r}, t) = \delta f(\mathbf{r}, t)\varphi_0 + \nu_x Z_{xx} + \nu_y Z_{yy} - K\nabla^2(\nabla^2 Z). \qquad \text{(J.5)}$$

The sign of ν_x and ν_y is determined by the Schwoebel effect. A positive value of ν_x would imply step bunching, which normally does not occur during growth as discussed in section 6.7. However, the Schwoebel effect is a cause for the step meandering instability as shown in section 10.2, and this instability requires $\nu_y < 0$. Rost *et al.* (1996) have found (J.5) as the linearization of a non-linear equation, which rules growth on a vicinal surface, and leads to the same kind of mound morphology obtained on a singular surface.

Appendix K
Molecular beam epitaxy, the KPZ model, the Edwards-Wilkinson model, and similar models

As argued in section 13.1, the local growth rate $\dot{z} = \partial z/\partial t$ is the sum of two terms: the first term $\dot{z}_1$ is the effect of the beam, and the second one, $\dot{z}_2$, is due to further reorganization of the surface by adatom diffusion and evaporation-condensation.

Here, we will discuss the beam contribution $\dot{z}_1$. In chapters 12, 13 and 14, the beam was assumed to be normal to the surface so that $\dot{z}_1$ was just $f(\mathbf{r}, t)$, the beam intensity. It will now be assumed that the beam makes an angle α with the z direction. The x axis may be chosen such that the unit vector in the beam direction is $\mathbf{b} = (\sin\alpha, 0, \cos\alpha)$. The unit vector normal to the surface is $\mathbf{n} = (-\partial_x z, -\partial_y z, 1)/\sqrt{1 + (\partial_x z)^2 + (\partial_y z)^2}$. The number $\dot{N}\mathrm{d}S$ of atoms collected by a surface element $\mathrm{d}S$ per unit time is

$$\dot{N}\mathrm{d}S = f(\mathbf{r}, t)\mathbf{n}\cdot\mathbf{b}\mathrm{d}S = f(\mathbf{r}, t)\frac{(-\partial_x z\sin\alpha + \cos\alpha)}{\sqrt{1 + (\partial_x z)^2 + (\partial_y z)^2}} .$$

On the other hand, in the absence of healing mechanisms such as diffusion, the local growth rate $\dot{z}_1$ would be such that

$$\dot{N}\mathrm{d}S = \dot{z}_1 n_z \mathrm{d}S = \frac{\dot{z}_1\mathrm{d}S}{\sqrt{1 + (\partial_x z)^2 + (\partial_y z)^2}} .$$

Equating these two relations, we obtain

$$\dot{z}_1 = (-\partial_x z\sin\alpha + \cos\alpha)f(\mathbf{r}, t) . \tag{K.1}$$

As in chapters 12, 13 and 14, we write the beam intensity as the sum $f(\mathbf{r}, t) = F + \delta f(\mathbf{r}, t)$ of a constant, uniform term F and a fluctuating term δf with vanishing average value. Formula (K.1) reads

$$\dot{z}_1 = F\cos\alpha - F\partial_x z\sin\alpha + \cos\alpha\,\delta f(\mathbf{r}, t) - \delta f(\mathbf{r}, t)\partial_x z\sin\alpha . \tag{K.2}$$

The first term corresponds to a uniform growth rate $F\cos\alpha$, instead of F as written in chapters 12, 13 and 14. This change is unimportant.

Similarly, the third term is the fluctuation term of chapters 12, 13 and 14 multiplied by a trivial factor $\cos\alpha$ which can easily be accounted for. It is usual in MBE to rotate the sample at a uniform velocity around the z axis. In that case, the average value of $\sin\alpha$ vanishes, so that the second and fourth terms of (K.2) vanish.

In certain cases, it is preferable not to rotate the sample. Then, it is tempting to neglect the last term of (K.2) because it is of higher order, but this procedure would require a better justification. The second term on the right-hand side of (K.2) is just a modification of the linear terms of (13.1) and can be eliminated by a Galilean transformation as explained in section 13.1. However, if there are two constituents or more (for instance if dopants are used) the various constituents necessarily impinge onto the surface at different angles α, and it is impossible to dispose of the linear term for all constituents. The case of GaAs is special because As on As evaporates. Thus, one can work with a large excess of As, and it is possible to consider Ga as if it were the only constituent.

Note that shadowing effects can take place if α is too large (Bruinsma *et al.* 1990).

Appendix L
Renormalization of the KPZ model

We shall try in this appendix to give the maximum number of details on the fundamental points, while skipping the worst algebraic complications, which are admittedly quite serious. The truly uninitiated reader, who is truly anxious of knowing it all, shall be resigned to spending a really huge time reading textbooks on critical phenomena such as Ma's or Le Bellac's (1988). On the subject of the computation of the important integrals for the KPZ model, our reader shall find some details in the fundamental papers by Forster *et al.* (1977) and by Medina *et al.* (1989).

Following Wolf (1991), we consider the anisotropic version of the KPZ equation obtained in appendix J:

$$\dot{z}(\mathbf{r},t) = \delta f(\mathbf{r},t) + \frac{1}{2} \sum_{\alpha=x,y} \lambda_\alpha (\partial_\alpha z)^2 + \sum_{\alpha=x,y} \nu_\alpha \partial^2_{\alpha\alpha} z \ . \tag{L.1}$$

The reason for this choice is that we want to allow λ_x and λ_y to have opposite signs. In fact, in this case the calculation can be succesfully done, while no really meaningful result is obtained in the symmetrical case $\lambda_x = \lambda_y$.

We will use the Wilson's renormalization group, whose basis are simpler than other more powerful, but more bothering, techniques (Le Bellac 1988). The goal is to obtain an equation for the field z, after its short-wavelength Fourier components have been eliminated:

$$\tilde{z}\left(k_c \mid \mathbf{r},t\right) = V^{-1/2} \sum_{k<k_c} \hat{z}_k(t) \exp\left(i\mathbf{k}\cdot\mathbf{r}\right) \ . \tag{L.2}$$

Our secret hope is that the equation will keep its form (L.1), at least within a good approximation. However, the coefficients λ and ν will be likely to be modified, or 'renormalized' as well as

$$\tilde{D} = \left\langle \delta f^2(\mathbf{r},t) \right\rangle \ . \tag{L.3}$$

Physically, the latter renormalization means that evaporation modifies fluctuations.

We will first exactly and simply–without any more or less horrible diagram!–prove that λ_x and λ_y are not renormalized! This is clearly a miraculous simplification. It even shows that the argument of section 13.5 is indeed correct, as well as the relation (13.8).

An invariance property (Medina *et al.* 1989, Wolf 1991)

The key to the problem is provided by the infinitesimal transformation

$$x \rightarrow x - \epsilon\lambda_x t \qquad z \rightarrow z + \epsilon x \tag{L.4}$$

where ϵ is infinitesimal. Note that Medina *et al.*'s paper contains a wrong sign corrected by Wolf (1991). Wolf proposed a general, non-infinitesimal form of transformation, but this complication is not needed here.

A small calculation with differential geometry, that we leave to the reader, shows that (L.1) is invariant with respect to (L.4). Now, this invariance property is certainly independent of k_c. Therefore, it follows that λ_x is independent of k_c, too. The same necessarily holds for λ_y.

One third of the work is already done! But we still have to renormalize $\tilde{D}$, ν_x and ν_y.

Solution of (L.1) by iteration

In order to renormalize what is left, the only way presently known starts from the iterative solution of (L.1), whose zero order is the linear problem solved in chapter 12. Resorting to an iterative solution reveals that we are powerless to face the non-linear problem we are tackling. However, experiments are performed in three dimensions, and three is–as we saw in chapter 13–the upper critical dimension of the problem. The non-linearity should start to be negligible there. An iterative solution has thus some chances of proving worth trying.

In view of this, it is preferable to take a space *and* time Fourier transform:

$$\zeta(\mathbf{k},\omega) = V^{-1/2}\int_{-\infty}^{\infty} \mathrm{d}t \int \mathrm{d}^{d'} r z(\mathbf{r},t)\exp(i\omega t - i\mathbf{k}\cdot\mathbf{r}) \tag{L.5}$$

where V is the volume. For the sake of simplicity, we wrote $z(\mathbf{r},t)$ in place of $\tilde{z}(k_c|\mathbf{r},t)$. In the same way, we introduce

$$\varphi(\mathbf{k},\omega) = V^{-1/2}\int_{-\infty}^{\infty} \mathrm{d}t \int \mathrm{d}^{d'} r \delta f(\mathbf{r},t)\exp(i\omega t - i\mathbf{k}\cdot\mathbf{r}) \ .$$

The initial conditions are assumed as in chapter 12:

$$z(\mathbf{r},t) = \delta f(\mathbf{r},t) = 0 \quad \text{if } t < 0 \ . \tag{L.6}$$

The time Fourier transform coincides with the Laplace transform:

$$\begin{aligned}\int_{-\infty}^{\infty} \mathrm{d}t \dot{z}(\mathbf{r},t) \exp(i\omega t) &= \int_{0}^{\infty} \mathrm{d}t \dot{z}(\mathbf{r},t) \exp(i\omega t) \\ &= \int_0^\infty \mathrm{d}t \frac{\mathrm{d}}{\mathrm{d}t}\left[z(\mathbf{r},t)\exp(i\omega t)\right] - \int_0^\infty \mathrm{d}t z(\mathbf{r},t) \frac{\mathrm{d}}{\mathrm{d}t}\exp(i\omega t) \\ &= -z(\mathbf{r},0) - i\omega \int_0^\infty \mathrm{d}t z(\mathbf{r},t)\exp(i\omega t) \\ &= -i\omega \int_0^\infty \mathrm{d}t z(\mathbf{r},t)\exp(i\omega t) \ .\end{aligned}$$

Multiplying both sides of (L.1) by $V^{-1/2}\exp(i\omega t - i\mathbf{k}\cdot\mathbf{r})$, and integrating over t and $\mathbf{r}$, one gets

$$\begin{aligned}-i\omega\zeta(\mathbf{k},\omega) &= \varphi(\mathbf{k},\omega) - \sum_\alpha v_\alpha k_\alpha^2 \zeta(\mathbf{k},\omega) \\ &+ V^{-1/2}\sum_\alpha \frac{1}{2}\lambda_\alpha \int_{-\infty}^{\infty} \mathrm{d}t \int \mathrm{d}^{d'} r\, (z_\alpha(\mathbf{r},t))^2 \exp(i\omega t - i\mathbf{k}\cdot\mathbf{r}) \ .\end{aligned}$$

The last term can be written as a convolution, according to the theory of the Fourier transform:

$$\begin{aligned}-i\omega\zeta(\mathbf{k},\omega) &= \varphi(\mathbf{k},\omega) - \sum_\alpha v_\alpha k_\alpha^2 \zeta(\mathbf{k},\omega) \\ &- \sum_{\alpha,q} \frac{1}{2}\lambda_\alpha q_\alpha (k_\alpha - q_\alpha) \int_{-\infty}^{\infty} \mathrm{d}\Omega \zeta(\mathbf{q},\Omega)\zeta(\mathbf{k}-\mathbf{q},\omega-\Omega) \ .\end{aligned} \tag{L.7}$$

If both λs vanish, the result of chapter 12 is easily re-obtained. This result is in fact the zeroth order of the iterative solution, which can be written as follows:

$$\zeta(\mathbf{k},\omega) = G_0(\mathbf{k},\omega)\varphi(\mathbf{k},\omega) \ , \tag{L.8}$$

with

$$G_0(\mathbf{k},\omega) = \left(\sum_\alpha v_\alpha k_\alpha^2 - i\omega\right)^{-1} \ . \tag{L.9}$$

To go to next order, equation (L.7) is rewritten in the exact form

$$\begin{aligned}\zeta(\mathbf{k},\omega) &= G_0(\mathbf{k},\omega)\varphi(\mathbf{k},\omega) \\ &+ G_0(\mathbf{k},\omega)\sum_{\alpha,q}\frac{1}{2}\lambda_\alpha q_\alpha(k_\alpha - q_\alpha)\int_{-\infty}^{\infty}\mathrm{d}\Omega\zeta(\mathbf{q},\Omega)\zeta(\mathbf{k}-\mathbf{q},\omega-\Omega) \ ;\end{aligned} \tag{L.10}$$

then (L.8) is inserted into (L.10), which yields

$$\zeta(\mathbf{k},\omega) = G_0(\mathbf{k},\omega)\varphi(\mathbf{k},\omega) + G_0(\mathbf{k},\omega)\sum_{\alpha,q}\frac{1}{2}\lambda_\alpha q_\alpha(k_\alpha - q_\alpha) \times \int_{-\infty}^{\infty} \mathrm{d}\Omega G_0(\mathbf{q},\omega)G_0(\mathbf{k}-\mathbf{q},\Omega-\omega)\varphi(\mathbf{q},\omega)\varphi(\mathbf{k}-\mathbf{q},\Omega-\omega)\,. \tag{L.11}$$

Suppose we want to decrease k_c by a quantity δk_c. We are only interested in the values of k smaller than $k_c' = k_c - \delta k_c$. In the non-linear term of (L.11), three contributions need to be distinguished. i) The terms where both q and $|\mathbf{k}-\mathbf{q}|$ are smaller than k_c'. They are perfectly harmless. ii) The terms where only one between q and $|\mathbf{k}-\mathbf{q}|$ is smaller than k_c'. Such terms correspond–alas!–to a change of form of the equation (L.1), and not only of its coefficients ν_α and $\tilde{D}$. We will neglect them, which requires a justification that will not be given. iii) The terms where q and $|\mathbf{k}-\mathbf{q}|$ are both included between k_c and k_c'. Such terms correspond to the following modification of the fluctuation (or, in the jargon, of the 'noise') $\varphi(\mathbf{k},\omega)$:

$$\delta\varphi(\mathbf{k},\omega) = \sum_\alpha \frac{1}{2}\lambda_\alpha \sum_q^{\star} q_\alpha(k_\alpha - q_\alpha) \times \int_{-\infty}^{\infty} \mathrm{d}\Omega G_0(\mathbf{q},\omega)G_0(\mathbf{k}-\mathbf{q},\Omega-\omega)\varphi(\mathbf{q},\omega)\varphi(\mathbf{k}-\mathbf{q},\Omega-\omega)\,.$$

The superscript '$\star$' means that summation is limited to the terms where q and $\mid \mathbf{k}-\mathbf{q} \mid$ are larger than k_c'. We see that (L.3) is modified, though unfortunately by a quantity which is not necessarily independent of k and ω. An easy calculation yields

$$\delta\tilde{D}(\mathbf{k},\omega) = \sum_{\gamma\alpha} \frac{\lambda_\alpha\lambda_\gamma}{4\pi} \sum_q^{\star} q_\alpha(k_\alpha - q_\alpha)q_\gamma(k_\gamma - q_\gamma) \int_{-\infty}^{\infty} \mathrm{d}\Omega \left| G_0(\frac{\mathbf{k}}{2}+\mathbf{q},\frac{\omega}{2}+\Omega)G_0(\frac{\mathbf{k}}{2}-\mathbf{q},\frac{\omega}{2}-\Omega)\right|^2 \times\tilde{D}(\frac{\mathbf{k}}{2}+\mathbf{q},\frac{\omega}{2}+\Omega)\tilde{D}(\frac{\mathbf{k}}{2}-\mathbf{q},\frac{\omega}{2}-\Omega)\,. \tag{L.12}$$

It is common to call this a one-loop approximation. In fact, rather than writing such complicated expressions, a diagrammatical form is often preferred. Unfortunately, the latter becomes quickly quite esoteric if a complete dictionary of the diagrams is not provided. But since this is a somewhat bothersome job, we only start to sketch it, and leave to the reader to finish it with the aid of the example (L.12).

$G_0(\mathbf{k},\omega)$ is represented by a thin continuous line; it is called 'bare propagator'.

$\tilde{D}(\mathbf{k},\omega)$ is represented by a circle with a dot at its centre $\odot$. The initial or 'bare' value will be represented simply by an empty circle 'o'.

The coefficients λ_α, together with all appropriate sums and integrals will be represented by a full dot •.

Equation (L.12) becomes then:

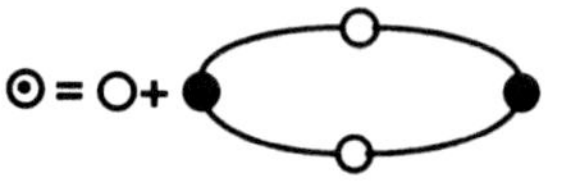

We also need to renormalize the propagator. This can be achieved by considering the iteration of (L.11) beyond first order. This can be cast in diagrammatic form by representing the bare function $\zeta(\mathbf{k},\omega)$ by a thin dashed line — — — — — —, and by a thick dashed line - - - - - - after renormalization. The noise function $\varphi(\mathbf{k},\omega)$ will be denoted by a cross $\times$. Equation (L.11) becomes

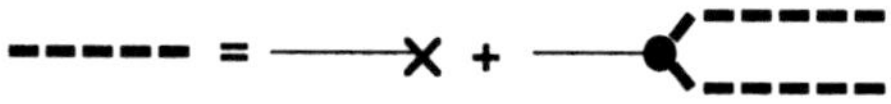

whose iterative solution is

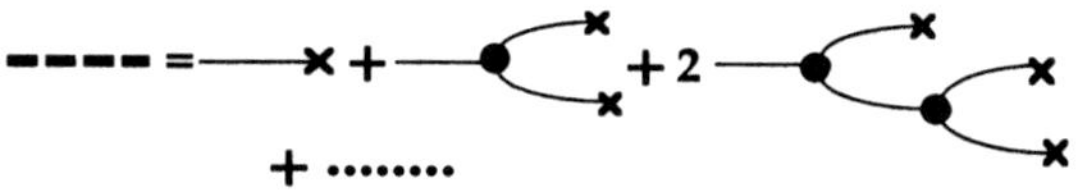

To see what is the effect of modifying k_c on the propagator, the product $\varphi(\mathbf{q},\omega)\varphi(\mathbf{q}',\omega')$ (where $k_c > q, q' > k_c'$) can be replaced by its average value. Translation invariance requires now $\mathbf{q} = -\mathbf{q}'$. The second term of the right-hand side gives a vanishing contribution to the renormalization of the propagator (although it provides the main contribution to the renormalization of the noise, as we accounted for previously). The result is

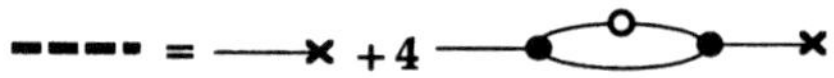

The bare propagator of (L.8) is thus replaced by the renormalized propagator

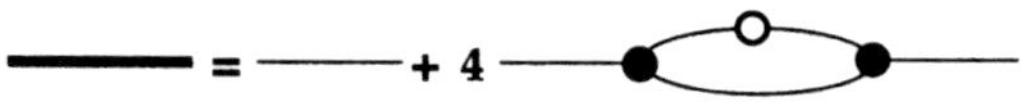

or, explicitly:

$$\delta G(\mathbf{k},\omega) = \sum_{\alpha\gamma} \frac{\lambda_\alpha \lambda_\gamma}{\pi} q_\alpha (k_\alpha - q_\alpha) q_\gamma k_\gamma \int_{-\infty}^{\infty} \mathrm{d}\Omega \sum_q^{\star} G_0(\mathbf{q},\Omega)$$
$$G_0(-\mathbf{q},-\Omega) G_0(\mathbf{k}-\mathbf{q},\omega-\Omega) \tilde{D}(\mathbf{q},\Omega)\,. \qquad \text{(L.13)}$$

The evaluation of the integrals is quite annoying (Medina *et al.* 1989). Assuming $k \ll k_c$, one can prove that renormalization maintains the form (L.1) and the local character in space and time of the fluctuations δf. Equations (L.12) and (L.13) have to produce three equations describing the evolution of $\tilde{D}$, v_x and v_y when $\ell \equiv \ln k_c$ varies. Wolf (1991) has preferred to introduce the parameters $g = \tilde{D}\lambda_x^2/v_x^3$ and $\tilde{r} = v_y/v_x$. Defining the constant quantity $r_0 = \lambda_y/\lambda_x$, and starting from (L.12) and (L.13),

Wolf finds the following two equations for g and $\tilde{r}$ (we do not give the third equation for ν_x, since we will not need it):

$$\frac{\mathrm{d}}{\mathrm{d}\ell}\tilde{r} = -A(\tilde{r})\left[1 - \left(\frac{r_0}{\tilde{r}}\right)^2\right] g\tilde{r} \tag{L.14a}$$

$$\frac{\mathrm{d}}{\mathrm{d}\ell}g = \frac{1}{2}A(\tilde{r})\left[8\frac{r_0}{\tilde{r}} - 3\left\{1 - (r_0/\tilde{r})^2\right\}\right] g^2 \,. \tag{L.14b}$$

The only property of the function $A(\tilde{r})$ we need to know about is that it is positive. The function has been investigated by Wolf (1991) and for the enormously interested readers we will write how it looks like, $A(\tilde{r}) = \left[\arctan\sqrt{\tilde{r}} + \arctan\sqrt{1/\tilde{r}}\right]/(16\pi^2\sqrt{\tilde{r}})$. Note that stability requires that ν_x and ν_y must be both positive, hence their ratio $\tilde{r}$ must also be positive.

Inspired by the theory of critical phenomena, we look for a self-similar behaviour. We know from the theory that this behaviour is connected to a fixed point, where, in particular, $\tilde{r}$ must be independent of ℓ. When this occurs, the right-hand side of equation (L.14a) vanishes. In turn, vanishing of the right-hand side of (L.14a) takes place for $\tilde{r} = r_0$, and for $\tilde{r} = -r_0$, too.

As we just recalled, $\tilde{r}$ must be positive. Thus, since $r_0 = \lambda_y/\lambda_x$ by definition, either the two λs have equal signs (Kardar *et al.* 1986), or opposite signs (Wolf 1991). Inserting these two solutions into (L.14b), we see that they possess very different properties.

i) If $\tilde{r} = r_0$, g is a monotonically increasing function of ℓ, which diverges with ℓ. Therefore, the starting assumption of a weak non-linearity is unacceptable.
ii) If $\tilde{r} = -r_0$, g is a decreasing function of ℓ, which goes to zero with increasing ℓ. In this case, the starting assumption of a weak non-linearity is acceptable.

The symmetrical KPZ equation corresponds to a positive (equal to 1, in fact) value of r_0. Since $\tilde{r}$ is positive, it is the fixed point i) which is selected. Our technique based on the assumption of a weak non-linearity fails.

On the contrary, if λ_x and λ_y have opposite signs, r_0 is negative, and the fixed point ii) is appropriate. Hence g vanishes for large ℓ, and the terms whose coefficients are λ_x and λ_y can be neglected. We are thus brought back to the linear problem of chapter 12, which gives, as we have seen, logarithmic roughness in three dimensions.

Appendix M
Elasticity in a discrete lattice

An atomistic presentation of elasticity may be more intuitive than the usual continuum approach. Moreover, certain features, such as surface relaxation, are accessible this way, while they are not so in continuum elasticity.

In the present appendix, the pressure is assumed to be zero for the sake of simplicity. Extension to non-vanishing pressure is straightforward.

The drawback of any microscopic model is that it can hardly be general. We shall assume a Bravais lattice with pair interactions between nearest and next nearest neighbours. This model is a satisfactory approximate description of rare gas crystals. It is not possible to consider nearest neighbours only, as will be seen, because most of the general features (surface relaxation, surface stress...) are absent.

Fig. M.1 shows what the surface relaxation is, and what its mechanism is in this model. The lattice is a two-dimensional, triangular one. The interaction potential $V(r)$ is assumed to be the same for nearest and next nearest neighbours. If there were no surface, the distance r between nearest

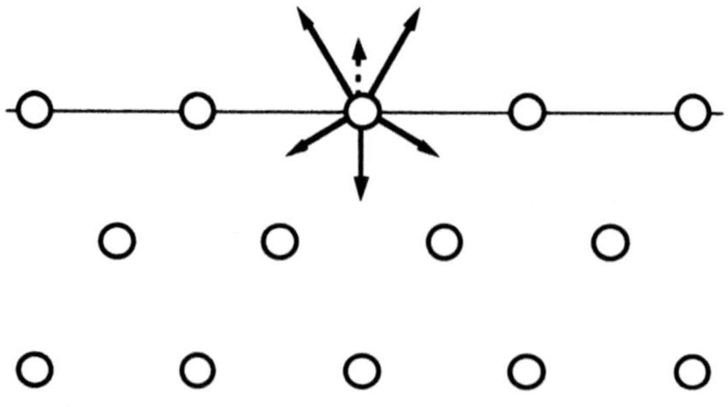

Fig. M.1. Forces responsible for surface relaxation in a triangular lattice with pair interactions between nearest and next nearest neighbours.

neighbours should minimize the energy, which is $3N\left[V(r)+V(r\sqrt{3})\right]$. Therefore, $V'(r)+V'(r\sqrt{3})\sqrt{3}=0$. But $V'(r)$ is the force between two atoms at a distance r. Therefore, each atom is subject to a force f_1 exerted by each nearest neighbour, and a force $f_2=f_1/\sqrt{3}$ exerted by each second neighbour. The force f_1 is repulsive ('hard core') while f_2 is attractive (Van der Waals force).

Now, if there is a planar surface (Fig. M.1), and if the change in the local atomic distance is neglected, each surface atom is seen to be subject to a force directed toward the outer side, normal to the surface and equal to $f_1\sqrt{3}-2f_2=-f_2$. Therefore, the surface atoms move out until the force acting on them vanishes (of course, at equilibrium, the force acting on each atom vanishes). This is the *surface relaxation*, and its result is that the atomic distance (normal to the surface) is larger at the surface than in the bulk. The experimental observation in transition metals is just the opposite! Indeed the pair interaction model is not appropriate to metals. Another simple approximation, the tight binding approximation, predicts the correct sign of the relaxation (Desjonquères & Spanjaard 1993).

While the force acting on each atom vanishes at equilibrium, two parts of a solid can exert upon each other a non-vanishing force. This fact has been used in chapter 16 where we defined the strain. An example, in the case of an infinite planar surface perpendicular to the z direction, is the force acting on the two parts of the outer atomic layer on both sides of the plane $x=0$ (Fig. M.2). If the relaxation is neglected, the tangential component of this force is easily seen to be $\left(3f_1-f_2\sqrt{3}\right)/2$, which is equal to f_1. There is also a normal component, but it vanishes after relaxation. The force acting on all half-atomic layers is easily seen to vanish, as it should.

The existence of a non-vanishing force acting on a part of a system at equilibrium is a paradox which disappears in a finite system. It also

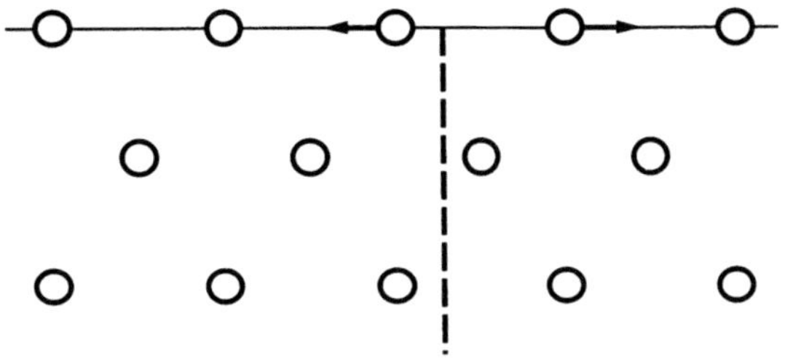

Fig. M.2. Surface stress. The vectors correspond to the forces acting on the two outer atomic half-layers. The force acting on all other horizontal half-layers is equal to zero.

disappears for any finite part of an infinite system. In Fig. M.2, the same calculation as above shows that a group of ℓ atoms in the outer atomic layer is subject to two opposite forces f_1 and $-f_1$ acting on its ends. This force system may be viewed as the sum of ℓ force dipoles of moment af_1, if a is the atomic distance. The non-vanishing force which has been evidenced is thus seen to correspond to the surface stress Π defined in section 16.2, this surface stress is tangential and its value is f_1. Its dimension is W/L rather than W/L^2 because the crystal of Fig. M.2 is only two-dimensional.

With interactions between nearest neighbours only, the atomic distance r would satisfy $V'(r) = 0$, so that $f_1 = f_2 = 0$. There would be no surface relaxation and the surface stress would vanish.

Appendix N
Linear response of a semi-infinite elastic, homogeneous medium

The most direct and regular method of solving this problem is to use Fourier's method. In that case, however, some fairly complicated integrals have to be calculated.

L. Landau, E. Lifshitz (*Theory of elasticity*)

In this appendix, the standard method to solve the equations of linear elasticity will be explained in the case of a homogeneous medium bounded by an infinite plane. Clever alternative methods have been proposed (Landau & Lifshitz 1959b, Nozières 1991). They are extremely efficient in certain problems with cylindrical symmetry (Ling 1948) but in the present case they are not much simpler than the standard method.

In order to obtain the essential results without complicated integrations, it is appropriate to assume that two adatoms interact according to formula (15.1), to deduce the effect of a sinusoidal distribution of surface force dipoles, and to compare with the result of elasticity theory.

Elasticity theory yields equations (16.10) and we want to solve them. We will first solve equation (16.10a), valid in the bulk, i.e. for $z < 0$ if the z axis is chosen perpendicular to the surface (Fig. 16.5). Equation (16.10a) can be alternatively written as

$$\nabla(\mathrm{div}\mathbf{u}) + (1 - 2\zeta)(\nabla^2\mathbf{u}) = 0 \quad (z < 0) \tag{N.1}$$

where $\zeta = \lambda/2(\lambda + \mu)$ is the Poisson coefficient.

The solid is assumed to be submitted to forces acting at its surface, which will be called 'external forces'. Possible origins of these forces are described in chapters 15 and 16. The force per unit area is assumed to be

$$f_\alpha^{\mathrm{ext}}(x) = \mathrm{e}^{ixq} f_\alpha^0 \, . \tag{N.2}$$

We will look for solutions of the form:

$$u_\alpha(x,z) = h_\alpha(z)\mathrm{e}^{ixq}\,. \tag{N.3}$$

The system of interest is an adsorbate on an infinite substrate. It will be sufficient to treat the case when no shear is present in the xy plane. If $\mathbf{q}$ is parallel to x, then $u_y = 0$ and $h_y = 0$.

Equation (N.3) implies then

$$\mathrm{div}\mathbf{u} = \left(\mathrm{d}h_z/\mathrm{d}z + iqh_x\right)\mathrm{e}^{iqx}\,,$$

whence

$$\nabla(\mathrm{div}\mathbf{u}) = \begin{pmatrix} iq[\mathrm{d}h_z/\mathrm{d}z + iqh_x] \\ h_z'' + iq\mathrm{d}h_x/\mathrm{d}z \end{pmatrix}\mathrm{e}^{iqx}$$

and

$$\nabla^2\mathbf{u} = (-q^2\mathbf{h} + \mathrm{d}^2\mathbf{h}/\mathrm{d}z^2)\mathrm{e}^{iqx}\,.$$

Equation (N.1) now reads

$$(\lambda+\mu)\left(iq\mathrm{d}h_z/\mathrm{d}z - q^2h_x\right) + \mu\left(\mathrm{d}^2h_x/\mathrm{d}z^2 - q^2h_x\right) = 0 \tag{N.4a}$$

$$(\lambda+\mu)\left(\mathrm{d}^2h_z/\mathrm{d}z^2 + iq\mathrm{d}h_x/\mathrm{d}z\right) + \mu\left(\mathrm{d}^2h_z/\mathrm{d}z^2 - q^2h_z\right) = 0 \tag{N.4b}$$

or

$$(\lambda+2\mu)q^2h_x - \mu\mathrm{d}^2h_x/\mathrm{d}z - iq\,(\lambda+\mu)\,\mathrm{d}h_z/\mathrm{d}z = 0 \tag{N.5a}$$

$$(\lambda+\mu)iq\mathrm{d}h_x/\mathrm{d}z + (\lambda+2\mu)\,\mathrm{d}^2h_z/\mathrm{d}z^2 - \mu q^2h_z = 0\,. \tag{N.5b}$$

We will now eliminate h_x. For this purpose, we differentiate (N.5b) and obtain

$$\mathrm{d}^2h_x/\mathrm{d}z^2 = i\frac{(\lambda+2\mu)\mathrm{d}^3h_z/\mathrm{d}z^3 - \mu q^2\mathrm{d}h_z/\mathrm{d}z}{q(\lambda+\mu)}\,.$$

Inserting this expression into (N.5a), one finds

$$h_x = i\frac{\mu\mathrm{d}^3h_z/\mathrm{d}z^3 + \lambda q^2\mathrm{d}h_z/\mathrm{d}z}{q^3(\lambda+\mu)}\,.$$

Differentiating this equation and inserting the result into (N.5b), one obtains an equation for h_z:

$$\mathrm{d}^4h_z/\mathrm{d}z^4 - 2q^2\mathrm{d}^2h_z/\mathrm{d}z^2 + q^4h_z = 0\,. \tag{N.6}$$

The solutions of this linear equation have the form

$$h_z(z) = \sum_n H_n\exp(k_nz)\,,$$

where the wavevectors k_n are determined by the equation

$$k^4 - 2q^2k^2 + q^4 = 0 .$$

The negative double root $k = -q$ is not acceptable since $h_z(-\infty)$ should vanish. The other root is also double, $k_1 = k_2 = q$. Writing

$$h_z = H_1 \left(e^{k_1 z} + \epsilon e^{k_2 z} \right) \simeq H_1 e^{k_1 z} \left[1 + \epsilon + \epsilon (k_2 - k_1) z \right] ,$$

it is seen that $h_z = z e^{qz}$ is also a solution. The general solution of (N.6) is therefore $(A + Bz)e^{qz}$. A similar discussion is valid for h_x, so that the solution of (N.5) has the form

$$\begin{cases} h_x(z) = (C + Dz)e^{|q|z} & \text{(N.7a)} \\ h_z(z) = (A + Bz)e^{|q|z} & \text{(N.7b)} \end{cases}$$

whence

$$\begin{cases} dh_x/dz = \left[C|q| + D(z|q| + 1) \right] e^{|q|z} \\ dh_z/dz = \left[A|q| + B(z|q| + 1) \right] e^{|q|z} \end{cases}$$

and

$$\begin{cases} d^2h_x/dz^2 = \left[Cq^2 + D(zq^2 + 2|q|) \right] e^{|q|z} \\ d^2h_z/dz^2 = \left[Aq^2 + B(zq^2 + 2|q|) \right] e^{|q|z} . \end{cases}$$

Insertion of these formulae into (N.5) yields

$$\begin{cases} (\lambda + \mu)q^2(C + Dz) - 2\mu D|q|) - iq(\lambda + \mu)(A|q| + B|q|z + B) = 0 \\ (\lambda + \mu)iq(C + Dz + \frac{D}{|q|}) + (\lambda + \mu)|q|(A + Bz + 2\frac{B}{|q|}) + 2\mu B = 0 . \end{cases}$$

Setting to zero all terms proportional to z, one obtains

$$D = iB\frac{q}{|q|} . \tag{N.8a}$$

Setting to zero all terms independent of z, one obtains

$$B = -\frac{\lambda + \mu}{\lambda + 3\mu} q \left(iC + A\frac{q}{|q|} \right) . \tag{N.8b}$$

On the surface, the solution (u_x, u_z) should verify equation (16.10b). Since $n_\alpha = \delta_{\alpha z}$, those equations can be written as

$$f_x^{\text{ext}} = \mu \partial_x u_z + \mu \partial_z u_x \tag{N.9a}$$

and

$$f_z^{\text{ext}} = 2\mu \partial_z u_z + \lambda \left(\partial_x u_x + \partial_z u_z \right) . \tag{N.9b}$$

Using (N.3) and (N.7), one finds

$$\begin{aligned}
\partial_x u_x(x,0) &= iqC\mathrm{e}^{ixq} \\
\partial_z u_x(x,0) &= |q|C\mathrm{e}^{ixq} + D\mathrm{e}^{ixq} \\
\partial_x u_z(x,0) &= iqA\mathrm{e}^{ixq} \\
\partial_z u_z(x,0) &= |q|A\mathrm{e}^{iqx} + B\mathrm{e}^{ixq}
\end{aligned}$$

so that equations (N.9) read

$$\begin{aligned}
f_x^{\text{ext}} &= i\mu qA\mathrm{e}^{iqx} + \mu|q|C\mathrm{e}^{iqx} + \mu D\mathrm{e}^{iqx} = f_x^0\mathrm{e}^{iqx} \\
f_z^{\text{ext}} &= 2\mu|q|A\mathrm{e}^{iqx} + 2\mu B\mathrm{e}^{iqx} + \lambda(iqC + |q|A + B)\mathrm{e}^{iqx} = f_z^0 e^{iqx}
\end{aligned}$$

or

$$i\mu qA + \mu|q|C + \mu D = f_x^0 \tag{N.10a}$$

and

$$2\mu|q|A + 2\mu B + \lambda(iqC + |q|A + B) = f_z^0\,. \tag{N.10b}$$

Using (N.8), equations (N.10a) and (N.10b) respectively read

$$i\frac{\mu}{\lambda+3\mu}Aq + \frac{\lambda+2\mu}{\lambda+3\mu}C|q| = f_x^0/(2\mu) \tag{N.11a}$$

and

$$\frac{\lambda+2\mu}{\lambda+3\mu}A|q| - i\frac{\mu}{\lambda+3\mu}Cq = f_z^0/(2\mu)\,. \tag{N.11b}$$

The solution of this set of two linear equations is

$$2\mu Aq = \frac{if_x^0\mu + f_z^0(\lambda+2\mu)q/|q|}{\lambda+\mu} \tag{N.12a}$$

$$2\mu Cq = \frac{-if_z^0\mu + f_x^0(\lambda+2\mu)q/|q|}{\lambda+\mu}\,. \tag{N.12b}$$

The other coefficients B and D follow from (N.8), namely

$$B = -iDq/|q| = -\frac{if_x^0 q/|q| + f_z^0}{2\mu}\,. \tag{N.12c}$$

It is usual to introduce the Poisson coefficient ζ. Using the relations

$$\frac{\mu}{\lambda+\mu} = 1 - 2\zeta$$

$$\frac{\lambda+2\mu}{\lambda+\mu} = 2(1-\zeta)\,,$$

insertion of (N.12) into (N.7) and (N.3) yields

$$u_x(\mathbf{q},z) = \frac{1}{2\mu|q|}\left\{[|q|z + 2(1-\zeta)]f_x^0 - i[qz + (1-2\zeta)\frac{q}{|q|}]f_z^0\right\} e^{z|q|}e^{ixq} \tag{N.13a}$$

$$u_z(\mathbf{q},z) = \frac{-1}{2\mu|q|}\left\{i[qz - (1-2\zeta)\frac{q}{|q|}]f_x^0 + [|q|z - 2(1-\zeta)]f_z^0\right\} e^{z|q|}e^{ixq}. \tag{N.13b}$$

It may be interesting to check formula (N.13b) in the limit $q = 0$, for instance for a uniform force density parallel to z and equal to -1. The displacement is infinite, but the strain is finite. In order to make the displacement finite, an opposite force should be applied to the solid at $z = -\infty$. The strain can be deduced from (N.13b) as

$$\begin{aligned}\partial_z u_z &= -\lim_{q\to 0} \partial_z \left[-qze^{qz} + 2(1-\zeta)e^{qz}\right]/(2\mu q) \\ &= -\lim_{q\to 0}\left[-qe^{qz} - q^2ze^{qz} + 2q(1-\zeta)e^{qz}\right]/(2\mu q) \\ &= -\frac{1}{2\mu}(1-2\zeta)\,.\end{aligned}$$

Our main goal is to establish formula (15.3), which gives the interaction free energy of two dipole moments on the surface. As seen from (16.26d), this interaction can be deduced from the strain produced by each force dipole at the site occupied by the other one. In this appendix, only the effect of components of the dipole moment $m_{\alpha\gamma}$ with $\gamma \neq z$ will be considered. Moreover, off-diagonal components will be assumed to vanish, $m_{xy} = m_{yx} = m_{zx} = m_{zy} = 0$. Finally, the symmetry relation $m_{xx} = m_{yy} = m$ will be assumed. These conditions are independent of the choice of the x and y axes in the surface plane, and are reasonable if the force dipoles are produced by atoms in a symmetric position. The effect of the m_{zz} component will be investigated in appendix O.

From (N.13) it is now possible to calculate the displacement produced by a localized force dipole moment $m_{xx} = m_{yy} = m$ at the origin. This dipole may be assumed to stem from a force $(f,0)$ situated at $(b,0)$, a force $(-f,0)$ at $(-b,0)$, a force $(0,f)$ at $(0,b)$, and a force $(0,-f)$ at $(0,-b)$. The limit $b \to 0$ should then be taken, keeping $2bf = m$ constant.

The surface force density corresponding to this dipole is given by

$$\begin{aligned}\frac{dF_x}{dS} &= f[\delta(x-b)\delta(y) - \delta(x+b)\delta(y)] \\ &= \frac{f}{4\pi^2}\int d^2q\left(e^{iq_xb} - e^{-iq_xb}\right)e^{i\mathbf{q}\cdot\mathbf{r}} = \frac{2ibf}{4\pi^2}\int d^2q\; q_x \exp(i\mathbf{q}\cdot\mathbf{r}) \\ &= \frac{mi}{4\pi^2}\int d^2q\; q_x\exp(i\mathbf{q}\cdot\mathbf{r})\,,\end{aligned}$$

and a similar formula for the y component. This implies

$$\mathbf{f} = \frac{mi}{4\pi^2} \int d^2q \mathbf{q} \exp(i\mathbf{q} \cdot \mathbf{r}) . \tag{N.14}$$

In order to compute the energy given by (16.26), one needs only the tangential component $\mathbf{u}_T(\mathbf{r}, 0)$ of the displacement on the surface. This displacement is given by (N.13) if $\mathbf{q}$ is parallel to x. If $\mathbf{q}$ has another direction, the response to a surface force density $\mathbf{f}_T(\mathbf{r}) = \mathbf{f}_T^0(\mathbf{q}) \exp(i\mathbf{q} \cdot \mathbf{r})$ parallel to $\mathbf{q}$ is easily deduced from (N.13a), namely

$$\mathbf{u}_T(\mathbf{r}, 0) = \frac{1}{\mu|q|}(1 - \zeta)\mathbf{f}_T^0(\mathbf{q}) \exp(i\mathbf{q} \cdot \mathbf{r}) .$$

The response to (N.14) is obtained by replacing $\mathbf{f}_T^0(\mathbf{q})$ by $mi\mathbf{q}/(4\pi^2)$ and integrating over q. One finds

$$\mathbf{u}_T(\mathbf{r}, 0) = mi\frac{1-\zeta}{4\pi^2} \int d^2q \frac{\mathbf{q}}{\mu|q|} \exp(i\mathbf{q} \cdot \mathbf{r}) . \tag{N.15}$$

As argued in section 15.1, we expect this to be proportional to $\mathbf{r}/r^3$ at long distance. In order to check this expectation, we will evaluate the Fourier transform of $\mathbf{r}/r^3$:

$$\begin{aligned}\int d^2r \frac{\mathbf{r}}{r^3} \exp(-i\mathbf{q} \cdot \mathbf{r}) &= -\frac{i\mathbf{q}}{|q|} \int_{-\infty}^{\infty} dx x \sin(qx) \int_{-\infty}^{\infty} \frac{dy}{(x^2 + y^2)^{3/2}} \\ &= -2i\frac{\mathbf{q}}{|q|} \int_{-\infty}^{\infty} dx \frac{\sin(qx)}{x} = -2i\pi \frac{\mathbf{q}}{|q|} .\end{aligned}$$

Inverting the Fourier transform, the elastic displacement (N.15) resulting from a force dipole m is found to be

$$\mathbf{u}_T(\mathbf{r}, 0) = \frac{m}{\mu} \frac{1-\zeta}{2\pi} \frac{\mathbf{r}}{r^3} = m \frac{1-\zeta^2}{\pi E} \frac{\mathbf{r}}{r^3} .$$

This expression can also be deduced from formulae (8.19) of Landau & Lifshitz (1959b).

We can now compute the interaction energy between a force dipole m at the origin and a force dipole m' at $(x, 0, 0)$. The non-vanishing components of the strain are:

$$\partial_x u_x = -\frac{2m}{\pi E} \frac{1-\zeta^2}{r^3}$$

and

$$\partial_y u_y = \frac{m}{\pi E} \frac{1-\zeta^2}{r^3} .$$

Insertion of these relations into (16.26d) yields the interaction energy of two tangential force dipoles as

$$W_{\text{int}} = \frac{mm'}{\pi E} \frac{1-\zeta^2}{r^3} \,.$$

This corresponds to the first term between brackets in (15.3).

Instead of localized force dipoles, it is often interesting to consider a sinusoidal distribution of external surface forces given by (N.2). Relations (N.3), (N.7) and (N.12) give the displacement, for instance

$$u_x(x,y,z) = \frac{e^{|q|z+iqx}}{2\mu(\lambda+\mu)} \left[\frac{\lambda+2\mu}{|q|} f_x^0 - \frac{\mu}{q} f_x^0 \right] + \frac{z e^{|q|z+iqx}}{2\mu} \left[f_x^0 - \frac{iq}{|q|} f_z^0 \right] \,.$$

It is easy to deduce the strain $\epsilon_{xx} = iqu_x(x,y,z)$ produced by an external surface stress (or surface density of force dipoles) $\pi_{xx}^{\text{ext}}(x,y,0) = \pi_{xx}^0 e^{iqx}$. Indeed, $f_x^0 = -iq\pi_{xx}^0$. For instance the strain at the surface produced by an external surface stress $\pi_{xx}^{\text{ext}}(x,y,0) = \pi_{xx}^0 \cos(qx)$ is

$$\begin{aligned} \epsilon_{xx}(x,y,0) &= \frac{\lambda+2\mu}{2\mu(\lambda+\mu)} |q| \pi_{xx}^0 \cos(qx) \\ &= 2\frac{1-\zeta^2}{E} |q| \pi_{xx}^0 \cos(qx) \,. \end{aligned} \tag{N.16}$$

The corresponding elastic free energy is readily deduced from (16.26 d), namely

$$\frac{\mathrm{d}\mathscr{F}}{\mathrm{d}S} = \frac{1-\zeta^2}{2E} |q| \left(\pi_{xx}^0\right)^2 \,. \tag{N.17}$$

Note that the response (N.16) vanishes with the wavevector q.

The response to a localized dipole moment with xz and yz components can easily be deduced from (N.13) by a Fourier transformation, as we have done for the xx and yy components.

Appendix O
Elastic dipoles in the z direction

As in appendix N, we wish to solve the equation of elasticity for a semi-infinite, homogeneous solid medium bounded by a plane, on which force dipoles are acting. As seen in chapter 16, these dipoles may be due to an adsorbate, to steps or to other defects. In appendix N, dipole moments components m_{xx} and m_{xz} were considered. In the present appendix, we calculate the strain of an elastic solid subject to a distribution of external force dipoles which have components m_{zz} and m_{zx} . Let

$$\pi_{zz}^{\text{ext}}(x, y) = \frac{\mathrm{d}m_{zz}^{\text{ext}}}{\mathrm{d}S}$$

be the surface density of dipole moment, i.e. the external surface stress. It is convenient to assume that this surface stress is produced by a volume density of external stress

$$p_{zz}^{\text{ext}}(x, y, z) = \pi_{zz}^{\text{ext}}(x, y)/b$$

acting inside a layer of thickness b of the solid, for $-b < z < 0$. The elastic free energy is

$$\mathscr{F} = \frac{\lambda}{2}\int \mathrm{d}^3r(\epsilon_{xx} + \epsilon_{yy} + \epsilon_{zz})^2 + \mu\int d^3r\left[\sum_{\alpha,\gamma}\epsilon_{\alpha\gamma}^2\right] - \int_b \mathrm{d}^3r p_{zz}^{\text{ext}}\epsilon_{zz}\,, \quad \text{(O.1)}$$

where the subscript b restricts the integration to the domain $-b < z < 0$ in the last term. The linear terms can be transformed by introducing the new displacement variable

$$\tilde{u}_z(x, y) = \begin{cases} u_z(x, y) - \dfrac{zp_{zz}^{\text{ext}}}{\lambda + 2\mu} & (-b < z < 0) \\ u_z(x, y) & (z < -b) \end{cases}$$

so that the strain is replaced by

$$\tilde{\epsilon}_{zz}(x,y) = \begin{cases} \epsilon_{zz}(x,y) - \dfrac{p_{zz}^{\text{ext}}}{\lambda + 2\mu} & (-b < z < 0) \\ \epsilon_{zz}(x,y) & (z < -b) \,. \end{cases}$$

Now, the free energy takes the form

$$\begin{aligned} \mathscr{F} &= \frac{\lambda}{2} \int \mathrm{d}^3 r \, (\epsilon_{xx} + \epsilon_{yy} + \tilde{\epsilon}_{zz})^2 \\ &+ \mu \int d^3 r \left(\epsilon_{xx}^2 + \epsilon_{yy}^2 + \tilde{\epsilon}_{zz}^2 + 2\epsilon_{xy}^2 + 2\epsilon_{yz}^2 + 2\epsilon_{zx}^2 \right) \\ &+ \frac{\lambda}{\lambda + 2\mu} \int_b \mathrm{d}^3 r p_{zz}^{\text{ext}} \, (\epsilon_{xx} + \epsilon_{yy}) - \frac{1}{2(\lambda + 2\mu)} \int_b \mathrm{d}^3 r \left[p_{zz}^{\text{ext}} \right]^2 \,. \end{aligned} \tag{O.2}$$

The last term does not depend on the displacement and can therefore be ignored when minimizing $\mathscr{F}$ with respect to the strain. The presence of the third term in (O.2) shows that the external stress $p_{zz}^{\text{ext}}(x,y) = p(x,y)$ produces at long distance the same strain as an external stress

$$p_{xx}^{\text{ext}}(x,y) = p_{yy}^{\text{ext}}(x,y) = -\lambda p(x,y)/(\lambda + 2\mu) = -\zeta p(x,y)/(1-\zeta) \,.$$

In the limit $b = 0$, one sees that external surface stress $\pi_{zz}^{\text{ext}}(x,y) = \pi(x,y)$ has the same effect at long distance as an external surface stress

$$\pi_{xx}^{\text{ext}}(x,y) = \pi_{yy}^{\text{ext}}(x,y) = -\zeta \pi(x,y)/(1-\zeta) \,.$$

Therefore, a localized force dipole m_{zz} at the surface, $m_{zz} = m$, has the same effect at long distance as a localized force dipole moment whose non-vanishing components are

$$m_{xx} = m_{yy} = -\zeta m/(1-\zeta) \,.$$

Relations (15.3) and (15.4) are easily deduced from this relation and the results of appendix N.

The effect of a surface density of dipole moment of component zx, namely

$$\pi_{zx}^{\text{ext}}(x,y) = \frac{\mathrm{d}m_{zx}^{\text{ext}}}{\mathrm{d}S} \,,$$

can be discussed in the same way. It can be replaced by a volume density of external dipole moments

$$p_{zx}^{\text{ext}}(x,y,x) = \pi_{zx}^{\text{ext}}(x,y)/b$$

acting inside a layer of thickness b of the solid, for $-b < z < 0$. The elastic free energy is

$$\mathscr{F} = \frac{\lambda}{2}\int \mathrm{d}^3r(\epsilon_{xx} + \epsilon_{yy} + \epsilon_{zz})^2 + \mu \int d^3r \left[\sum_{\alpha,\gamma} \epsilon_{\alpha\gamma}^2\right] - \int_b \mathrm{d}^3r p_{zx}^{\mathrm{ext}} \partial_z u_x \, . \tag{O.3}$$

It is convenient to introduce the variable defined by

$$\tilde{u}_x(x, y) = \begin{cases} u_x(x, y) - \dfrac{z p_{zx}^{\mathrm{ext}}}{\mu} & (-b < z < 0) \\ u_x(x, y) & (z < -b) \, . \end{cases}$$

so that (O.3) takes the form

$$\begin{aligned} \mathscr{F} = &\frac{\lambda}{2}\int \mathrm{d}^3r(\epsilon_{xx} + \epsilon_{yy} + \epsilon_{zz})^2 \\ &+ \mu \int d^3r \left[\epsilon_{xx}^2 + \epsilon_{yy}^2 + \tilde{\epsilon}_{zz}^2 + 2\epsilon_{xy}^2 + 2\epsilon_{yz}^2 + 2\tilde{\epsilon}_{zx}^2\right] \\ &+ \int_b \mathrm{d}^3r p_{zx}^{\mathrm{ext}} \partial_x u_z - \frac{1}{4\mu}\int_b \mathrm{d}^3r \left[p_{zx}^{\mathrm{ext}}\right]^2 \, , \end{aligned} \tag{O.4}$$

where

$$\tilde{\epsilon}_{zx}(x, y) = \begin{cases} \epsilon_{zx}(x, y) - \dfrac{p_{zx}^{\mathrm{ext}}}{\mu} & (-b < z < 0) \\ \epsilon_{zx}(x, y) & (z < -b) \, . \end{cases}$$

The last term of (O.4) does not depend on the displacement and can therefore be ignored. The third term has the same form as in (O.3) except that $\partial_z u_x$ has been replaced by $\partial_x u_z$ and the sign has been changed. Therefore, external force dipole moments of the zx type produce the same strain as external force dipole moments of the xz type with identical absolute value and opposite sign.

Appendix P
Elastic constants of a cubic crystal

The concept of isotropic solid is widely used in the current literature as well as in our chapters 15 and 16. However, it is not an adequate description of a crystal. The simplest crystals are cubic. For a cubic crystal, the tensor Ω of formula (16.5) has the following non-vanishing components (Bacon *et al.* 1979).

$$c_{11} = \Omega_{xx}^{xx} = \Omega_{yy}^{yy} = \Omega_{zz}^{zz} ,$$

$$c_{12} = \Omega_{xx}^{yy} = \Omega_{yy}^{xx} = \Omega_{yy}^{zz} = \Omega_{zz}^{yy} = \Omega_{zz}^{xx} = \Omega_{xx}^{zz} ,$$

$$c_{44} = \Omega_{xy}^{xy} = \Omega_{yz}^{yz} = \Omega_{zx}^{zx} = \Omega_{xy}^{yx} = \Omega_{yx}^{yx} = \dots$$

How isotropic a cubic crystal is? This information is contained in the quantity

$$A = \frac{2c_{44}}{c_{11} - c_{12}}$$

which would be equal to 1 in an isotropic solid. It is indeed close to 1 in tungsten, but can be very different from 1 in other elements as seen from table P.1.

It is possible to define average Lamé coefficients as follows (Hirth & Lothe 1982):

$$\lambda = \frac{1}{15}\left(2\sum_{\alpha\gamma}\Omega_{\alpha\alpha}^{\gamma\gamma} - \sum_{\alpha\gamma}\Omega_{\alpha\gamma}^{\alpha\gamma}\right) ,$$

$$\mu = \frac{1}{30}\left(3\sum_{\alpha\gamma}\Omega_{\alpha\gamma}^{\alpha\gamma} - \sum_{\alpha\gamma}\Omega_{\alpha\alpha}^{\gamma\gamma}\right) .$$

Table P.1. *Elastic coefficients for selected elements (GPa).*

Element	μ	λ	ζ	A
Al	26.5	59.3	0.347	1.21
Ag	33.8	81.1	0.354	3.01
Au	31.0	146.0	0.412	2.90
Cu	54.6	100.6	0.324	3.21
Pb	10.1	34.8	0.387	3.90
W	160.0	201.0	0.278	1.00
Si	68.1	52.4	0.218	1.56
Diamond	536.0	85.0	0.068	1.21

The Poisson ratio ζ can then be defined as in the isotropic case by

$$\lambda = \frac{2\zeta\mu}{1 - 2\zeta} .$$

Values of λ and μ (in Gigapascals) and of ζ and A are given for selected elements in table P.1.

References

Nous sommes comme des nains assis sur les épaules de géants.
Nous voyons donc plus de choses que les anciens et de plus éloignées,
non par la pénétration de notre propre vue
ou par l'élévation de notre taille,
mais parce qu'ils nous soulèvent et nous exhaussent
de toute leur hauteur gigantesque.

Bernard de Chartres (XII Century)

If I have seen further it is by standing
on the shoulders of giants.

Sir Isaac Newton

ABELMANN, L. (1994). *Oblique evaporation of thin* $Co_{80}NI_{20}$ *films for magnetic recording*. Ph. D. Thesis, Twente University.

ABRAHAM, D.B. (1983) *Phys. Rev. Lett.* **51**, 1279. Surface reconstruction in crystals.

ADAMS, D.P., YALISOVE, S.M., EAGLESHAM, D.J. (1993) *Appl. Phys. Lett.* **63**, 3571. Effect of hydrogen on surface roughening during Si homoepitaxial growth.

ADDA, Y., PHILIBERT, J. (1966) *La diffusion dans les solides* (I.N.S.T.N., Saclay).

AKUTSU, Y., AKUTSU, N., YAMAMOTO, T. (1988) *Phys. Rev. Lett.* **61**, 424. Universal jump of gaussian curvature at the facet edge of a crystal.

ALERHAND, O.L., VANDERBILT, D., MEADE, R.D., JOANNOPOULOS, J.D. (1988) *Phys. Rev. Lett.* **61**, 1973. Spontaneous formation of stress domains on crystal surfaces.

ALERHAND, O.L., BERKER, A.N., JOANNOPOULOS, J.D., VANDERBILT, D., HAMERS, R.J., DEMUTH, J.E. (1990) *Phys. Rev. Lett.* **64**, 1990. Finite temperature phase diagram of vicinal Si(100) surfaces.

ALFONSO, C., BERMOND, J.M., HEYRAUD, J.C., MÉTOIS, J.J. (1992) *Surface Sci.* **262**, 371. The meandering of steps and the terrace width distribution on clean Si(111).

ALFONSO, C., HEYRAUD, J.C., MÉTOIS, J.J. (1993) *Surface Sci.* **291**, L745. About the sublimation of Si surfaces vicinal of $\{111\}$.

AMAR, J.G., FAMILY, F. (1990) *Phys. Rev. Lett.* **64**, 543. Phase transition in a restricted solid-on-solid surface-growth model in 2+1 dimensions.

ANDREEV, A.F. (1981) *Zh. Exp. Teor. Fiz.* **80**, 2042 and *Sov. Phys. J.E.T.P.* **53**, 1063. Faceting phase transitions of crystals.

ANDREEV, A.F., KOSEVICH, YU.A. (1981) *Zh. Exp. Teor. Fiz.* **81**, 1435 and *Sov. Phys. J.E.T.P.* **54**, 761. Capillary phenomena in the theory of elasticity.

ANDREEVA, O.A., KESHISHEV, K.O., KOGAN A.G., MARCHENKOV, A.N. (1992) *Europhys. Lett.* **19**, 683. Growth kinetics of vicinal surfaces of ^{4}He crystals.

ANDRIEU, S., ARNAUD D'AVITAYA, F., PFISTER, J.C. (1989) *J. Appl. Phys.* **65**, 2681. Surface segregation mechanism during two-dimensional epitaxial growth: the case of dopants in Si and GaAs molecular-beam epitaxy.

ARBEL, E., CAHN, J.W. (1975) *Surface Sci.* **51**, 305. On invariances in surface thermodynamic properties and their applications to low symmetry crystals.

ARMAND, G., MANSON, J.R. (1988) *Phys. Rev. B* **37**, 4363. Thermal attenuation in atom-surface scattering: a useful approximation.

ARTHUR, J.R. (1974) *Surface Sci.* **43**, 449. Surface stoichiometry and structure of GaAs.

ASARO, R.J., TILLER, W.A. (1972) *Metall. Trans.* **3**, 1789. Interface morphology development during stress corrosion cracking.

AVIGNON, M. (1971) *J. Cryst. Growth* **11**, 265. Role of interface kinetics on the shape stability of a two-dimensional nucleus on a substrate.

AVIGNON, M., CHAKRAVERTY, B.K. (1969) *Proc. Roy. Soc. A* **310**, 277. Morphological stability of two-dimensional nucleus.

BACON, D.J., BARNETT, D.M., SCATTERGOOD, R.O. (1979) in *Progress in Mat. Science* **23**, p. 51. Anisotropic continuum theory of lattice defects.

BAK, P, CHEN K. (1991) *Scientific American*, January issue, p. 46. Self-organized criticality.

BALES, G.S., CHRZAN, D.C. (1994) *Phys. Rev. B* **50**, 6057. Dynamics of irreversible island growth during submonolayer epitaxy.

BALES, G.S., ZANGWILL, A. (1990a) *Phys. Rev. B* **41**, 5500. Morphological instability of a terrace edge during step flow growth.

BALES, G.S., ZANGWILL, A. (1990b) *J. Vac. Sc. Technol. A* **9**, 145. Macroscopic model for columnar growth of amorphous films by sputter deposition.

BALIBAR, S., BOUCHAUD, J.P. (1992) *Phys. Rev. Lett.* **69**, 862. Kardar-Parisi-Zhang equation and the dynamic roughening of crystal surfaces.

BALIBAR, S., CASTAING, B. (1985) *Surface Sci. Reports* **5**, 87. Helium: solid-liquid interfaces.

BALIBAR, S., GALLET, J.F., ROLLEY, E., WOLF, P.E. (1987) *La Recherche*, **18**, 1454. La transition rugueuse.

BALIBAR, S., GUTHMANN, C., ROLLEY, E. (1993) *J. de Physique I* **3**, 1475. From vicinal to rough crystal surfaces.

BARABÁSI, A.-L., STANLEY, H.E. (1995) *Fractal concepts in surface growth* (Cambridge University Press).

BARDOTTI, L., JENSEN, P., TREILLEUX, M., HOAREAU, A., CABAUD, B. (1995) *Phys. Rev. Lett.* **74**, 4694. Experimental observation of fast diffusion of antimony clusters on graphite surfaces.

BARTELT, M.C., EVANS, J.W. (1992) *Phys. Rev. B* **46**, 12675. Scaling analysis of diffusion-mediated island growth in surface adsorption processes.

BARTELT, M.C., EVANS, J.W. (1993) *Europhys. Lett.* **21**, 99. Crossover from anisotropic-to-isotropic diffusion-mediated island growth on surfaces.

BARTELT, M.C., EVANS, J.W. (1994a) *Surface Sci. Lett.* **314**, L829. Dendritic island in metal-on-metal epitaxy: I. Shape transitions and diffusion at island edge.

BARTELT, M.C., EVANS, J.W. (1994b) *Surface Sci. Lett.* **314**, L835. Dendritic island in metal-on-metal epitaxy: II. Coalescence and mutilayer growth.

BARTELT, M.C., PERKINS, L.S., EVANS, J.W. (1995) *Surface Sci. Lett.* **344**, L1193. Transition in critical size for metal(100) homoepitaxy.

BARTELT, N.C., GOLDBERG, J.L., EINSTEIN, T.L., WILLIAMS, E.D. (1992) *Surface Sci.* **273**, 252. The equilibration of terrace width distributions on stepped surfaces.

BARTELT, N.C., TROMP, R.M., WILLIAMS, E.D. (1994) *Phys. Rev. Lett.* **73**, 656. Step capillary waves and equilibrium island shapes on Si(001).

BARTOLINI, A., ERCOLESSI, F., TOSATTI, E. (1989) *Phys. Rev. Lett.* **63**, 872. *Magic* vicinal surfaces stabilized by reconstruction.

BATEMAN, H. and Staff of the Bateman Manuscript Project (1953) *Higher transcendental functions*, Vol. II (McGraw-Hill, New York).

BAUER, E. (1992) *Appl. Surface Sci.* **60/61**, 350. Low-energy electron microscopy of surface processes.

BECKER, R., DÖRING, G. (1935) *Ann. Phys.* (Leipzig) **24**, 719. Kinetische Behandlung der Keimbildung in übersättigten Dämpfen.

BENA, I., MISBAH, C., VALANCE, A. (1993) *Phys. Rev. B* **47**, 7408. Nonlinear evolution of a terrace edge during step-flow growth.

BEN AMAR, M. (1991a) *Phys. Rev. A* **43**, 5724. Exact self-similar shapes in viscous fingering.

BEN AMAR, M. (1991b) *Phys. Rev. A* **44**, 3673. Viscous fingering in a wedge.

BENNEMA, P. (1990) in *Kinetics of Ordering and Growth at Surfaces* M.G. Lagally ed. (Plenum, New York) p. 415. Morphology of crystals based on crystallography, statistical mechanics and thermodynamics.

BERGE, B., FAUCHEUX, L., SCHWAB, K., LIBCHABER, A. (1991) *Nature* **350**, 322. Faceted crystal growth in two dimensions.

BERMOND, J.M., MÉTOIS, J.J., ÉGEA, X., FLORET, F. (1995) *Surface Sci.* **330**, 48. The equilibrium shape of silicon.

Berrehar, J.C., Caroli, C., Lapersonne-Meyer, C., Schott, M. (1992) *Phys. Rev. B* **46**, 13487. Surface patterns on single-crystal films under uniaxial stress: experimental evidence for the Grinfeld instability.

Berrehar, J.C., C., Lapersonne-Meyer, C., Schott, M., Villain, J. (1989) *J. de Physique* **50**, 923. Formation of periodic cracks in polyacethylene single crystal thin films.

Bethge, H. (1990) in *Kinetics of Ordering and Growth at Surfaces*, M.G. Lagally ed. (Plenum, New York) p. 125. Molecular kinetics of steps.

Bienfait, M. (1987) in *Dynamics of Molecular Crystals*, J. Lascombe ed. (Elsevier, Amsterdam) p. 353. Mobility of 2D adsorbed phases of methane and ethane. A quasi-elastic neutron scattering study.

Bienfait, M., Gay, J.M., Blank, H. (1988) *Surface Sci.* **204**, 331. Surface premelting of thin films of methane.

Binnig, G., Rohrer, H., Gerber, CH., Weibel, E. (1983) *Phys. Rev. Lett.* **50**, 120. 7×7 reconstruction on Si(111) resolved in real space.

Binnig, G., Rohrer, H. (1987) *Rev. Mod. Phys.* **59**, 615. Scanning tunneling microscopy–from birth to adolescence.

Blatter, G. (1984) *Surface Sci.* **145**, 419. Scattering of atomic beams off rough stepped surfaces.

Boistelle, R. (1985) in *Actualités néphrologiques 1985*, J. Crosnier, J.L. Funck-Brentano, J.F. Bach, J.P. Grunfeld eds., p. 159-202 (Flammarion Médecine Sciences). Concepts de la cristallisation en solution.

Boisvert, G., Lewis, L.J., Puska, M.J., Nieminen, R.M. (1995) *Phys. Rev. B* **52**, 9078. Energetics of diffusion on the (100) and (111) surfaces of Ag, Au, and Ir from first principles.

Bonzel, H.P. (1975) in *Surface Physics of Materials* Vol. 2, J.M. Blakely ed., p. 279. Transport of matter at surfaces.

Bonzel, H.P. (1990) in *Diffusion in Solid Metals and Alloys*, Landolt-Börnstein, New Series III-26 (Springer-Verlag, Berlin) p. 717. Surface diffusion on metals.

Bonzel, H.P., Mullins, W.W. (1996) *Surface Sci.* **350**, 285. Smoothing of perturbed vicinal surfaces.

Bonzel, H.P., Preuss, E., Steffen B. (1984a) *Appl. Phys. A* **35**, 1. The dynamical behaviour of periodic surface profiles on metals under the influence of anisotropic surface energy.

Bonzel, H.P., Preuss, E., Steffen B. (1984b) *Surface Sci.* **145**, 20. Periodic surface profiles under the influence of anisotropic surface energy: a steady-state solution.

Bostanov, V., Staikov, G., Roe, D.K. (1975) *J. Electrochem. Soc.* **122**, 1301.

Bott, M., Michely, T., Comsa, G. (1992) *Surface Sci.* **272**, 161. The homoepitaxial growth of Pt on Pt(111) studied with STM.

Bourdin, J.P., Ganachaud, J.P., Jardin, J.P., Spanjaard, D., Desjonquères, M.C. (1988) *J. Phys. F* **18**, 1801. Influence of the ledge orientation on the adsorption of transition atoms on stepped BCC transition metal surfaces.

Bowley, R., Caroli, B., Caroli, C., Graner, F., Nozières, P., Roulet, B. (1989) *J. de Physique* **50**, 1377. On directional solidification of a faceted crystal.

Braun, O.M., Medvedev, V.K. (1989) *Uspekhi fizicheskikh nauk* **157**, 631. *Soviet Phys. Uspekhi* **32**, 328. Interaction between particles adsorbed on metal surfaces.

Bray, A.J. (1994) *Adv. Phys.* **43**, 357. Theory of phase ordering kinetics.

Brener, E., Müller-Krumbhaar, H., Temkin, D. (1991) *Europhys. Lett.* **17**, 535. Kinetic phase diagram and scaling relations for stationary diffusional growth.

Brommer, K.D., Needels, M., Larson, B.E., Joannopoulos, J.D. (1992) *Phys. Rev. Lett.* **68**, 1355. Ab initio theory of the Si(111)-(7 × 7) surface reconstruction: a challenge for massively parallel computation.

Bruinsma, R., Karunasiri, R.P.U., Rudnick, J. (1990) in *Kinetics of Ordering and Growth at Surfaces*, M.G. Lagally ed. (Plenum, New York) p. 395. Growth and erosion of thin solid films.

Bruinsma, R., Zangwill, A. (1986) *J. de Physique* **47**, 2055. Structural transitions in epitaxial overlayers.

Brune, H., Romainczyk, C., Röder, H., Kern, K. (1994) *Nature* **369**, 469. Fractal and dendritic growth patterns in two dimensions.

Bucher, J.-P., Hahn, E., Fernandez P. Massobrio, C., Kern, K. (1994), *Europhys. Lett.* **27**, 473. Transition from one- to two-dimensional growth of Cu on Pd(110) promoted by cross-exchange migration.

Burgers, J.M. (1974) *The Nonlinear Diffusion Equation* (Reidel, Boston).

Burton, W.K., Cabrera, N., Frank, F. (1951) *Phil. Trans. Roy. Soc.* **243**, 299. The growth of crystals and the equilibrium structure of their surfaces.

Buschow, K.H.J., Long, G.J., Grandjean, F. eds. (1993) *High density digital recording* (Kluwer, Dordrecht).

Cadoret, R. (1980) in *Current Topics in Materials Science* Vol. 5, E. Kaldis ed., p. 221 (North Holland). Application of the theory of rate processes in the CVD of GaAs.

Cadoret, R., Laporte, J.L., Harrous, M. (1977) *Mat. Res. Bull.* **21**, 1259. Approach of the interplay between kinetics and diffusion in hot wall reactors used in V.P.E. of III-V compounds.

Cahn, J.W. (1977) *J. Chem. Phys.* **66**, 3667. Critical point wetting.

Cahn, J.W. (1982) *J. de Physique Suppl. C6* **43**, 199. Transitions and phase equilibria among grain boundary structures.

Carmi, Y., Lipson, S.G., Polturak, E. (1987) *Phys. Rev. B* **36**, 1894. Critical behavior of vicinal surfaces of ^{4}He.

Caroli, B., Caroli, C., Roulet, B. (1991) in *Solids far from equilibrium*, C. Godreche ed. (Cambridge University Press). Instabilities of planar solidification fronts.

Chadi, D.J. (1987) *Phys. Rev. Lett.* **59**, 1691. Stabilities of single-layer and bilayer steps on Si(001) surfaces.

Chadi, D.J. (1994) *Surface Sci.* **299/300**, 311. Energy-minimization approach to the atomic geometry of semiconductor surfaces.

Chakraverty, B.K. (1967) *J. Phys. Chem. Solids* **28**, 2401. Grain size distribution in thin films.

CHERNOV, A.A. (1984) in *Modern crystallography III-Crystal growth*, p. 1 (Springer Verlag, Berlin).

CHEVRIER, J., LE THANH, V., BUYS, R., DERRIEN, J. (1991) *Europhys. Lett.* **16**, 737. A RHEED study of epitaxial growth of iron on a silicon surface: experimental evidence for kinetic roughening.

CHUI, S.T., WEEKS, J.D. (1976) *Phys. Rev. B* **14**, 4978. Phase transition in the two-dimensional Coulomb gas, and the interfacial roughening transition.

CLARKE, S., VVEDENSKY, D. (1987) *Phys. Rev. Lett.* **58**, 2235. Origin of reflection high-energy electron diffraction intensity oscillations during molecular-beam epitaxy: a computational modeling approach.

COLAS, E., KAPON, E., SIMHONY, S., COX, H.M., BHAT, R., KASH, K., LIN, P.S.D. (1989) *Appl. Phys. Lett.* **55**, 867. Generation of macroscopic steps on patterned (100) vicinal GaAs surfaces.

COMBESCOT, R., BEN AMAR, M. (1991) *Phys. Rev. Lett.* **67**, 453. Selection of Saffman-Taylor fingers.

COMSA, G., DAVID, R. (1985) *Surface Sci. Reports* **5**, 145. Dynamical parameters of desorbing particles.

COPEL, M., REUTER, M.C., KAXIRAS, E., TROMP, R.M. (1989) *Phys. Rev. Lett.* **63**, 632. Surfactants in epitaxial growth.

COPEL, M., REUTER, M.C., HORN VON HOEGEN, M., TROMP, R.M. (1990) *Phys. Rev. B* **42**, 11682. Influence of surfactants in Ge and Si epitaxy on Si(001).

COUSTY, J., PEIX, R., PERRAILLON, B. (1981) *Surface Sci.* **107**, 586. Diffusion superficielle sur le cuivre: influence des marches.

DASH, G. (1989) *Contemporary Physics* **30**, 89. Surface melting.

DAS SARMA, S. (1992) *J. Vac. Sci. Tech. B* **10**, 1695. MBE growth as a self-organized critical phenomenon.

DAS SARMA, S., TAMBORENEA, P.I. (1991) *Phys. Rev. Lett.* **66**, 325. A new universality class for kinetic growth: one-dimensional molecular beam epitaxy.

DAS SARMA, S., TAMBORENEA, P.I. (1992) *Phys. Rev. B* **46**, 1925. Limiting thickness versus epitaxial-growth temperature in molecular-beam epitaxy.

DAVIS, C.I. (1974) *Microphysics of clouds and precipitation* (Dordrecht: Reidel).

DE GENNES, P.G. (1968) *J. Chem. Phys.* **48**, 2257. Soluble model for fibrous structures with steric constraints.

DE GENNES, P.G. (1985) *Rev. Mod. Phys.* **57**, 827. Wetting: statics and dynamics.

DE LORENZI, G., JACUCCI, G., PONTIKIS, V. (1982) *Surface Sci.* **116**, 391. Diffusion of adatoms and vacancies on otherwise perfect surfaces: a molecular dynamics study.

DE MIGUEL, J.J., SANCHEZ, A., CEBOLLADA, A., GALLEGO, J.M., FERRON, J., FERRER, S. (1987) *Surface Sci.* **189/190**, 1062. The surface morphology of a growing crystal studied by TEAS.

DEN NIJS, M. (1990) *Phys. Rev. Lett.* **64**, 435. Preroughening of crystal surfaces and energy differences between inside and outside corners.

DEN NIJS, M. (1991) *Phys. Rev. Lett.* **66**, 907. Roughening-induced deconstruction in (110) facets of fcc crystals.

Den Nijs, M., Rommelse, K. (1989) *Phys. Rev. B* **40**, 4709. Preroughening transitions in crystal surfaces and valence-bond phases in quantum spin chains.

Desjonquères, M.C., Spanjaard, D. (1993), *Concepts in Surface Physics*, Springer Series in Surface Science, Vol. 30 (Springer Verlag).

Diehl, H.W. (1986) in *Phase Transitions and Critical Phenomena* Vol. 10, C. Domb and J. Lebowitz eds. (Acad. Press, London). Field-theoretical approach to critical behaviour at surfaces.

Dietrich, S. (1989) in *Phase Transitions and Critical Phenomena* Vol. 12, C. Domb and J. Lebowitz eds. (Acad. Press, London). Wetting phenomena.

Dinghas, A. (1944) *Z. Krist.* **105**, 304. Über einen geometrischen Satz von Wulff für die Gleichgewichtsform von Kristallen.

Duport, C. (1996) Ph. D. Thesis (Université Joseph Fourier, Grenoble).

Duport, C., Nozières, P., Villain, J. (1995a) *Phys. Rev. Lett.* **74**, 134. New instability in molecular beam epitaxy.

Duport, C., Politi, P., Villain, J. (1995b) *J. de Physique I*, **5**, 1317. Growth instabilities induced by elasticity in a vicinal surface.

Dürr, H., Wendelken, J.F., Zuo, J.-K. (1995), *Surface Sci.* **328**, L527. Island morphology and adatom energy barriers during homoepitaxy on Cu(001).

Eaglesham, D.J., Gossmann, H.J., Cerullo, M. (1990) *Phys. Rev. Lett.* **65**, 1227. Limiting thickness for epitaxial growth and room temperature Si growth on Si(100).

Eaglesham, D.J., Cerullo, M. (1990) *Phys. Rev. Lett.* **64**, 632. Dislocation-free Stranski-Krastanov growth of Ge on Si(100).

Eaglesham, D.J., Gossmann, H.J., Cerullo, M., Pfeiffer, L.N., West, K.W. (1991), *J. Cryst. Growth* **111**, 833. Limited thickness epitaxy of semiconductors and Si MBE down to room temperature.

Eaglesham, D.J., Gilmer, G.H. (1992) in *Surface disordering: growth, roughening and phase transitions*, R. Jullien, J. Kertèsz, P. Meakin and D.E. Wolf eds. (Nova Science, New York). Roughening during Si deposition at low temperature.

Eaglesham, D.J., Unterwald, F.C., Jacobson, D.C. (1993a) *Phys. Rev. Lett.* **70**, 966. Growth morphology and the equilibrium shape: the role of *surfactants* in Ge/Si island formation.

Eaglesham, D.J., Unterwald, F.C., Lufftman, H., Adams, D.P., Yalisove, S.M. (1993b) *J. Appl. Phys.* **74**, 6615. Effect of H on Si molecular-beam epitaxy.

Eaglesham, D.J., White, A.E., Feldman, L.C., Moriya, N., Jacobson, D.C. (1993c) *Phys. Rev. Lett.* **70**, 1643. Equilibrium shape of Si.

Eden, M. (1961) in *Proceedings of the fourth Berkeley Symposium on Mathematical Statistics and probability* Vol. IV, H.P. Neyman ed. (University of California Press, Berkeley) p. 223. A two-dimensional growth process. Reprinted in Family & Vicsek (1991).

Edwards, S.F., Wilkinson, D.R. (1982) *Proc. Roy. Soc. A* **381**, 17. The surface statistics of a granular aggregate.

EHRLICH, G. (1977) *Surface Sci.* **63**, 422, Direct observation of individual atoms on metals.

EHRLICH, G., HUDDA, F.G. (1966) *J. Chem. Phys.* **44**, 1039. Atomic view of surface diffusion: tungsten on tungsten.

ELKINANI, I., VILLAIN, J. (1994) *J. de Physique I* **4**, 949. Growth roughness and instability due to the Schwoebel effect: a one-dimensional model.

ELWENSPOEK, M., VAN DER EERDEN, J.P. (1987) *J. Phys. A* **20**, 669. Kinetic roughening and step free energy in the solid-on-solid model and on naphtalene crystals.

ERNST, H.J., FABRE, F., LAPUJOULADE, J. (1992a) *Phys. Rev. Lett.* **69**, 458. Observation of dynamical scaling in *spinodal decomposition* in two dimensions.

ERNST, H.J., FABRE, F., LAPUJOULADE, J. (1992b) *Phys. Rev. B* **46**, 1929. Nucleation and diffusion of Cu adatoms on Cu(100): a helium-atom-beam scattering study.

ERNST, H.J., FABRE, F., LAPUJOULADE, J. (1993) in *Surface disordering: growth, roughening and phase transitions*, R. Jullien, J. Kertèsz, P. Meakin and D.E. Wolf eds. (Nova Science, New York). Kinetics of growth and ordering of Cu on Cu(100) investigated with helium atom beam scattering.

ERNST, H.J., FABRE, F., FOLKERTS, R., LAPUJOULADE, J. (1994) *Phys. Rev. Lett.* **72**, 112. Observation of a growth instability during low temperature molecular beam epitaxy.

ERNST, H.J., FOLKERTS, R., SCHWENGER, L. (1995) *Phys. Rev. B* **52**, 8461. Thermal roughening of Cu(115): An energy-resolved helium-atom-beam scattering study.

EVANS, J.W. (1991) *Phys. Rev. B* **43**, 3897. Factors mediating smoothness in epitaxial film growth.

EVANS, J.W., BARTELT, M.C. (1993) *Surface Sci.* **284**, L437. Irreversible island formation during deposition: separation distributions and diffraction profiles.

FABRE, F., GORSE, D., SALANON, B., LAPUJOULADE, J. (1986) *Surface Sci.* **175**, L693. Low temperature behaviour of surface roughness of Cu(115).

FABRE, F., GORSE, D., LAPUJOULADE, J., SALANON, B. (1987a) *Europhys. Lett.* **3**, 737. An experimental study of the roughening transition on a stepped surface: Cu(115).

FABRE, F., GORSE, D., SALANON, B., LAPUJOULADE, J. (1987b) *J. de Physique* **48**, 1017. Observation of surface roughening on Cu(115).

FAMILY, F., VICSEK, T. (1985) *J. Phys. A* **18**, 75. Scaling of the active zone in the Eden process on percolation networks and the ballistic deposition model.

FAMILY, F., VICSEK, T. (1991) *Dynamics of fractal surfaces* (World Scientific, Singapore).

FERRANDO, R., SCALAS, E. (1993) *Surface Sci.* **281**, 178. Self-diffusion in a 2D lattice gas with lateral interactions.

FINK, H.W., EHRLICH, G. (1986) *Surface Sci.* **173**, 128. Atom binding at overlayers: Re on W(211).

FOLLSTAEDT, D.M. (1993) *Appl. Phys. Lett.* **62**, 1116. Relative free energies of Si surfaces.

FONTES, E., PATEL, J.R., COMIN, F. (1993) *Phys. Rev. Lett.* **70**, 2790. Direct measurement of the asymmetric dimer buckling of Ge on Si(001).

FORSTER, D., NELSON, D.R., STEPHEN, M.J. (1977) *Phys. Rev. A* **16**, 732. Large-distance and long-time properties of a randomly stirred fluid.

FORREST, B.M., TANG, L.-H. (1990) *Phys. Rev. Lett.* **64**, 1405. Surface roughening in a hypercube-stacking model.

FOXON, C.T., JOYCE, B.A. (1975) *Surface Sci.* **50**, 434 and **64**, 293. Interaction kinetics of As_4 (*resp.* As_2) on (100) GaAs surfaces.

FOURCADE, B., MUTZ, M., BENSIMON, D. (1992) *Phys. Rev. Lett.* **68**, 2551. Experimental and theoretical studies of toroidal vesicles.

FRANK, F.C. (1958) in *Growth and Perfection of Crystals* (John Wiley & sons, New York) p. 411. On the kinematic theory of crystal growth and dissolution processes.

FRANK, F.C., IVES, M.B. (1960) *J. Appl. Phys.* **31**, 1996. Orientation-dependent dissolution of germanium.

FRAXEDAS, J., FERRER, S., COMIN, F. (1994) *Europhys. Lett.* **25**, 119. Incomplete melting of the Si(001) surface: a photoelectron diffraction study.

FRENKEL, J. (1924) *Z. Physik* **26**, 117.

FRENKEN, J.W.M., MARÉE, P.M.J., VAN DER VEEN, J.F. (1986) *Phys. Rev. B* **34**, 7506. Observation of surface-initiated melting.

FRENKEN, J.W.M., TOENNIES, J.P., WÖLL, CH. (1988) *Phys. Rev. Lett.* **60**, 1727. Self-diffusion at a melting surface observed by He scattering.

FREYER, N. (1985) Ph. D. Thesis (Universität Aachen) and Report Jül-2003 of the KFA Jülich.

FRIEDEL, J. (1956) *Les dislocations* (Gauthier-Villars, Paris) and its English translation (1964) (Pergamon, Oxford).

GALILEI, G. (1632) *Dialogo sopra i due massimi sistemi del mondo*, (Florence).

GALLET, F., BALIBAR, S., ROLLEY, E. (1987) *J. de Physique* **48**, 369. The roughening transition of crystal surfaces. II. Experiments on static and dynamic properties near the first roughening transition of hcp ^{4}He.

GEHRING, G.A., KESKIN, M. (1993) *J. Phys: Condens. Matter* **5**, L581. The temperature dependence of the domain spacing in ultrathin magnetic films.

GHAISAS, S.V., DAS SARMA, S. (1992) *Phys. Rev. B* **46**, 7308. Surface diffusion length under kinetic growth conditions.

GIESEN-SEIBERT, M., JENTJENS, R., POENSGEN, M., IBACH, H. (1993) *Phys. Rev. Lett.* **71**, 3521. Time dependence of step fluctuations on vicinal Cu(1,1,19) surfaces investigated by tunneling microscopy.

GIRARD, J.C., GAUTHIER, S., ROUSSET, S., SACKS, W., DE CHEVEIGNE, S., KLEIN, J. (1994) *Surface Sci.* **301**, 245. Dynamics of high index step equilibrium fluctuations as observed by scanning tunneling microscopy.

GLASSTONE, S., LAIDLER, K., EYRING, H. (1941) *The theory of rate processes* (McGraw Hill, New York).

GOLUBOVIČ, L., BRUINSMA, R. (1991) *Phys. Rev. Lett.* **66**, 321. Surface diffusion and fluctuations of growing interfaces.

GOMER, R. (1990) *Rep. Prog. Phys.* **53**, 917. Diffusion on metal surfaces.

Gómez-Rodríguez, J.M., Saénz, J.J., Baró, A.M., Veuillen, J.Y., Cinti, R.C. (1996) *Phys. Rev. Lett.* **76**, 799. Real-time observation of the dynamics of single Pb atoms on Si(111)-(7 × 7) by scanning tunneling microscopy.

Gossmann, H.J., Sinden, F.W., Feldman, L.C. (1990a) *J. Appl. Phys.* **67**, 745. Evolution of terrace size distributions during thin-film growth by step mediated epitaxy.

Gossmann, H.J., Sinden, F.W., Feldman, L.C. (1990b) *J. Vac. Sci. Technol. A* **8**, 5316. Growth of multidimensional superlattices using step array templates: evolution of the terrace size distribution.

Gradshteyn, I.S., Ryzhik, I.M. (1980), *Table of integrals, series, and products* (Academic Press).

Grinfeld, M. Ya. (1986) *Dokl Akad. Nauk SSSR* **290**, 1358; *Sov. Phys. Dokl.* **31**, 831. Instability of the separation boundary between a nonhydrostatically stressed elastic body and a melt.

Grinfeld, M. Ya. (1993) *J. Intell. Mat. Syst. and Struc.* **4**, 76. The stress driven *rearrangement* instability in crystalline films.

Gross, E.K.U. (1990) *Density functional theory.* Springer Verlag, Heidelberg.

Gruber, E.E., Mullins, W.W. (1967) *J. Phys. Chem. Solids* **28**, 875. On the theory of anisotropy of crystalline surface tension.

Haldane, F. D. M., Villain J. (1981) *J. de Physique* **42**, 1673. Commensurate-incommensurate transitions of physisorbed monolayers on anisotropic substrates.

Halpin-Healy, T. (1989) *Phys. Rev. Lett.* **62**, 442. Diverse manifolds in random media.

Hansen, L., Stoltze, P., Jacobsen, K.W., Nørskov, J.K. (1991) *Phys. Rev. B* **44**, 6523. Self-diffusion on copper surfaces.

Hardy, J.R., Bullough, R. (1966) *Phil. Mag.* **237**. Point defect interactions in harmonic cubic lattices.

Harris, J.J., Joyce, B.A., Dobson, P.J. (1981a), *Surface Sci.* **103**, L90. Oscillations in the surface structure of Sn-doped GaAs during growth by MBE.

Harris, J.J., Joyce, B.A., Dobson, P.J. (1981b), *Surface Sci.* **108**, L444. Comments on *RED intensity oscillations during MBE growth of* GaAs.

Harris, S. (1993) *Phys. Rev. B* **48**, 8286. Step-motion-imposed asymmetry during molecular-beam epitaxy on vicinal surfaces.

Hegeman, P.E., Zandvliet, H.J.W., Kip, G.A.M., van Silfhout, A. (1994) *Surface Sci.* **311**, L655. Kinetic roughening of vicinal Si(001).

Held, G.A., Jordan-Sweet, J.L., Horn, P.M., Mak, A., Birgeneau, R.J. (1987) *Phys. Rev. Lett.* **59**, 2075. X-ray scattering study of the thermal roughening of Ag(110).

Hentschel, H.G.E., Family, F. (1991) *Phys. Rev. Lett.* **66**, 1982. Scaling in open dissipative systems.

Henyey, F.S., Seshadri, V. (1982) *J. Chem. Phys.* **76**, 5530. On the number of distinct sites visited in 2D lattices.

Herring, C. (1951) *Phys. Rev.* **82**, 87. Some theorems on the free energy of crystal surfaces.

HERRING, C. (1953) in *Structure and properties of solid surfaces*, R. Gomer and C.S. Smith eds. (University of Chicago Press, Chicago) pp. 5-81. The use of classical macroscopic concepts in surface energy problems.

HEYRAUD, J.C., MÉTOIS, J.J. (1986) *Surface Sci.* **177**, 213. Surface free energy anisotropy measurement of indium.

HEYRAUD, J.C., MÉTOIS, J.J. (1987) *J. Cryst. Growth* **82**, 269. Growth shapes of metallic crystals and roughening transition.

HIRTH, J.P., LOTHE, J. (1982) *Theory of Dislocations* (Mc Graw Hill).

HOBBS, P.V. (1974) *Ice physics* (Clarendon Press, Oxford).

HU, S.M. (1979) *J. Appl. Phys.* **50**, 4661. Film-edge-induced stress in substrates.

HUNT, A.W., ORME, C., WILLIAMS, D.R.M., ORR, B.G., SANDER, L.M. (1994) *Europhys. Lett.* **27**, 611. Instabilities in MBE growth.

HWA, T., KARDAR, M., PACZUSKI M. (1991) *Phys. Rev. Lett.* **66**, 441. Growth-induced roughening of crystalline facets.

HWANG, R.Q., SCHRÖDER, J. GÜNTHER, C., BEHM, R.J. (1991) *Phys. Rev. Lett.* **67**, 3279. Fractal growth of two-dimensional islands: Au on Ru(0001).

HWANG, R.Q., BEHM, R.J. (1992) *J. Vac. Sci. Technol. B* **10**, 256. Scanning tunneling microscopy studies on the growth and structure of thin metallic films on metal substrates.

ICHE, G., NOZIÈRES, P. (1976) *J. de Physique* **37**, 1313. A simple stochastic description of desorption rates.

IRISAWA, T., ARIMA, Y., KURODA, T. (1990) *J. Cryst. Growth* **99**, 491. Periodic changes in the structure of a surface growing under MBE conditions.

ITZYKSON, C., DROUFFE, J.-M. (1989) *Théorie statistique des champs* Vol.1 (InterEditions/Editions du CNRS, Paris).

IVANTSOV, G.P. (1947) *Dokl. Akad. Nauk SSSR* **58**, 567. Temperature field around a spherical, cylindrical and acicular crystal growing in a supercooled melt. English translation in Pelcé (1988).

IVES, M.B. (1961) *J. Appl. Phys.* **32**, 1534. Orientation-dependent dissolution of lithium fluoride.

JACCODINE, R.J. (1962) *J. Appl. Phys.* **33**, 2643. Use of modified free energy theorems to predict equilibrium growing and etching shapes.

JACOBSEN, K.W., NØRSKOV, J.K., PUSKA, M.J. (1987) *Phys. Rev. B* **35**, 7423. Interatomic interactions in the effective-medium theory.

JACOBSEN, K.W. (1988) *Comments. Cond. Matt. Phys.* **14**, 129. Bonding in metallic systems: an effective-medium approach.

JAMGOTCHIAN, H., TRIVEDI, R., BILLIA, B. (1993a) *Phys. Rev. E* **47**, 4313. Array of doublets: A branch of cellular solutions in directional solidification.

JAMGOTCHIAN, H., TRIVEDI, R., BILLIA, B. (1993b) *J. Cryst. Growth* **134**, 181. Interface dynamics and coupled growth in directional solidification in presence of bubbles.

JAYAPRAKASH, C., SAAM, W.F., TEITEL, S. (1983) *Phys. Rev. Lett.* **50**, 2017. Roughening and facets formation in crystals.

JENSEN, P., BARABÁSI, A.-L., LARRALDE, H., HAVLIN, S., STANLEY, H.E. (1994) *Phys. Rev. E* **50**, 618. Model incorporating deposition, diffusion, and aggregation in submonolayer nanostructures.

JENSEN, P., LARRALDE, H., PIMPINELLI, A. (1997) *Phys. Rev. B* **55**, 2556. Effect of monomer evaporation on a simple model of submonolayer growth.

JESSER, W,A., VAN DER MERWE, J.H. (1988) *J. Appl. Phys.* **63**, 1509 and 1928. An exactly solvable model for calculating misfit and thickness in epitaxial superlattices. See also *J. Appl. Phys.* **64**, 4968 (1988)

JOHNSON, H.T., FREUND, L.B. (1997) Brown University preprint. The mechanics of coherent and dislocated island morphologies in strained epitaxial material systems. See also *J. Appl. Phys.* **64**, 4968

JOHNSON, M.D., SUDIJONO, J., HUNT, A.W., ORR, B.G. (1993) *Surface Sci.* **298**, 392. The dynamical transition to step-flow growth during molecular-beam epitaxy of GaAs(001).

JOHNSON, M.D., ORME, C., HUNT, A.W., GRAFF, D., SUDIJONO, J. SANDER, L.M., ORR, B.G. (1994) *Phys. Rev. Lett.* **72**, 116. Stable and unstable growth in molecular beam epitaxy.

JOHNSON, R.A. (1991) *Phil. Mag. A* **63**, 865. Many-body effects on calculated defect properties in hcp metals.

JULLIEN, R., KOLB, M. BOTET, R. (1984) *J. de Physique* **45**, 395. Diffusion limited aggregation with directed and anisotropic diffusion.

JULLIEN, R., BOTET, R. (1985) *J. Phys. A* **18**, 2279. Scaling properties of the surface of the Eden model in $d = 2, 3, 4$.

KANDEL, D., WEEKS, J. D. (1992) *Phys. Rev. Lett.* **69**, 3758. Step bunching as a chaotic pattern formation process.

KANDEL, D., WEEKS, J. D. (1995) *Phys. Rev. Lett.* **74**, 3632. Simultaneous bunching and debunching of surface steps: theory and relation to experiments.

KARDAR, M., PARISI, G., ZHANG, Y.C. (1986) *Phys. Rev. Lett.* **56**, 889. Dynamic scaling of growing interfaces.

KARIOTIS, R. (1989) *J. Phys. A* **22**, 2781. Non-equilibrium roughening transitions.

KAXIRAS, E., ALERHAND, O.L., JOANNOPOULOS, J.D., TURNER, G.W. (1989) *Phys. Rev. Lett.* **62**, 2484. Microscopic model of heteroepitaxy of GaAs on Si(100).

KAWAMURA, T., WILBY, M.R. (1993) *Surface Sci.* **283**, 360. Transient growth in molecular beam epitaxy of Si on Si(100) vicinal surfaces.

KEHR, K.W., ARGYRAKIS, P. (1986) *J. Chem. Phys.* **84**, 5816. Mean number of distinct sites visited by correlated walks.

KELLER, K.W. (1975) in *Crystal Growth and Characterization*, R. Ueda and J.B. Mullin eds. (North Holland). Surface microstructures and processes of crystal growth observed by electron microscopy.

KEPLER, J. (1611) *The six-cornered snowflakes*, english translation by C. Hardie (Oxford University Press 1966).

KERN, K. (1991) in *Phase transitions in Surface Films 2*, H. Taub et al. eds., Plenum (New York) p. 269. Thermal dynamics of vicinal metal surfaces.

KERN, R., MÉTOIS, J.J., MASSON, A. (1971) *Surface Sci.* **27**, 483. Migration brownienne de cristallites sur une surface et relation avec l'épitaxie.

KERN, R. (1987) in *Morphology of Crystals* I. Sunagawa ed. (Terra Scientific Publishing Company, Tokyo) pp. 77-206. The equilibrium form of a crystal.

KESSLER, D.A., LEVINE, H., SANDER, L.M. (1992) *Phys. Rev. Lett.* **69**, 100. Molecular-beam epitaxial growth and surface diffusion.

KIM, J.M., KOSTERLITZ, J.M. (1989) *Phys. Rev. Lett.* **62**, 2289. Growth in a restricted solid-on-solid model.

KIM, J.M., ALA-NISSILA, T., KOSTERLITZ, J.M. (1990) *Phys. Rev. Lett.* **64**, 2333. Comment on *Phase transition in a restricted solid-on-solid surface-growth model in* $2+1$ *dimensions.*

KNOPS, H.J.F. (1977) *Phys. Rev. Lett.* **39**, 766. Exact relation between the solid-on-solid model and the XY model.

KOHN, W., LAU, K. H. (1976) *Solid State Comm.* **18**, 553. Adatom dipole moments on metals and their interactions.

KOJIMA, T., KAWAI, N.J., NAKAGAWA, T., OHTA, K., SAKAMOTO, T., KAWASHIMA, M. (1985) *Appl. Phys. Lett.* **47**, 286. Layer-by-layer sublimation observed by reflection high-energy electron diffraction intensity oscillations in a molecular beam epitaxy system.

KOSTERLITZ, J.M., THOULESS, D.J. (1973) *J. Phys. C* **6**, 1181. Metastability and phase transitions in two-dimensional systems.

KOSTERLITZ, J.M. (1974) *J. Phys. C* **7**, 1046. The critical properties of the two-dimensional xy model.

KOTRLA, M., LEVI, A.C., ŠMILAUER, P. (1992) *Europhys. Lett.* **20**, 25. Roughness and nonlinearities in $2+1$-dimensional growth models with diffusion.

KRUG, J. (1987) *Phys. Rev. A* **36**, 5465. Scaling relation for a growing interface.

KRUG, J., PLISCHKE, M., SIEGERT, M. (1993) *Phys. Rev. Lett.* **70**, 3271. Surface diffusion currents and the universality classes of growth.

KRUG, J., SPOHN, H. (1991) in *Solids far from equilibrium*, C. Godreche ed. (Cambridge University Press). Kinetic roughening of growing surfaces.

KRUG, J., SCHIMSCHAK, M. (1995) *J. de Physique I* **5**, 1065. Metastability of step flow growth in $1+1$ dimension.

KUIPERS, L. (1994) Ph. D. Thesis (Amsterdam). *Surface dynamics studied with a high temperature high-speed scanning microscope.*

KUIPERS, L., HOOGEMAN, M.S., FRENKEN, J.W.M. (1993) *Phys. Rev. Lett.* **71**, 3517. Step dynamics on Au(110) studied with a high-temperature, high speed scanning tunneling microscope.

KUNKEL, R., POELSEMA, B., VERHEIJ, L.K., COMSA, G. (1990) *Phys. Rev. Lett.* **65**, 733. Reentrant layer by layer growth during molecular-beam epitaxy of metal-on-metal surfaces.

LAGALLY, M., MO, Y.-W., KARIOTIS, R., SWARTZENTRUBER, B.S., WEBB, M.B. (1990) in *Kinetics of Ordering and Growth at Surfaces*, M.G. Lagally ed. (Plenum, New York) p. 145. Microscopic aspects of the initial stages of epitaxial growth: a scanning tunneling microscopy study of Si on Si(001).

LAI, Z.W., DAS SARMA, S. (1991) *Phys. Rev. Lett.* **66**, 2348. Kinetic growth with surface relaxation: continuum versus atomistic models.

LAMELAS, F.J. (1992) *Appl. Phys. Lett.* **60**, 2610. Gallium arsenide surface reconstruction during organometallic vapor-phase epitaxy.

LANÇON, F., VILLAIN, J. (1990) in *Kinetics of Ordering and Growth at Surfaces*, M.G. Lagally ed. (Plenum, New York) p. 369. Dynamics of a crystal surface below its roughening transition.

LANDAU, L., LIFSHITZ, E. (1959a) *Statistical Physics*, Pergamon Press, London.

LANDAU, L., LIFSHITZ, E. (1959b) *Theory of Elasticity*, Pergamon Press, London.

LANGER, J. (1980) *Rev. Mod. Phys.* **52**, 1. Instabilities and pattern formation in crystal growth.

LANNOO, M., ALLAN, G. (1978) *Solid State Comm.* **28**, 715. Band effects and the dipole interaction between two adatoms on a metal surface.

LANNOO, M., FRIEDEL, P. (1991) *Atomic and electronic structure of surfaces*, Springer series in surface sciences Vol. 16 (Springer Verlag, Heidelberg).

LAPUJOULADE, J. (1990) in *Interaction of atoms and molecules with solid surfaces*, V. Bortolani, N.H. March, M.P. Tosi eds. (Plenum) p. 381. Defect structures at surfaces.

LAPUJOULADE, J. (1994) *Surf. Sci. Rep.* **20**, 191. The roughening of metal surfaces.

LAPUJOULADE, J., SALANON, B., FABRE, F., LOISEL, B. (1990) in *Kinetics of Ordering and Growth at Surfaces*, M. Lagally ed. (Plenum, New York) p. 355. The roughening transition at vicinal surfaces.

LATYSHEV, A.V., ASEEV, A.L., KRASILNIKOV, A.B., STENIN, S.I. (1989) *Surface Sci.* **213**, 157. Transformations on clean Si(111) stepped surface during sublimation.

LATYSHEV, A.V., ASEEV, A.L., KRASILNIKOV, A.B., STENIN, S.I. (1990) *Surface Sci.* **227**, 24. Reflection electron microscopy study of structural transitions on a clean silicon surface in sublimation phase transition and homoepitaxy.

LATYSHEV, A.V., KRASILNIKOV, A.B., ASEEV, A.L. (1993) *Ultramicroscopy* **48**, 377. In-situ REM study of monatomic step behaviour on Si(111) surface during sublimation.

LAU, K.H., KOHN, W. (1977) *Surface Sci.* **65**, 607. Elastic interaction of two atoms adsorbed on a solid surface.

LAU, K.H., KOHN, W. (1978) *Surface Sci.* **75**, 69. Indirect long range interaction between adsorbed atoms.

LE BELLAC, M. (1988) *Des phénomènes critiques aux champs de jauge* (InterEditions/Editions du CNRS, Paris).

LEE, J.K., AARONSON, H.I., RUSSELL K.C. (1975) *Surface Sci.* **51**, 302. On the equilibrium shape for a non-centro-symmetric γ-plot.

LEONARD, D., KRISHNAMURTHY, M., REAVES, C.M., DENBAARS, S.P., PETROFF, P.M. (1993) *Appl. Phys. Lett.* **63**, 3203. Direct formation of quantum dots from uniform coherent islands of InGaAs on GaAs surfaces.

LEVI, A.C. (1984) *Surface Sci.* **137**, 385. Scattering of atoms from a rough surface: non-markovian effects.

LEVI, A.C., KOTRLA, M. (1997) *J. Phys.: Condens. Matter* **9** 299. Theory and simulation of crystal growth.

Lifshitz, I.M., Slyozov, V.V. (1961) *J. Phys. Chem. Solids* **19**, 35. The kinetics of precipitation from supersaturated solid solutions.

Ling, C.-B. (1948) *J. of Mathematics and Physics* **26**, 284. On the stresses in a notched plane under tension.

Liu, C.I., Cohen, J.M., Adams, J.B., Voter, A.F. (1991) *Surface Sci.* **253**, 334. EAM study of surface self-diffusion of single adatoms of fcc metals Ni, Cu, Al, Ag, Au, Pd and Pt.

Liu. F., Metiu, H. (1993) *Phys. Rev. B* **48**, 5808. Dynamics of phase separation of crystal surfaces.

Liu, X.Y., Bennema, P. (1994) *Phys. Rev. B* **49**, 765. Morphology of crystals: internal and external controlling factors.

Lodder, A. (1989) *Physica A* **158**, 723. The driving force in electromigration.

Ma, S.K. (1976) *Modern theory of critical phenomena* (Benjamin, Reading, Massachusetts).

Madhukar, A. (1983) *Surface Sci.* **132**, 344. Far from equilibrium vapour phase growth of lattice matched III-V compound semi-conductor interfaces: some basic concepts and Monte-Carlo computer simulations.

Madhukar, A., Ghaisas, S.V. (1988) *CRC Crit. Rev. Sol. State and Matter Sci.* **14**, 1.

Mak, A., Evans-Lutterodt, K.W., Blum, K., Noh, D.Y., Brock, J.D., Held, G.A., Birgeneau, R.J. (1991) *Phys. Rev. Lett.* **68**, 2002. Synchrotron X-ray diffraction study of the disordering of the Ge(111) surface at high temperature.

Marchenko, V.I. (1981) *Zh. Eksp. Teor. Fiz.* **81**, 1141. *Sov. Phys. J.E.T.P.* **54**, 605. Theory of the equilibrium shape of crystals.

Marchenko, V.I., Parshin, A. Ya. (1980) *Zh. Eksp. Teor. Fiz.* **79**, 257. *Sov. Phys. J.E.T.P.* **52**, 129. Elastic properties of crystal surfaces.

Mazenko, G., Banavar, J.R., Gomer, R. (1981) *Surface Sci.* **107**, 459. Diffusion coefficients and the time auto-correlation function of density fluctuations.

Mason, B.J. (1992) *Contemporary Physics* **72**, 227. Snow crystals, natural and man made.

Masson, A., Kern, R., Métois, J.J. (1971) *Surface Sci.* **27**, 463 and 483. Migration brownienne de cristallites sur une surface et relation avec l'épitaxie.

Meakin, P., Ramanlal, P., Sander, L., Ball, R.C. (1986) *Phys. Rev. A* **34**, 5091. *Ballistic deposition on surfaces.*

Meakin, P., Ramanlal, P., Sander, L. (1987) *Phys. Rev. A* **35**, 5233. Structure of large two-dimensional square-lattice diffusion-limited aggregates: approach to asymptotic behavior.

Medina, E., Hwa, T., Kardar, M., Zhang, Y.C. (1989) *Phys. Rev. A* **39**, 3053. Burgers equation with correlated noise: renormalization-group analysis and applications to directed polymers and interface growth.

Men, F.K., Packard, W.E., Webb, M.B. (1988) *Bull. Am. Phys. Soc.* **33**, 472. Strain effects on Si(100) surfaces.

Merikoski, J., Ala-Nissila, T. (1995) *Phys. Rev. B* **52**, R8715. Diffusion processes and growth on stepped metal surfaces.

MÉTOIS, J.J. HEYRAUD, J.C. (1987) *Surface Sci.* **180**, 647. Analysis of the critical behaviour of curved regions in equilibrium shapes of In crystals.

MÉTOIS, J.J. WOLF, D.E. (1993) *Surface Sci.* **298**, 71. Kinetic surface roughening of Si(001) during sublimation.

MEYER, J.A., VRIJMOETH, J., VAN DER VEGT, H.A., VLIEG, E., BEHM, R.J. (1995) *Phys. Rev. B* **51**, 14790. Importance of the additional step-edge barrier in determining film morphology during epitaxial growth.

MIAO, L., FOURCADE, B., RAO, M., WORTIS, M. (1991) *Phys. Rev. A* **43**, 6843. Equilibrium budding and vesiculation in the curvature model of fluid vesicles.

MICHELY, T., HOHAGE, M., BOTT, M., COMSA, G. (1993) *Phys. Rev. Lett.* **70**, 3943. Inversion of growth speed anisotropy in two dimensions.

MIKHEEV, L.V., POKROVSKY, V.L. (1991) *J. de Physique I*, **1**, 373. Free-fermion solution for overall equilibrium crystal shape.

MISBAH, C., RAPPEL, W.J. (1993) *Phys. Rev. B* **48**, 12193. Cellular structures in step flow growth.

MISBAH, C., PIERRE-LOUIS, O., PIMPINELLI, A. (1995) *Phys. Rev. B* **51**, 17283. Advacancy-induced step bunching on vicinal surfaces.

MO, Y.-W., KLEINER, J., WEBB, M.B., LAGALLY, M.G. (1991) *Phys. Rev. Lett.* **66**, 1998, Activation energy for surface diffusion of Si on Si(001): a scanning-tunnelling-microscopy study.

MO, Y.-W., KLEINER, J., WEBB, M.B., LAGALLY, M.G. (1992) *Surface Sci.* **268**, 275, Surface self diffusion of Si on Si(001).

MOISON, J.M., HOUZAY, F., BARTHE, F., LEPRINCE, L., ANDRÉ, E., VATEL, O. (1994) *Appl. Phys. Lett.* **64** 196. Self-organized growth of regular nanometer-scale InAs dots on GaAs.

MONTROLL, E.W., WEISS, G. (1965) *J. Math. Phys.* **6**, 178. Random walks on lattices. II.

MORSE, P.M., FESHBACH, H. (1953) *Methods of theoretical physics*, part II (McGraw Hill, New York).

MOSER, K., WOLF, D.E., KERTESZ, J. (1991) *Physica A* **178**, 215. Numerical solution of the Kardar-Parisi-Zhang equation in one, two and three dimensions.

MUKHERJEE, S., PEHLKE, E., TERSOFF, J. (1994) *Phys. Rev. B* **49**, 1919. Calculation of temperature effects on the equilibrium crystal shape of Si near (001).

MÜLLER-KRUMBHAAR, H. (1978) in *Current Topics in Materials Science* Vol. 1, E. Kaldis ed. (North Holland, Amsterdam, New York) p. 1. Kinetics of crystal growth.

MULLINS, W.W. (1957) *J. Appl. Phys.* **28**, 333. Theory of thermal grooving.

MULLINS, W.W. (1959) *J. Appl. Phys.* **30**, 77. Flattening of a nearly plane solid surface due to capillarity.

MULLINS, W.W. (1961) *Phil. Mag.* **6**, 1313. Theory of linear facet growth during thermal etching.

MULLINS, W.W. (1963) in *Metal Surfaces: Structure, Energetics and Kinetics*, Am. Soc. Metals, Metals Park, Ohio, p. 17. Solid surface morphologies governed by capillarity.

MULLINS, W.W., SEKERKA, R.F. (1963) *J. Appl. Phys.* **34,** 323. Morphological stability of a particle growing by diffusion or heat flow.

MULLINS, W.W., SEKERKA, R.F. (1964) *J. Appl. Phys.* **35**, 444. Stability of a planar interface during solidification of a dilute binary alloy.

MUNDSCHAU, M., BAUER, E., TELIEPS, W., ŚWIĘCH, W. (1989) *Surface Sci.* **223**, 413. Atomic steps on Si(100) and step dynamics during sublimation studied by low energy electron microscopy.

MUTAFTSCHIEV, B. (1988) in *Epitaxial growth of electronic materials*, A. Baldereschi and C. Paorici eds. (World Scientific, Singapore) p. 1. Surface thermodynamics, nucleation and crystal growth of thin films.

MUTAFTSCHIEV, B. (1990) in *Kinetics of Ordering and Growth at Surfaces*, M.G. Lagally ed. (Plenum, New York) p. 169. Surface kinetics and growth morphologies: theoretical models and physical realizations.

NATTERMANN, T., TANG, L.-H. (1992) *Phys. Rev. A* **45**, 7156. Kinetic surface roughening: I. The Kardar-Parisi-Zhang equation in the weak coupling regime.

NEAVE, J.H., DOBSON, P.J., JOYCE, B.A., ZHANG, J. (1985) *Appl. Phys. Lett.* **47**, 100. Reflection high-energy electron diffraction oscillations from vicinal surfaces-a new approach to surface diffusion measurements.

NEAVE, J.H., JOYCE, B.A., DOBSON, P.J., NORTON, N. (1983) *Appl. Phys. A* **31**, 1. Dynamics of film growth of GaAs by MBE from RHEED observations.

NISHINAGA, T., CHO, K. (1988) *Jap. J. Appl. Phys.* **27**, L12. Theoretical study of the mode transition between 2-D nucleation and step flow in MBE growth of GaAs.

NOETZEL, R., TEMMYO, J., TAMAMURA, T. (1994) *Nature* **369**, 132. Self-organized growth of strained InGaAs quantum discs.

NOGUERA, C. (1995) *Physique et chimie des surfaces d'oxydes* (Alea-Saclay, Eyrolles, Paris).

NOH, D.Y., BLUM, K.I., RAMSTAD, M.J., BIRGENEAU, R.J. (1993) *Phys. Rev. B* **48**, 1612. Faceting, roughness and step disorder of vicinal Si(111) surfaces: an X-ray scattering study.

NOVACO, A.D., MCTAGUE, J.P. (1977) *Phys. Rev. Lett.* **38**, 100. Orientational epitaxy: the orientational ordering of incommensurate structures.

NOVACO, A.D., MCTAGUE, J.P. (1979) *Phys. Rev. B* **20**, 2469. Effect of monolayer-substrate dynamical coupling: Argon on graphite.

NØRSKOV, J.K., JACOBSEN, K.W., STOLTZE, P., HANSEN, L.B. (1993) *Surface Sci.* **283**, 277. Many-atom interactions in metals.

NOZIÈRES, P., GALLET, F. (1987) *J. de Physique* **48,** 353. The roughening transition of crystal surfaces. I. Static and dynamic renormalization theory, crystal shape and facet growth.

NOZIÈRES, P. (1989) *J. de Physique* **50**, 2541. Surface melting and crystal shape.

NOZIÈRES, P. (1991) in *Solids far from equilibrium*, C. Godrèche ed. (Cambridge University Press). Shape and growth of crystals.

NOZIÈRES, P. (1993) *J. de Physique I* **3**, 681. Amplitude expansion for the Grinfeld instability due to uniaxial stress at a solid surface.

OHTSUKA, M., MIYAZAWA, S. (1988) *J. Appl. Phys.* **64**, 3522. Model for molecular-beam-epitaxy growth over nonplanar surfaces.

ORR, B.G., KESSLER, D., SNYDER, C.W., SANDER, L.M. (1992) *Europhys. Lett.* **19**, 33. A model for strain-induced roughening and coherent island growth.

OSTWALD, W. (1908) *Foundation of Analytical Chemistry*, p. 22 (Mc Millan, London).

OZDEMIR, M., ZANGWILL, A. (1990) *Phys. Rev. B* **42**, 5013. Morphological equilibration of a corrugated crytalline surface.

PARKER, G. (1994) *Semiconductor Device Physics* (Prentice Hall, New York).

PAVLOVSKA, A., NENOW, D. (1977) *J. Cryst. Growth* **39**, 346. Experimental study of the surface melting of tetrabrommethane.

PELCÉ, P. (1988) *Dynamics of curved fronts* (Academic Press, New York).

PELLEGRINI, Y.P., JULLIEN, R. (1990) *Phys. Rev. Lett.* **64**, 1745. Roughening transition and percolation in random ballistic deposition.

PHANEUF, R.J., WILLIAMS, E.D., BARTELT, N.C. (1988) *Phys. Rev. B* **38**, 1984. Temperature dependence of vicinal Si(111) surfaces.

PHILIBERT, J. (1990) *Diffusion et transport de matière dans les solides* (Editions de Physique, Les Ulis). See also the english translation, (1991) (same publisher).

PIDDUCK, A.J., ROBBINS, D.J., CULLIS, A.G., LEONG, W.Y., PITT, A.M. (1992) *Thin Solid Films* **222**, 78. Evolution of surface morphology and strain during SiGe epitaxy.

PIETRONERO, L., ERZAN, A., EVERTSZ, C. (1988) *Physica A* **151**, 207. Theory of laplacian fractals: diffusion limited aggregation and dielectric breakdown model.

PIETRONERO, L., SCHNEIDER, W.R. (1990) *Physica A* **170**, 81. Invasion percolation as a fractal growth problem.

PIETRONERO, L., TOSATTI, E. eds. (1986) *Fractals in Physics* (North Holland).

PIMPINELLI, A., VILLAIN, J., WOLF, D.E. (1992) *Phys. Rev. Lett.* **69**, 985. Surface diffusion and island density.

PIMPINELLI, A., VILLAIN, J., WOLF, D.E. (1993a) *J. de Physique I* **3**, 447. Fractal terraces in M.B.E. growth.

PIMPINELLI, A., VILLAIN, J., WOLF, D.E., MÉTOIS, J.J., HEYRAUD, J.C., ELKINANI, I., UIMIN, G. (1993b) *Surface Sci.* **295**, 143 Equilibrium step dynamics on vicinal surfaces.

PIMPINELLI, A., ELKINANI, I. KARMA, A., MISBAH, C., VILLAIN, J. (1994) *J. Phys.: Condens. Matter* **6**, 2661. Step motions on high temperature vicinal surfaces.

PIMPINELLI, A., MÉTOIS, J.J. (1994) *Phys. Rev. Lett.* **72**, 3566. Macrovacancy nucleation on evaporating Si(001).

PIMPINELLI, A., VILLAIN, J. (1994) *Physica A* **204**, 521. What does an evaporating surface look like?.

PLUIS, B., VAN DER GON, A.W.D., FRENKEN, J.W.M., VAN DER VEEN, J.F. (1987), *Phys. Rev. Lett.* **59**, 2678. Crystal-face dependence of surface melting.

POLITI, P., VILLAIN, J. (1996) *Phys. Rev. B* **54**, 5114. Ehrlich-Schwoebel instability in Molecular-Beam-Epitaxy: a minimal model.

POMEAU, Y., BEN AMAR, M. (1991) in *Solids far from equilibrium*, C. Godreche ed. (Cambridge University Press). Dendritic growth and related topics.

POKROVSKI, V.L., TALAPOV, A.L. (1979) *Phys. Rev. Lett.* **42**, 65. Ground state, spectrum and phase diagram of two-dimensional incommensurate crystals.

PONCHET, A., ROCHER, A., EMERY, J.Y., STARCK, C., GOLDSTEIN, L. (1993) *J. Appl. Phys.* **74**, 3778. Lateral modulation in zero net strained GaInAsP multilayers grown by gas source molecular beam epitaxy.

PRIESTER, C., LANNOO, M. (1995) *Phys. Rev. Lett.* **75**, 93. Origin of self-assembled quantum dots in highly mismatched heteroepitaxy.

PUMIR, A. (1985) *J. de Physique* **46**, 511. Statistical properties of an equation describing fluid interfaces.

RAMANLAL, P., SANDER, L. (1985) *Phys. Rev. Lett.* **54**, 1828. Theory of ballistic aggregation.

RAMSTAD, M.J., BIRGENEAU, R.J., BLUM, K., NOH, D.Y. (1993) *Europhys. Lett.* **21**, 653. D.C. current-dependent faceting of vicinal Si(111).

RATSCH, C., ZANGWILL, A., ŠMILAUER, P., VVEDENSKY, D.D. (1994) *Phys. Rev. Lett.* **72**, 3194. Saturation and scaling of epitaxial island densities.

RETTORI, A., VILLAIN, J. (1988) *J. de Physique* **49**, 257. Flattening of grooves on a crystal surface: a method of investigation of surface roughness.

ROBINSON, I.K., MAC DOWELL, A.A., ALTMAN, M.S., ESTRUP, P.J., EVANS-LUTTERODT, K., BROCK, J.D., BIRGENEAU, R.J. (1989) *Phys. Rev. Lett.* **62**, 1294. Order-disorder transition of the W(001) surface.

ROBINSON, I.K., CONRAD, E.H., REED, D.S. (1990) *J. de Physique* **51**, 103. X-ray study of the roughening transition of the Ni(113) surface.

ROBINSON, I.K., WHITEAKER, K.L., WALKO, D.A. (1996) *Physica B* **221**, 70. Cu island growth on Cu(110).

RODER, H., HAHN, E., BRUNE, H., BUCHER, J.-P. KERN K. (1993) Nature **366**, 141. Building one- and two-dimensional nanostructures by diffusion-controlled aggregation at surfaces.

ROLAND, C., GILMER, G.H. (1992a) *Phys. Rev. B* **46**, 13428. Epitaxy on surfaces vicinal to Si(001) I. Diffusion of silicon adatoms over the terraces.

ROLAND, C., GILMER, G.H. (1992b) *Phys. Rev. B* **46**, 13437. Epitaxy on surfaces vicinal to Si(001). II. Growth properties of Si(001) steps.

ROLLEY, E., GUTHMANN, C., CHEVALIER, E., BALIBAR, S. (1995) *J. of Low Temp. Phys.* **99**, 85. The static and dynamic properties of vicinal surfaces on helium 4 crystals.

ROSENFELD, G., SERVATY, R., TEICHERT, C., POELSEMA, B., COMSA, G. (1993) *Phys. Rev. Lett.* **71**, 895. Layer-by-layer growth of Ag on Ag(111) induced by enhanced nucleation: a model study for surfactant-mediated growth.

ROST, M., SPOHN, H. (1994) *Phys. Rev. B* **49**, 3709. Renormalization of the driven sine-Gordon equation in $2+1$ dimensions.

ROST, M., ŠMILAUER, P., KRUG. J. (1996) *Surface Sci.* **369**, 393. Unstable epitaxy on vicinal surfaces.

RUGGERONE, P., RATSCH, C. SCHEFFLER, M. (1997) in *Growth and properties of ultrathin epitaxial layers. The chemical physics of solid surfaces*, Vol. 8,

D.A. King and D.P. Woodruff Eds. (Elsevier Science, Amsterdam). Density functional theory of epitaxial growth of metals.

RYS, F. (1986) *Phys. Rev. Lett.* **56**, 624. Critical and noncritical roughening of surfaces.

SAFFMAN, P.G., TAYLOR, G.I. (1958) *Proc. Roy. Soc. London A* **245**, 312. The penetration of a fluid into a porous medium or Heele-Shaw cell containing a more viscous liquid.

SAKAMOTO, Y., KANAMORI, J. (1993) *J. Phys. Soc. Jpn.* **62**, 563. Lattice gas model of Si(111).

SALDITT, T., SPOHN, H. (1993) *Phys. Rev. E* **47**, 3524. Kinetic roughening of a terrace ledge.

SCHROEDER, M., SIEGERT, M., WOLF, D.E., SHORE, J.D., PLISCHKE, M. (1993) *Europhys. Lett.* **24**, 563. Scaling of growing surfaces with large local slopes.

SCHROEDER, M., WOLF, D.E. (1995) *Phys. Rev. Lett.* **74**, 2062. Magic islands and submonolayer scaling in molecular beam epitaxy.

SCHWOEBEL, R.L. (1968) *J. Appl. Phys.* **40**, 614. Step motion on crystal surfaces II.

SEIFERT, U., BERNDL, K., LIPOWSKY, R. (1991) *Phys. Rev. A* **44**, 1182. Shape transformations of vesicles: phase diagram for spontaneous curvature and bilayer-coupling models.

SELKE, W., BICKER, T. (1993) *Surface Sci.* **281**, 163. Morphological changes of periodic surface profiles.

SELKE, W., DUXBURY, P.M. (1994) *Acta Physica Slovaca* **226**, 215. Surface profile evolution above and below the roughening transition.

SHAW, D.W. (1979) *J. Cryst. Growth* **47**, 509. *Morphology analysis in localized crystal growth and dissolution.*

SHCHUKIN, V.A., LEDENTSOV, N.N., KOP'EV, P.S., BIMBERG, D. (1995) *Phys. Rev. Lett.* **75**, 2968. Spontaneous ordering of arrays of coherent strained islands.

SHIFLET, G.J., VAN DER MERWE, J. (1991) *Surface Sci.* **256**, 187. The elastostatics of atomic steps on crystal surfaces 1.

SHITARA, T., VVEDENSKY, D.D., WILBY, M.R., ZHANG, J., NEAVE, J.H., JOYCE, B.A. (1992a) *Phys. Rev. B* **46**, 6825. Misorientation dependence of epitaxial growth on vicinal GaAs(001).

SHITARA, T., ZHANG, J., NEAVE, J.H., JOYCE, B.A. (1992b) *J. Appl. Phys.* **71**, 4229. Ga adatom incorporation kinetics at steps on vicinal GaAs(001) surfaces during growth.

SHLESINGER, M.F. (1974) *J. Stat. Phys.* **10**, 421. Asymptotic solutions of continuous-time random walks.

SHOLL, D.S., SKODJE, R.T. (1995) *Phys. Rev. Lett.* **75**, 3158. Diffusion of clusters of atoms and vacancies on surfaces and the dynamics of diffusion-driven coarsening.

SIEGERT, M., PLISCHKE, M. (1992) *Phys. Rev. Lett.* **68**, 2035. Instability in surface growth with diffusion.

SIEGERT, M., PLISCHKE, M. (1994) *Phys. Rev. Lett.* **73**, 1517. Slope selection and coarsening in molecular beam epitaxy.

SIEMS, R. (1968) *Phys. Stat. Sol.* **30**, 645. Mechanical interaction of point defects.

ŠMILAUER, P., HARRIS, S. (1995) *Phys. Rev. B* **51**, 14798. Determination of step-edge barriers to interlayer transport from surface morphology during the initial stages of homoepitaxial growth.

ŠMILAUER, P., VVEDENSKY, D.D. (1993) *Phys. Rev. B* **48**, 17603. Step edge barriers on GaAs(001).

ŠMILAUER, P., VVEDENSKY, D.D. (1995) *Phys. Rev. B* **52**, 14263. Coarsening and slope evolution during unstable epitaxial growth.

ŠMILAUER, P., WILBY, M.R., VVEDENSKY, D.D. (1993a) *Phys. Rev. B* **47**, 4119. Reentrant layer-by-layer growth: a numerical study.

ŠMILAUER, P., WILBY, M.R., VVEDENSKY, D.D. (1993b) *Surface. Sci. Lett.* **291**, L733. Morphology of singular and vicinal metal surfaces sputtered at different temperatures.

SNYDER, C.W., SUDIJONO, J. LAM, C. H., JOHNSON, M. D., ORR, B.G. (1994) *Phys. Rev. B* **50**, 18194. Surface transformation on annealed GaAs(001).

SONG, S., MOCHRIE, S.G.J. (1994) *Phys. Rev. Lett.* **73**, 995. Tricriticality in the orientational phase diagram of stepped Si(113) surfaces.

SOUCHIÈRE, J.L., VU THIEN BINH (1986) *Surface Sci.* **168**, 52. On the evaporation rate of silicon.

SOMMERFELD, A. (1964) *Partial differential equations in physics*, Lectures in Theor. Phys. vol. VI (Academic Press, New York).

SPENCER, D.S., VORHEES, P.W., DAVIS, S.H. (1991) *Phys. Rev. Lett.* **67**, 3696. Morphological instability in epitaxially strained dislocations-free solid films.

SPOHN, H. (1993) *J. de Physique I* **3**, 69. Surface dynamics below the roughening transition.

SROLOVITZ, D.J. (1989) *Acta Metall.* **17**, 621. On the stability of surfaces of stressed solids.

STEIGERWALD, D.A., JACOB, I., EGELHOFF, JR., W.F. (1988) *Surface Sci.* **202**, 472. Structural study of the epitaxial growth of fcc-Fe films, sandwiches, and superlattices on Cu(100).

STEWART, J., GOLDENFELD, N. (1992) *Phys. Rev. A* **46**, 6505. Spinodal decomposition of a crystal surface.

STEYER, A., GUENOUN, P., BEYSENS, D. (1992) *Phys. Rev. Lett.* **68**, 1869. Spontaneous jumps of a droplet.

ŠTICH, I., PAYNE, M.C., KING-SMITH, R.D., LIN, J.-S., CLARKE, L.J. (1992) *Phys. Rev. Lett.* **68**, 1351. Ab initio total-energy calculations for extremely large systems: application to the Takayanagi reconstruction of Si(111).

STOLTZE, P. (1994) *J. Phys.: Condens. Matter* **6**, 9495. Simulation of surface defects.

STOYANOV, S. (1988) *Surface Sci.* **199**, 226. Layer growth of epitaxial films and superlattices.

STOYANOV, S. (1990) *Appl. Phys. A* **50**, 349. On the interpretation of the RHEED intensity oscillations during the growth of vicinal faces.

STOYANOV, S. (1991) *Jap. J. Appl. Phys.* **30**, 1. Electromigration induced step bunching on Si surfaces–How does it depend on the temperature and heating current direction?

STOYANOV, S., MICHAILOV, M. (1988) *Surface Sci.* **202**, 109. Non-steady state effects in MBE: oscillations of the step density at the crystal surface.

STOYANOV, S., KASHCHIEV, D. (1981) in *Current Topics in Materials Science* Vol. 7, E. Kaldis ed. (North Holland), p. 69. Thin film nucleation and growth theories: a confrontation with experiment.

STROSCIO, J.A., PIERCE, D.T., DRAGOSET, R.A. (1993) *Phys. Rev. Lett.* **70**, 3615. Homoepitaxial growth of iron and a real space view of reflection-high-energy-electron diffraction.

STROSCIO, J.A., PIERCE, D.T. (1994) *Phys. Rev. B* **49**, 8522. Scaling of diffusion-mediated island growth in iron-on-iron homoepitaxy.

STUMPF, R., SCHEFFLER, M. (1994) *Phys. Rev. Lett.* **72**, 254. Theory of self-diffusion and growth of Al(111).

SUDIJONO, J., JOHNSON, M.D., SNYDER, C.W., ELOWITZ, M.B., ORR, B.G. (1992) *Phys. Rev. Lett.* **69**, 2811. Surface evolution during molecular beam epitaxy deposition of GaAs.

SUN, T., GUO, H., GRANT, M. (1989) *Phys. Rev. A* **40**, 6763. Dynamics of driven interfaces with a conservation law.

SUN, T., MORIN, B., GUO, H., GRANT, M. (1992) in *Surface disordering: growth, roughening and phase transitions*, ed. R. Jullien (Nova Science Publishers, Commack NY). Roughening transition of a driven interface with a conservation law.

SURNEV, S., VOIGTLÄNDER, B., H.P. BONZEL, W.W. MULLINS (1996) *Surface Sci.* **350**, 285 (1996) Anisotropic profile decay on perturbed Au(111) vicinal surfaces.

SUZUKI, T., MÉTOIS, J.J., YAGI, K. (1995) *Surface Sci.* **339**, 105. REM observations of the birth stage of the flat facets on an inner cylindrical silicon surface.

SWARTZENTRUBER, B.S., MO, Y.-W., KARIOTIS, R., LAGALLY, M.G., WEBB, M.B. (1990) *Phys. Rev. Lett.* **65**, 1913. Direct determination of step and kink energies on vicinal Si(001).

TAKAYANAGI K., TANISHIRO, Y., TAKAHASHI, S., TAKAHASHI, M. (1985) *Surface Sci.* **164**, 367. Structure analysis of Si(111)-7×7 reconstructed surface by transmission electron diffraction.

TAKAYANAGI K., TANISHIRO, Y., KOBAYASHI, K., AKIYAMA, K., YAGI, K. (1987) *Jpn. J. Appl. Phys.* **26**, L957. Surface structures observed by high-resolution UHV microscopy of atomic layers.

TAKEUCHI, N., SELLONI, A., TOSATTI, E. (1994) *Phys. Rev. Lett.* **72**, 2227. Metallization and incomplete melting of a semiconductor surface at high temperature.

TANG, L.-H. (1993) *J. de Physique I* **3**, 935. Island formation in submonolayer epitaxy.

TANG, L.-H., NATTERMANN, T. (1991) *Phys. Rev. Lett.* **66**, 2899. Kinetic roughening in molecular-beam epitaxy.

Taylor, E.J. (1974) *Symp. Mathem.* **14**, 499. Existence and structure of solutions to a class of nonelliptic variational problems.

Taylor, J., Cahn, J.W., Handwerker, C.A. (1992) *Acta Met. Mater.* **40**, 1443. I – Geometric models of crystal growth.

Taylor, J. (1992) *Acta Met. Mater.* **40**, 1475. II – Mean curvature and weighted mean curvature.

Tersoff, J., Denier van der Gon, A.W., Tromp, R.M. (1994) *Phys. Rev. Lett.* **72**, 266. Critical island size for layer-by-layer growth.

Tersoff, J., Legoues, F.K. (1994) *Phys. Rev. Lett.* **73**, 3570. Competing relaxation mechanisms in strained layers.

Tersoff, J., Phang, Y.H, Zhang, Z., Lagally, M.G. (1995) *Phys. Rev. Lett.* **75**, 2730. Step-bunching instability of vicinal surfaces under stress.

Tersoff, J., Tromp, R.M. (1993) *Phys. Rev. Lett.* **70**, 2782. Shape transition in growth of strained islands: spontaneous formation of quantum wires.

Thiel, M., Willibald, A., Evers, P., Levchenko, A., Leiderer, P., Balibar, S. (1992) *Europhys. Lett.* **20**, 707. Stress-induced melting and surface instability of ^{4}He crystals.

Thompson, M.D., Huntington, H.B. (1982) *Surface Sci.* **116**, 522. Adatom binding at the surface ledges of a jellium metal.

Thomy, A., Duval, X., Regnier, J. (1981) *Surface Sci. Reports* **1**, 1. Two-dimensional phase transitions as displayed by adsorption isotherms on graphite and other lamellar solids.

Tian, Z.J., Rahman, T.S. (1993) *Phys. Rev. B* **45**, 9751. Energetics of stepped Cu surfaces.

Tiller, W. A. (1992) *The science of crystallization: microscopic interfacial phenomena.* Cambridge University Press.

Tromp, R.M., (1993) *Phys. Rev. B* **47**, 7125. Surface stress and interface formation.

Tromp, R.M., Reuter, M.C. (1992a) *Phys. Rev. Lett.* **68**, 954. Local dimer exchange in surfactant-mediated epitaxial growth.

Tromp, R.M., Reuter, M.C. (1992b) *Phys. Rev. Lett.* **68**, 820. Wavy steps on Si(001).

Tromp, R.M., Reuter, M.C. (1993) *Phys. Rev. B* **47**, 7598. Step morphologies on small-miscut Si(001) surfaces.

Tsushiya, M., Petroff, P.M., Coldren, L.A. (1989) *Appl. Phys. Lett.* **54**, 1690. Spontaneous growth of coherent tilted superlattices on vicinal (100) GaAs substrates.

Uwaha, M. (1988) *J. Phys. Soc. Jpn.* **57**, 1681. Relaxation of crystal shapes caused by step motion.

Uwaha, M., Saito, Y. (1992) *Phys. Rev. Lett.* **68**, 225. Kinetic smoothening and roughening of a step with surface diffusion.

van Beijeren, H. (1977) *Phys. Rev. Lett.* **38**, 993. Exactly solvable model for the roughening transition of a crystal surface.

VAN BEIJEREN, H., NOLDEN, I. (1987) in *Structure and Dynamics of Surfaces* II, W. Schommers and P. von Blanckenhagen eds., Topics in Current Physics **43** (Springer, Berlin).

VAN DER EERDEN, J.P., MÜLLER-KRUMBHAAR, H. (1986) *Phys. Rev. Lett.* **57**, 2431. Dynamic coarsening of crystal surfaces.

VAN DER MERWE, J., SHIFLET, G.J. (1991) *Surface Sci.* **256**, 171. The elastostatics of atomic steps on crystal surfaces 2.

VAN DER VEEN, J.F., PLUIS, B., DENIER VAN DER GON, A.W. (1990) in *Kinetics of Ordering and Growth at Surfaces*, M.G. Lagally ed. (Plenum, New York) p. 343. Ion scattering studies of surface melting.

VAN DER VEGT, H.A., VAN PINXTEREN, H.M., LOHMEIER, M., VLIEG, E., THORNTON, J.M.C. (1992) *Phys. Rev. Lett.* **68**, 3335. Surfactant-induced layer-by-layer growth of Ag on Ag(111).

VAN HOVE, J.M., LENT, C.S., PUKITE, P.R., COHEN, P.I. (1983), *J. Vac. Sci. Technol. B* **1**, 741. Damped oscillations in reflection high energy electron diffraction during GaAs MBE.

VAN HOVE J.M., COHEN, P.I. (1985), *Appl. Phys. Lett.* **47**, 726. Mass-action control of AlGaAs and GaAs growth in molecular beam epitaxy.

VAN HOVE, J.M., COHEN, P.I. (1987) *J. Cryst. Growth* **81**, 13. Reflection high energy electron diffraction measurements of surface diffusion during the growth of gallium arsenide by MBE.

VAN SICLEN C. DEW. (1995) *Phys. Rev. Lett.* **75**, 1574. Single jump diffusion mechanisms for large cluster diffusion on metal surfaces.

VENABLES, J.A., SPILLER, G.D., HANBÜCKEN, M. (1984) *Rep. Prog. Phys.* **47**, 399. Nucleation and growth of thin films.

VENABLES, J.A. (1973) *Phil. Mag.* **27**, 697. Rate equation approach to thin film nucleation kinetics.

VERON, M.B., ARNOULT, A., DAUDIN, B., TATARENKO, S. (1996) *Phis. Rev. B* **54**, R5267. Reversibility of the elementary mechanisms of atomic-layer epitaxy and sublimation of (001) CdTe.

VICSEK, T. (1989) *Fractal growth phenomena* (World Scientific, Singapore).

VILFAN, I., VILLAIN, J. (1991) *Surface Sci.* **257**, 368. Phase transitions on reconstructed (110) surfaces.

VILLAIN, J. (1978) *Phys. Rev. Lett.* **41**, 36. Pseudoharmonic theory of orientational instabilities in physisorbed layers.

VILLAIN, J. (1980) in *Ordering in strongly fluctuating condensed matter systems*, T. Riste ed. (Plenum, New York). Two-dimensional solids and their interaction with substrates.

VILLAIN, J. (1986) *Europhys. Lett.* **2**, 531. Healing of a rough surface.

VILLAIN, J. (1991) *J. de Physique I* **1**, 19. Continuum models for crystal growth from atomic beams with and without desorption.

VILLAIN, J., GORDON, M.B. (1983) *Surface Sci.* **125**, 1. Static properties of physisorbed monolayers in their solid phases.

VILLAIN, J., GREMPEL, D., LAPUJOLADE, J. (1985) *J. Phys. F* **15**, 809. Roughening transition of high index crystal phases: the case of copper.

VILLAIN, J., LANÇON, F. (1989) *Comptes rendus de l'Académie des Sciences*, **309 II**, 647. Dynamique du lissage d'une surface cristalline en dessous de sa transition rugueuse.

VILLAIN, J., PIMPINELLI, A., WOLF, D.E. (1992a) *Comments Cond. Matt. Phys.* **16**, 1. Layer by layer growth in molecular beam epitaxy.

VILLAIN, J., PIMPINELLI, A., TANG, L.-H., WOLF, D. (1992b) *J. de Physique I* **2**, 2107. Terrace sizes in molecular beam epitaxy.

VINEYARD, G. (1957) *J. Phys. Chem. Solids* **3**, 121. Frequency factors and isotope effects in solid state rate processes.

VOIGTLÄNDER, B., ZINNER, A. (1993) *Appl. Phys. Lett.* **63**, 3055. Simultaneous molecular beam epitaxy growth and scanning tunneling microscopy imaging during Ge/Si epitaxy.

VOIGTLÄNDER, B., ZINNER, A., T. WEBER, H.P. BONZEL (1995) *Phys. Rev. B* **51**, 7583. Modification of growth kinetics in surfactant-mediated epitaxy.

VOTER, A.F. (1989) *Phys. Rev. Lett.* **63**, 167. Computing classically exact diffusion constants using short-time trajectories.

WANG, X.S., GOLDBERG, J.L., BARTELT, N.C., EINSTEIN, T.L., WILLIAMS, E.D. (1990) *Phys. Rev. Lett.* **65**, 2430. Terrace width distribution on vicinal Si(111).

WANG, S.C., EHRLICH, G. (1991) *Phys. Rev. Lett.* **67**, 2509. Atom incorporation at surface clusters: an atomic view.

WEEKS, J.D. (1980) in *Ordering in Strongly Fluctuating Condensed Matter Systems*, T. Riste ed. (Plenum, New York) p. 293. The roughening transition.

WEEKS, J.D., VAN SAARLOOS, W. (1991) *J. Cryst. Growth* **112**, 244. Stability and shapes of cellular profiles in directional solidification: expansion and matching methods.

WIDOM, B. (1965) *J. Chem. Phys.* **43**, 3892. Surface tension and molecular correlations near the critical point.

WIEGMAN, P. B. (1978) *J. Phys. C* **11**, 1583. One-dimensional Fermi system and plane xy model.

WILLIAMS, E.D. (1994) *Surface Sci.* **299/300**, 502. Surface steps and surface morphology: understanding macroscopic phenomena from atomic observations.

WILLIAMS, E.D., PHANEUF, R.J., WEI, J., BARTELT, N.C., EISNTEIN, T.L. (1993) *Surface Sci.* **294**, 219. Thermodynamics and statistical mechanics of the faceting of stepped Si(111).

WILSON, K.G. (1983) *Rev. Mod. Phys.* **55**, 585. Renormalization group and critical phenomena.

WITTEN, T.A., SANDER, L.M. (1981) *Phys. Rev. Lett.* **47**, 1400. Diffusion-limited aggregation, a kinetic critical phenomenon.

WOLF, D.E. (1990) in *Kinetics of Ordering and Growth at Surfaces* M.G. Lagally ed. (Plenum, New York) p. 383. Growth of rough and faceted surfaces.

WOLF, D.E. (1991) *Phys. Rev. Lett.* **67**, 1784. Kinetic roughening of vicinal surfaces.

WOLF, D.E., KERTESZ, J. (1987) *Europhys. Lett.* **4**, 651. Surface width exponents for three- and four-dimensional Eden growth.

WOLF, D.E., VILLAIN, J. (1990a) *Phys. Rev. B* **41**, 2434. Shape fluctuations of crystal bars.

WOLF, D.E., VILLAIN, J. (1990b) *Europhys. Lett.* **13**, 389. Growth with surface diffusion.

WOLF, J.F., IBACH, H. (1991) *Appl. Phys. A* **52**, 218. Dislocations on Ag(111).

WOLF, P.E., GALLET, F., BALIBAR, S., ROLLEY, E., NOZIÈRES, P. (1985) *J. de Physique* **46**, 1987. Crystal growth and crystal curvature near roughening transitions in hcp ^{4}He.

WORTIS, M. (1988) in *Chemistry and Physics of Solid Surfaces VII*, Springer Series in Surface Sciences 10, R. Vanselow ed. (Springer Verlag, Heidelberg). Equilibrium crystal shapes and interfacial phase transitions.

WULFF, G. (1901) *Z. Kristallogr.* **34**, 449. Zur Frage der Geschwindigkeit des Wachstums und der Auflösung der Krystallflächen.

WYNBLATT, P., GJOSTEIN, N.A. (1968) *Surface Sci.* **12**, 109. A calculation of relaxation, migration and formation energies for surface defects in copper.

XIAO, R.F., IWAN, I., ALEXANDER, D., ROSENBERGER, F. (1988) *Phys. Rev. Lett.* **38**, 2447. Morphological evolution of growing crystals: a Monte Carlo simulation.

XIE, Y.H., GILMER, G.H., ROLAND, C., SILVERMANN, P.J., BURATTO, S.K., GHENG, J.Y., FITZGERALD, E.A., KORTAN, A.R., SCHUPPLER, S., MARCUS, M.A., CITRIN, P.H. (1994) *Phys. Rev. Lett.* **73**, 3006. Semiconductor surface roughness: dependence on sign and magnitude of bulk strain.

YAMAMOTO, T., AKUTSU, N., AKUTSU, Y. (1990) *J. Phys. Soc. Jpn.* **59**, 3831. Facet shape and crystal profile of complex crystals.

YAMAMOTO, T., AKUTSU, Y., AKUTSU, N. (1994) *J. Phys. Soc. Jpn.* **63**, 915. Fluctuation of a single step on the vicinal surface–universal and non-universal behaviors.

YAMASHITA, K., BONZEL, H.P., IBACH, H. (1981) *Appl. Phys.* **25**, 231. Steady state shapes of periodic profiles on (110), (100) and (111) surfaces of nickel.

YANG, Y.-N., WILLIAMS, E.D., PARK, R.L., BARTELT, N.C., EINSTEIN, T.L. (1990) *Phys. Rev. Lett.* **64**, 2410. Disordering of the (3×1) reconstruction on Si(113) and the chiral three-state Potts model.

YOKOYAMA, E., SEKERKA, R. (1992) *J. Cryst. Growth* **125**, 389. A numerical study of the combined effect of anisotropic surface tension and interface kinetics on pattern formation during the growth of two-dimensional crystals.

YOUNG, A.P. (1978) *J. Phys. C* **11**, L453. On the theory of the phase transition in the two-dimensional planar spin model.

ZANGWILL, A. (1988) *Physics at surfaces* (Cambridge University Press, Cambridge).

ZANGWILL, A., LUSE, C.N., VVEDENSKY, D., WILBY, M.R. (1992) *Surface Sci.* **274**, L529. Equations of motion for epitaxial growth.

ZHANG, ZH., DETCH, J., METIU, H. (1993) *Phys. Rev. B* **48**, 4972. Surface roughness in thin-film growth: the effect of mass transport between layers.

ZHDANOV, V.P., NORTON, P.R. (1994) *Appl. Surf. Sci.* **81**, 109. Nucleation during molecular beam epitaxy.

ZINSMEISTER, G. (1968) *Thin solid films* **2**, 497. Theory of thin film condensation. Part B: solution of the simplified condensation equation.

ZINSMEISTER, G. (1971) *Thin solid films* **7**, 51. Theory of thin film condensation. Part D: influence of a variable collision factor.

ZUMOFEN, G., BLUMEN, A. (1982) *J. Chem. Phys.* **76**, 3713. Energy transfer as a random walk. II. Two-dimensional regular lattices.

ZUO, J.-K., WENDELKEN, J.F., DÜRR, H., LIU, C.-L. (1994) *Phys. Rev. Lett.* **72**, 3064. Growth and coalescence in submonolayer homoepitaxy on Cu(100) studied with high-resolution low-energy electron diffraction.

Index

Printed in the United Kingdom
by Lightning Source UK Ltd.
134634UK00001BA/11-12/A

9 780521 558556